MAGNETISM AND MAGNETIC RESONANCE IN SOLIDS

MAGNETISM AND MAGNETIC RESONANCE IN SOLIDS

A. P. Guimarães
Centro Brasileiro de Pesquisas Físicas
Rio de Janeiro

I. S. Oliveira
Exercises
Centro Brasileiro de Pesquisas Físicas
Rio de Janeiro

A Wiley-Interscience Publication
JOHN WILEY & SONS, INC.
New York / Chichester / Weinheim / Brisbane / Singapore / Toronto

This book is printed on acid-free paper. ∞

Copyright ©1998 by John Wiley & Sons, Inc. All rights reserved.

Published simultaneously in Canada.

No part of this publication may be reproduced, stored in a retrieval system or transmitted in any form or by any means, electronic, mechanical, photocopying, recording, scanning or otherwise, except as permitted under Sections 107 or 108 of the 1976 United States Copyright Act, without either the prior written permission of the Publisher, or authorization through payment of the appropriate per-copy fee to the Copyright Clearance Center, 222 Rosewood Drive, Danvers, MA 01923, (978) 750-8400, fax (978) 750-4744. Requests to the Publisher for permission should be addressed to the Permissions Department, John Wiley & Sons, Inc., 605 Third Avenue, New York, NY 10158-0012, (212) 850-6011, fax (212) 850-6008, E-Mail: PERMREQ@WILEY.COM.

Library of Congress Cataloging-in-Publication Data:

Guimarães, Alberto Passos.
   Magnetism and magnetic resonance in solids / by A. P. Guimarães.
      p.   cm.
   Includes bibliographical references and index.
   ISBN 0-471-19774-2 (cloth : alk. paper)
   1. Solids–Magnetic properties. 2. Magnetism. 3. Magnetic materials. 4. Nuclear magnetic resonance. I. Title.
   QC176.8.M3G85   1998
   530.4'12–dc21   97–35865

Printed in the United States of America

10 9 8 7 6 5 4 3 2 1

To the memory of my father

CONTENTS

Preface   xi

1 Introduction   1

    1.1 Magnetism of Matter / 1
    1.2 Magnetic Quantities and Units / 2
    1.3 Types of Magnetism / 12
    1.4 Magnetic Properties of Some Magnetic Materials / 20
    1.5 Permanent Magnets / 21

2 Atomic Magnetic Moments   27

    2.1 Diamagnetism / 28
    2.2 Electrons in Atoms / 29
    2.3 Magnetic Moment of an Assembly of Atoms / 40
    2.4 Langevin Paramagnetism / 47
    2.5 Nuclear Magnetism / 48
    2.6 Ferromagnetism / 49
    2.7 Crystal Fields / 55

3 Interaction between Two Spins   63

    3.1 Exchange Interaction / 63
    3.2 The Mean Field / 69
    3.3 Indirect Interactions in Metals / 70
    3.4 Pair of Spins in the Molecular Field (Oguchi Method) / 73
    3.5 Spin Waves: Introduction / 79

4 Magnetism Associated with the Itinerant Electrons — 91

4.1 Introduction / 91
4.2 Paramagnetic Susceptibility of Free Electrons / 93
4.3 Ferromagnetism of Itinerant Electrons / 100
    4.3.1 Magnetism at $T = 0\,\text{K}$: The Stoner Criterion / 100
    4.3.2 Magnetism at $T \neq 0\,\text{K}$ / 102
4.4 Coupled Localized Itinerant Systems / 107
4.5 Magnetic Phase Transitions: Arrott Plots / 109

5 The Magnetization Curve — 115

5.1 Ideal Types of Magnetic Materials / 115
5.2 Contributions to the Energy in Magnetic Materials / 119
    5.2.1 Magnetostatic Energy / 119
    5.2.2 Magnetic Anisotropy / 121
    5.2.3 Exchange Interaction / 126
    5.2.4 Magnetoelastic Energy and Magnetostriction / 127
5.3 Magnetic Domains / 132
5.4 Reversible and Irreversible Effects in the Magnetization / 135
5.5 The Magnetization Process / 137
5.6 Dynamic Effects in the Magnetization Process / 148

6 Hyperfine Interactions — 159

6.1 Introduction / 159
6.2 Electrostatic Interactions / 160
6.3 Magnetic Dipolar Interaction / 165
    6.3.1 Contribution of the Electronic Spin to the Magnetic Hyperfine Field / 168
    6.3.2 Orbital Contribution to the Magnetic Hyperfine Field / 170
6.4 Contributions to B_{hf} in the Free Ion / 174
6.5 Hyperfine Fields in Metals / 175
    6.5.1 Intraionic Interactions in the Metals / 175
    6.5.2 Extraionic Magnetic Interactions / 176
    6.5.3 Hyperfine Fields Observed Experimentally / 178
    6.5.4 The Knight Shift / 181
6.6 Electrostatic Interactions in Metals / 182
6.7 Combined Magnetic and Electrostatic Interactions / 183

7 Nuclear Magnetic Resonance — 189

7.1 The Phenomenon of Magnetic Resonance / 189
7.2 Equations of Motion: Bloch Equations / 191

7.3 Magnetization from Rotating Axes / 196
  7.4 Relaxation / 203
    7.4.1 Longitudinal Relaxation / 207
    7.4.2 Transverse Relaxation / 210
    7.4.3 Nuclear Magnetic Relaxation Mechanisms / 211
  7.5 Diffusion / 214
  7.6 Pulsed Magnetic Resonance / 214
  7.7 Quadrupole Oscillations / 221

8 Magnetic Resonance in Magnetic Materials   **227**

  8.1 Nuclear Magnetic Resonance / 227
    8.1.1 NMR Studies of Magnetically Ordered Solids / 229
  8.2 Resonance in a Coupled Two-Spin System / 230
  8.3 NMR Enhancement Factor: Domains and Domain Walls / 237
  8.4 Ferromagnetic Resonance / 242

Appendix A Table of NMR Nuclides   **251**

Appendix B Solutions to the Exercises   **255**

Appendix C Physical Constants   **283**

Index   **285**

PREFACE

Two fascinating subjects have been put together in this book: the magnetism of matter and magnetic resonance phenomena. The first of these has captured the imagination of humankind for 3000 years; the second theme has a much shorter history, started half a century ago.

Both subjects have grown to become major technical forces: the world market for magnetic media and recording equipment reaches about $100 billion per year (Simonds 1995); also, the value of the magnetic materials produced today is larger than that of semiconductor materials, which are the basis of the present technological revolution. The application of magnetic resonance to medicine is revolutionizing diagnosis and changing the image of the human body.

The present text provides a succinct presentation of the properties of magnetic materials, hyperfine interactions in condensed matter, and the phenomenon of nuclear magnetic resonance. Nuclear magnetic resonance in magnetic materials is included as an application of these subjects.

This book is intended for final-year undergraduate courses, or graduate courses in magnetism, magnetic materials, and magnetic resonance. With the growth in the applications of magnetism in permanent magnets, soft magnetic materials, magnetic recording, and magnetic resonance imaging (MRI), the text will be useful to materials scientists, physicists, and other specialists.

The text was organized from lecture notes on the introductory part of the course "Introduction to Magnetism," taught to graduate students at the Brazilian Center for Physical Research (CBPF). The exercises and solutions were written by Dr. Ivan S. Oliveira; I owe him also the stimulus for writing this book and the conversion of the original manuscript to Latex.

I have used SI units throughout; the units, symbols, and nomenclature recommended by the General Conference on Weights and Measures, and the International Organization of Standardization (ISO), are contained in a publication of the National Institute of Standards and Technology (NIST) (1995) (available from www.nist.gov). At each point I have included the conversion factors to the centimeter–gram–second (CGS) system. Some plots of experimental results are in CGS units, reflecting the present state of the literature on magnetism, in which the two systems coexist. On the use of the magnetic induction **B** or magnetic field intensity **H**, see Shadowitz (1975) and Crooks (1979).

In the choice of references presented along the text, I have preferred to indicate review articles and textbooks, with the intention of reinforcing the didactic function of the book.

I have tried to keep the text short; the books of Smart (1966) and McCausland and Mackenzie (1980) are two very different and successful examples of short texts, and were used as references in some chapters.

Acknowledgements are due to Dr. R. C. O'Handley for the hospitality during the period 1993–94, spend at the Massachusetts Institute of Technology, where some parts on macroscopic magnetism were first written.

I thank especially J. S. Helman, H. Micklitz, and X. A. da Silva for reading the manuscript and for many suggestions. We also acknowledge the comments of W. Baltensperger, V. M. T. Barthem, G. J. Bowden, W. D. Brewer, M. A. Continentino, H. Figiel, E. Gratz, D. Guenzburger, L. Iannarella, Cz. Kapusta, M. Knobel, H. R. Rechenberg, H. Saitovitch, L. C. Sampaio, J. Terra, and M. Wójcik.

I am especially grateful to my students V. L. B. de Jesus, R. Sarthour, Jr., and C. V. B. Tribuzy for their help in the trial of the first version of the text. I acknowledge the work of L. Baltar, with the figures.

Finally, suggestions and comments are welcome.

A. P. GUIMARÃES

Centro Brasileiro de Pesquisas Físicas
R. Xavier Sigaud 150
22290-180 Rio de Janeiro, RJ, Brazil

apguima@cat.cbpf.br

REFERENCES

Crooks, J. E., *J. Chem. Educ.* **56**, 301 (1979).
McCausland, M. A. H. and I. S. Mackenzie, *Nuclear Magnetic Resonance in Rare Earth Metals*, Taylor & Francis, London, 1980.
Shadowitz, A., *The Electromagnetic Field*, McGraw-Hill, New York, 1975.
Smart, J. S., *Effective Field Theories of Magnetism*, Saunders, Philadelphia, 1996.
Taylor, B. N., *Guide for the Use of the International System of Units (SI)*, NIST Special Publication 811, 1995 edition, Gaithersburg, MD, 1995.

MAGNETISM AND MAGNETIC RESONANCE IN SOLIDS

1

INTRODUCTION

1.1 MAGNETISM OF MATTER

The discipline of *magnetism* studies the magnetic fields, the magnetic properties of matter, and the interactions between matter and the fields. The magnetic field is a fundamental concept of magnetism: it is a field of forces that describes a property of space in the neighborhood of either charges in motion, or of magnets. Its presence can be detected, for example, through the force exerted on a probe consisting of a wire traversed by a current.

Materials exhibit different behaviors in the presence of a magnetic field. The most evident differences may be observed through the changes in the magnetic field itself in the neighborhood of the samples under study, or through the forces exerted on them by a distribution of magnetic fields. The three traditional classes of materials, in terms of magnetic behavior, are diamagnets, paramagnets, and ferromagnets. Materials of the first type are repelled from a region of more intense field, those of the second type are attracted, and the last type are strongly attracted; except for the last effect, these observations require very sensitive instruments. Inside these three classes of materials, a property known as *magnetic induction* also behaves in a differentiated way: it is, respectively, reduced, increased, or greatly increased in relation to its value in vacuum.

Before examining the classes of magnetic materials, we will briefly survey the magnetic quantities. For a revision of this topic, see, for example, Grant and Phillips (1990). Definitions and recommendations on the use of magnetic units can be found in Taylor (1995) and Cohen and Giacomo (1987).

1.2 MAGNETIC QUANTITIES AND UNITS

A magnet creates a magnetic field in the space around it; a magnetic field can also be created by an electric current. The physical quantity that describes the effect of a magnet or current in its neighborhood is the magnetic induction or magnetic flux density **B**. A measure of the magnetic induction may be given by the Lorentz force, the force on a charge q in motion. The Lorentz force acting on a charge that moves with a velocity **v** is given by

$$\mathbf{F} = q\mathbf{v} \times \mathbf{B} \qquad (1.1)$$

The unit of magnetic induction in the SI (Système International d'Unités) system of units is the *tesla* (T), defined as the magnetic induction that produces a force of 1 newton on a charge of 1 coulomb, moving with a velocity of 1 m s^{-1} in the direction perpendicular to that of **B**. The unit of **B** in the CGS (centimeter–gram–second) system is the gauss (G), which corresponds to 10^{-4} T.

The fundamental equations of classical electromagnetism, which involve its main physical quantities, are Maxwell equations, given in Table 1.I (SI) (e.g., Grant and Phillips 1990).

In the equations shown in Table 1.I, **H** is the magnetic field intensity, **D** is the displacement vector, ρ is the electric charge density and **j** is the current density. In **j** are included the conduction currents proper and also convection currents (those where there is motion of matter) (Shadowitz 1975).

The effect of a magnetic field may be characterized by the magnetic induction (or flux density) **B** or the magnetic field intensity **H**. The tendency in recent decades has been to emphasize **B** as a more fundamental quantity (e.g., Crooks 1979). One justification for this choice is the higher degree of generality of **B**; the curl of **B** is equal to μ_0 times the total current density $\mathbf{j}_t$, including convection currents and currents associated to the magnetization (Shadowitz 1975); μ_0 is the vacuum magnetic permeability, a constant equal to $4\pi \times 10^{-7}$ H m^{-1} (henries per meter) (SI).

As is usual in the literature, we will normally employ the word *field* when referring to the induction; the notation **B** should remove any ambiguity with the magnetic field intensity (**H**).

Before defining *magnetization*, we will define *magnetic moment*. Let us imagine an infinitesimal closed circuit of area dS through which a current I flows. The

Table 1.I Maxwell equations

curl **H** = **j** + $\partial \mathbf{D}/\partial t$	Generalized Ampère law
curl **E** = $-\partial \mathbf{B}/\partial t$	Faraday law
div **D** = ρ	Gauss law
div **B** = 0	(Nonexistence of magnetic monopoles)

magnetic dipole moment, or magnetic moment associated with this circuit is

$$d\mathbf{m} = I\, d\mathbf{S} \tag{1.2}$$

where $d\mathbf{S}$ is the oriented element of area, defined by the sense of the current, and given by the right-hand rule. The magnetic moment is measured in joules per tesla $J\,T^{-1}$; ($\equiv A\,m^2$) (in SI), and in ergs per gauss (erg/G) [also referred to as *emu* (electromagnetic unit), although strictly speaking, this is not a unit] (in CGS).

The magnetization $\mathbf{M}$ of a body is its total magnetic moment divided by its volume, a quantity that is numerically equivalent to

$$\mathbf{M} = n\, \mathbf{m} \tag{1.3}$$

where n is the number of magnetic moments $\mathbf{m}$ per unit volume. The magnetization $\mathbf{M}$ is measured in amperes per meter ($A\,m^{-1}$) or webers per square meter ($Wb\,m^{-2}$) in SI units and in oersteds (Oe) in the CGS system. Table 1.II presents values of the magnetization M measured for the elements iron, cobalt, and nickel.

The magnetic field intensity $\mathbf{H}$ is defined as

$$\mathbf{H} = \frac{\mathbf{B}}{\mu_0} - \mathbf{M} \tag{1.4}$$

and has no specific unit in the SI, being measured in amperes per meter ($A\,m^{-1}$); note that it has the same dimension as $\mathbf{M}$. In the CGS, it is measured in oersteds (Oe). The constant μ_0 is the vacuum permeability [$\mu_0 = 4\pi \times 10^{-7}\,H\,m^{-1}$ (SI)]. In the CGS system, μ_0 has a value of 1 gauss per oersted; therefore, the magnetic field intensity (in oersted) and the magnetic induction (in gauss) have the same numerical value.

A magnetic current that flows through a conductor produces a magnetic field; $1\,A\,m^{-1}$ is the intensity of the field inside a solenoid of infinite length, with N turns per meter, traversed by a current of $1/N$ amperes. For practical purposes, the magnetic fields are produced either by the action of electrical currents flowing through coils, or by means of permanent magnets. Table 1.III

Table 1.II Magnetization M and polarization J for the elements Fe, Co, and Ni at low temperature

Element	M^a ($10^3\,A\,m^{-1}$)	$J (\equiv \mu_0 M)$ (T)	T (K)
Fe	1766	2.219	4.2
Co	1475	1.853	4.21
Ni	528	0.663	4.21

[a] M values from Landolt-Börnstein, *Magnetic Properties of 3d Elements*, New Series III/19a, Springer-Verlag, Berlin, 1986, p. 37.

Table 1.III Typical values of the magnetic induction B in tesla, associated with different sources, including the surface of some astronomical bodies

Origin or Site	B (in T)
Brain	10^{-12}
Galactic disk	10^{-10}
Heart	10^{-10}
Earth	10^{-4}
Sunspots	10^{-1}
Permanent magnets	10^{-1}
Ap-type stars	1
Electromagnets	1
Superconducting coil	1–10
Nucleus of metallic Fe	30
Nucleus of metallic Ho (4.2 K)	737
White dwarf stars	$\leq 10^4$
Neutron stars	10^8

presents some values of magnetic induction obtained either through these means, or observed in Nature.

The vectors **B** and **H** obey different boundary conditions in the frontier of material media:

$\mathbf{B}_\perp$ is continuous.
$\mathbf{H}_\parallel$ is continuous.

We can read the definition of **H** [Eq. (1.4)] in the following way: there are two contributions to the magnetic induction **B** in a material medium; one arising from the magnetic field **H** (equal to $\mu_0 \mathbf{H}$), and another contribution proportional to the magnetization, equal to $\mu_0 \mathbf{M}$. The total induction in the medium becomes

$$\mathbf{B} = \mu_0(\mathbf{H} + \mathbf{M}) \tag{1.5}$$

Whenever we refer to material media, we will mean by **H** the internal magnetic field intensity.

In the CGS system, the fields **B** and **H** are related through

$$\mathbf{B} = \mathbf{H} + 4\pi \mathbf{M} \quad \text{(CGS)} \tag{1.6}$$

with **B** in gauss (G) and **H** in oersted (Oe); **M** is measured in ergs G^{-1} cm^{-3}, or sometimes in emu cm^{-3}.

Another form of characterizing magnetic materials is through the use of the polarization **J** (also known as intensity of magnetization **I**), defined by

$$\mathbf{J} = \mu_0 \mathbf{M} \tag{1.7}$$

and measured, as **B**, in tesla.[1]

The measure of the magnetic response of a medium to the action of a magnetic field of intensity **H** is given by its magnetic susceptibility χ. The magnetic susceptibility (volume susceptibility) χ is dimensionless and given by

$$\chi = \frac{M}{H} \tag{1.8}$$

or in differential form, by $\chi = \partial M / \partial H$. The mass susceptibility (or specific susceptibility) χ_g is the total magnetic moment divided by the field H, divided by the mass, and is measured in cubic meters per kilogram; it is related to the volume susceptibility through $\chi_g = \chi/\rho$, where ρ is the density.

The magnetic response of a medium can also be measured by its magnetic permeability, denoted by μ. If the proportionality of **M** and **H** is observed, then

$$\mathbf{B} = \mu \mathbf{H} \tag{1.9}$$

where μ is the magnetic permeability. A more general definition of μ is given by $\mu = B/H$. The magnetic permeability of a material is in general not a constant, but depends on the value of the field **H**.

In anisotropic media, μ, and also χ, depend on the direction of the applied field, and are second-rank tensors.

We can define the relative magnetic permeability μ_r of a medium, in terms of the vacuum magnetic permeability μ_0:

$$\mu_r = \frac{\mu}{\mu_0} \tag{1.10}$$

The relative magnetic permeability μ_r is related to the susceptibility χ, and it follows, from Eqs. (1.8) and (1.9)

$$\mu_r = 1 + \chi \tag{1.11}$$

The relative permeability in the SI is numerically equal to the CGS permeability. From Eq. (1.6), we obtain $\mu_r = 1 + 4\pi\chi$ (CGS), and therefore, the SI susceptibility is a number 4π times larger than χ_{CGS}.

---

[1]This form of describing the effects of the magnetization is part of the Kennelly convention; it is different, but it is not incompatible with the Sommerfeld convention, which we have adopted, and which is the one most frequently used in physics.

Table 1.IV Magnetic (volume) susceptibility χ and relative magnetic permeability μ_r of some elements at room temperature[a]

Element	$\chi \times 10^6$	μ_r
Na	8.09	1.000008
K	5.76	1.000006
Al	20	1.000020
Ti	182	1.000182
Cr	286	1.000286
Mn	830	1.000830
Cu	−9.7	0.999990
Zn	−12	0.999988
Ge	−7.14	0.999993
Pd	789	1.000789
Ag	−25.2	0.999975
Sb	−68	0.999932
La	56	1.000056
Pt	261	1.000261
Au	−34.6	0.999965
Tl	−36.4	0.999964

[a] Derived from values of mass susceptibility χ_g (Landolt-Börnstein, *Magnetic Properties of 3d Elements*, Springer-Verlag, Berlin, 1962, p. 1–5); μ_r was computed from χ using Eq. (1.11).

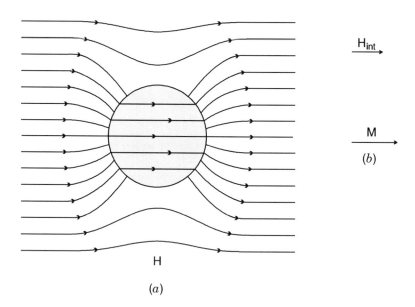

Figure 1.1 (*a*) Lines of magnetic field intensity **H** in the neighborhood of a sphere of material with relative permeability $\mu_r > 1$ (or, equivalently, $\chi > 0$) introduced into an originally uniform field. Note the smaller density of lines inside the sphere, representing a reduction in the modulus of **H**; (*b*) the internal field **H**$_{int}$ and the magnetization **M** induced in the sphere.

Table 1.IV shows room-temperature values of the susceptibility χ and relative permeability μ_r for some elements.

The directions and intensities of the field **H** and of the magnetic induction **B** may be represented by lines of force; these are lines that have at every point of space the same direction as **H** or **B**, with density per unit area proportional to the intensity of the corresponding field. The lines of **H** and **B** reflect the different properties of these fields; the lines of force of **H** start from the north (N) pole of each magnetic dipole and end at the south (S) pole; the lines of **B** are closed. The lines of **H** are analogous to the lines of force of the electric field **E**, with the charges substituted by (fictitious) magnetic poles (Fig. 1.1).

In Fig. 1.1 one can see the lines of force of the field **H** in an (initially uniform) field where a sphere of paramagnetic material (therefore a material with relative permeability $\mu_r > 1$) was introduced; Fig. 1.2 shows the lines of **B** in this case.

The lines of **H** inside a magnet and in the adjacent space initiate on the N pole and end on the S pole; this can be seen in Fig. 1.3, which shows a magnet in the absence of external applied fields.

The lines of induction **B** inside a magnetized medium point in the opposite sense, specifically, from S to N (Fig. 1.4). In empty space, the lines of **H** and **B** coincide, since $\mathbf{B} = \mu_0 \mathbf{H}$.

A magnetized body has at its surface "free poles" that arise where the normal component of the magnetization **M** goes through a discontinuity (Fig. 1.5). In

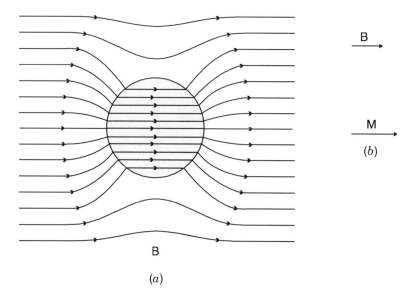

Figure 1.2 (a) Lines of magnetic induction **B** around a sphere of relative permeability $\mu_r > 1$ (or, equivalently, $\chi > 0$) introduced into an originally uniform field; (b) directions of **B**, and magnetization **M** inside the sphere.

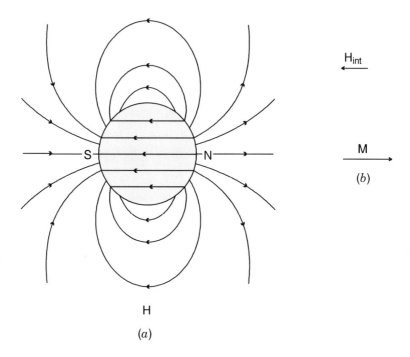

Figure 1.3 (a) Lines of magnetic field intensity **H** inside and outside a magnet; (b) the internal field H_{int} (equal to the demagnetizing field H_d in the absence of external field) and magnetization **M** inside a magnet.

the interior of the body, the opposite poles of the individual magnetic moments compensate each other. The density of free poles is given by

$$\rho_s = \mathbf{M} \cdot \mathbf{n} \qquad (1.12)$$

where **n** is the unit vector normal to the surface. The free poles produce an additional field $\mathbf{H}_d$, opposed to **M**. If the magnetization originates from the action of an external magnetic field $\mathbf{H}_0$, the field intensity **H** inside the body therefore differs from $\mathbf{H}_0$: $\mathbf{H} = \mathbf{H}_0 - \mathbf{H}_d$. The field $\mathbf{H}_d$ is called the *demagnetizing field*, its intensity is proportional to the value of the magnetization **M**, and its direction is opposite to **M**. The intensity of the internal magnetic field in the material, under an applied field $\mathbf{H}_0$, is therefore:

$$\mathbf{H} = \mathbf{H}_0 - N_d \mathbf{M} \qquad (1.13)$$

where N_d is the demagnetizing factor.

The demagnetizing factor N_d depends on the shape of the body and the angle between its axes of symmetry and the field **H**; it varies between 0 and 1 (or

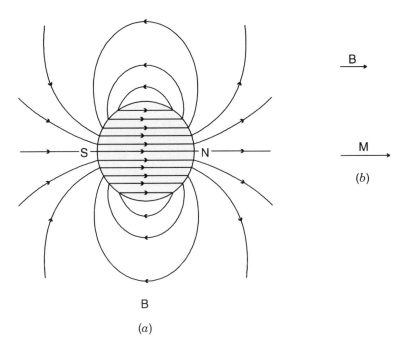

Figure 1.4 (a) Lines of the magnetic induction **B** inside and outside a magnet; (b) directions of **B**, and magnetization **M** inside a magnet.

between 0 and 4π in the CGS system). For example, it is zero for an infinite cylinder under the action of a field **H** parallel to its axis, and is equal to $\frac{1}{3}$ for a sphere ($4\pi/3$ in CGS). To obtain the value of the demagnetizing factor in the SI, the value in CGS has to be divided by 4π (see Table 1.V).

As an illustration, we can obtain the value of N_d for a flat plate, under a field H_0 applied perpendicular to its surface, noting simply that due to the continuity of $\mathbf{B}_\perp$, we have $\mathbf{B}_0 = \mathbf{B}$. Therefore

$$\mathbf{B} = \mu_0(\mathbf{H} + \mathbf{M}) = \mu_0(\mathbf{H}_0 - N_d\mathbf{M} + \mathbf{M}) = \mu_0\mathbf{H}_0 \qquad (1.14)$$

and it follows that $N_d = 1$ for a plane plate, with a perpendicular field.

In samples of arbitrary shape, the demagnetizing field varies from point to point; the field is homogeneous only inside ellipsoidal samples. This includes limiting case ellipsoids, with $a = b = c$ (sphere), $a = 0$ (plane), or $b = c = 0$ (wire). The demagnetizing factors along the three axes of an ellipsoid are related: $N_d^a + N_d^b + N_d^c = 1$.

In the most general case the demagnetizing field does not point along the same axis as **M**, and the demagnetizing factor is a tensor N_d.

The conversion factors between the SI and CGS systems for the units of the

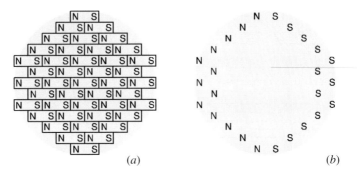

Figure 1.5 (a) Elementary magnetic dipoles in a magnetized sphere; (b) "free poles" at the surface of the sphere, arising from the uncompensated poles of these dipoles.

most important magnetic quantities are given in Table 1.VI. For a discussion of the units and quantities of magnetism, see Grandjean and Long (1990) and Evetts (1992, p. 253).

One can analyze the behavior of the magnetic flux inside samples of magnetic material using the properties of the fields **B** and **H**. Material media may be used to form magnetic circuits, where the lines of force may be studied. In the case of a closed magnetic circuit formed of a toroidal piece of magnetic material, there are no free poles, and consequently no demagnetizing field ($N_d = 0$). To be of any use, however, a magnet normally has to have a gap, and the presence of a gap opens the magnetic circuit.

As an example of a magnetic circuit, we may examine the case of a toroidal magnetized sample, of cross section A, with a gap of length l_g (Fig. 1.6). From Ampère's law, the line integral of **H** around the magnetic circuit is zero, because there is no current. Calling H_m the intensity of the magnetic field inside the magnetic material, and H_g the field in the gap, one has

$$H_g l_g - H_m l_m = 0 \qquad (1.15)$$

where l_m is the length of the magnet. The magnetic flux across an area A of unit

Table 1.V Demagnetizing factors N_d (SI)[a]

Shape	Direction	N_d
Plane	$\perp$	1
Plane	$\parallel$	0
Cylinder ($l/d = 1$)	$\parallel$	0.27
Cylinder ($l/d = 5$)	$\parallel$	0.04
Long cylinder	$\parallel$	0
Sphere	—	$\frac{1}{3}$

[a] To obtain the values in the CGS system, divide by 4π.

Table 1.VI Magnetic quantities and units[a]

Quantity	Symbol	CGS	SI	Conversion Factor
Magnetic induction	**B**	G	T	10^{-4}
Magnetic field intensity	**H**	Oe	$A\,m^{-1}$	$10^3/4\pi$
Magnetization	**M**	$erg\,G^{-1}\,cm^{-3}$ or $emu\,cm^{-3}$	$A\,m^{-1}$	10^3
Magnetic polarization	**J**	—	T	—
Magnetic moment	m	$erg\,G^{-1}\,(\equiv emu)$	$J\,T^{-1}\,(\equiv A\,m^2)$	10^{-3}
Specific magnetization	σ	$emu\,g^{-1}$	$A\,m^2\,kg^{-1}(\equiv J\,T^{-1}\,kg^{-1})$	1
Magnetic flux	ϕ	Mx (maxwell)	Wb (weber)	10^{-8}
Magnetic energy density	E	$erg\,cm^{-3}$	$J\,m^{-3}$	10^{-1}
Demagnetizing factor	N_d	—	—	$1/4\pi$
Susceptibility (volume)	χ	—	—	4π
Mass susceptibility	χ_g	$erg\,G^{-1}\,g^{-1}\,Oe^{-1}$ or $emu\,g^{-1}\,Oe^{-1}$	$m^3\,kg^{-1}$	$4\pi \times 10^{-3}$
Molar susceptibility	χ_{mol}	$emu\,mol^{-1}\,Oe^{-1}$	$m^3\,mol^{-1}$	$4\pi \times 10^{-6}\,m^3\,mol^{-1}$
Magnetic permeability	μ	$G\,Oe^{-1}$	$H\,m^{-1}$	$4\pi \times 10^{-7}$
Relative permeability	μ_r	—	1	1
Vacuum permeability	μ_0	$G\,Oe^{-1}$	$H\,m^{-1}$	$4\pi \times 10^{-7}$
Anisotropy constant	K	$erg\,cm^{-3}$	$J\,m^{-3}$	10^{-1}
Gyromagnetic ratio	γ	$s^{-1}\,Oe^{-1}$	$m\,A^{-1}\,s^{-1}$	$4\pi \times 10^{-3}$

[a] To obtain the values of the quantities in the SI, the corresponding CGS values should be multiplied by the conversion factors.

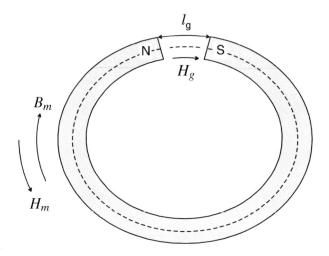

Figure 1.6 Lines of magnetic field intensity **H** and **B** inside a sample of toroidal shape, with a gap.

normal $\hat{\mathbf{n}}$ is $\phi = \mathbf{B} \cdot \hat{\mathbf{n}} A$. The continuity of ϕ along the circuit gives

$$\phi = B_g A = \mu_0 H_g A = B_m A \qquad (1.16)$$

where B_m is the field B inside the magnet. Combining these two equations, it follows

$$H_g^2 = \frac{1}{\mu_0} \frac{V_m}{V_g} (B_m H_m) \qquad (1.17)$$

where V_m and V_g are the volumes of the magnet and gap, respectively. This shows that the intensity of the magnetic field in the gap (H_g) increases with the product $B_m H_m$ (or BH), called the *energy product*; the square of H_g is directly proportional to (BH).

Magnetic circuits are analogous in many respects to electrical circuits; one important difference arises from the fact that the ratio of permeabilities in the magnetic material and in vacuum is much smaller than the corresponding ratio of conductivities between an electric conductor and the vacuum. This implies that the flux density across an open magnetic circuit is significant, whereas the charge flow in an open electrical circuit can be neglected.

1.3 TYPES OF MAGNETISM

The magnetic properties of matter originate essentially from the magnetic moments of electrons in incomplete shells in the atoms (see Chapter 2), and from unpaired electrons in the conduction band (in the case of metals, see Chapter 4). The incomplete shell may be, for example, the $3d$ shell—in the case of the elements of the iron group, or the $4f$ shell—in the rare earths.

Magnetic materials are those that present permanent magnetic moments, with spontaneous long range order; this order is due to an interaction of electrostatic origin and quantum nature, called *exchange interaction* (see Chapter 3). The interaction responsible for the magnetic order may be of short range—direct exchange interaction—of long range, or indirect interaction.

A sample of magnetic material is generally formed of ordered regions, called *domains*, inside which the magnetization points along the same direction, which varies from one such region to the other (see Section 5.3). An external magnetic field alters the structure of domains, but leaves practically unaltered the magnetization inside each domain, that remains equal to the saturation magnetization.

The degree of structural order is important for the magnetism of matter; the materials can be (1) crystalline, in which the atomic sites have translation symmetry; (2) disordered, with atoms occupying randomly the sites of a crystalline lattice; and (3) amorphous, where there are no equivalent atomic sites.

In sequence, we will enumerate very briefly the main classes of magnetic materials (Hurd 1982); although some concepts used in this classification are defined in more detail in later chapters, the reader may benefit from this survey by exposure to the wealth of magnetic properties of the substances.

1. *Diamagnetism.* Type of magnetism characterized by a small and negative susceptibility, independent of temperature (Fig. 1.7). The susceptibility of every substance presents a diamagnetic component; its origin lies in the shielding effect due to the motion of atomic electrons. In diamagnetic materials this component is dominant. Conduction electron currents in metals are responsible for Landau diamagnetism, an effect of larger magnitude. Examples of diamagnetic substances are the compound NaCl (sodium chloride) and copper oxide (CuO).

2. *Paramagnetism.* Magnetism characterized by a positive susceptibility whose inverse varies linearly with temperature (Fig. 1.8). This type of temperature dependence (called the *Curie law*) is found at any temperature in the paramagnetic materials, or above a certain temperature of magnetic order, in ferromagnetic and antiferromagnetic materials (called in these cases the *Curie–Weiss law*). The fall in the susceptibility with temperature originates in the increase in the ratio of thermal energy to the energy of the atomic magnetic moments in the presence of the external magnetic field. One type of paramagnetism—Pauli paramagnetism—is due to the magnetic moments of the conduction electrons and in this case the susceptibility is practically constant with temperature.

3. *Ferromagnetism.* Type of magnetism characterized by an spontaneous parallel alignment of atomic magnetic moments, with long range order (Fig. 1.9). Examples of ferromagnets are the elements iron, nickel, and gadolinium. This order disappears above a certain temperature, called the *Curie temperature* (T_C).

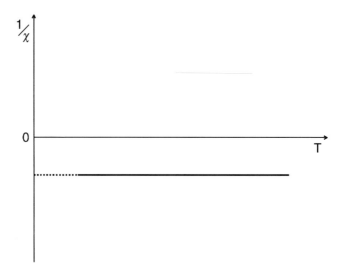

Figure 1.7 Temperature dependence of the inverse of the susceptibility ($1/\chi$) of a diamagnetic material. [Adapted from C. M. Hurd, *Contemp. Phys.*, **23**, 479 (1982), with permission from Taylor & Francis, Bristol, PA.]

4. *Antiferromagnetism.* Magnetism in which the atomic moments align antiparallel, with zero resulting magnetization (Fig. 1.10). Above the ordering temperature—the Néel temperature (T_N)—the inverse of the susceptibility follows a linear dependence. Examples are FeO and Fe$_3$Mn; α-Mn is an itinerant antiferromagnet, and does not obey the Curie–Weiss law.

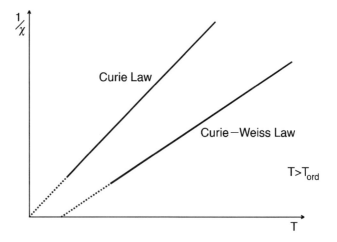

Figure 1.8 Temperature dependence of the inverse of the susceptibility ($1/\chi$) of a paramagnetic material (Curie law) and of a ferromagnetic material above the ordering temperature (Curie–Weiss law). [Adapted from C. M. Hurd, *Contemp. Phys.*, **23**, 479 (1982), with permission from Taylor & Francis, Bristol, PA.]

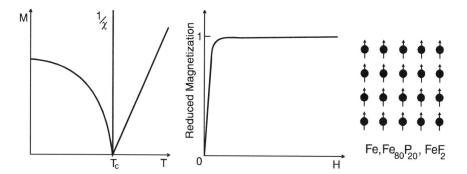

Figure 1.9 Temperature dependence of the magnetization M of a ferromagnetic material, dependence of the inverse susceptibility ($1/\chi$), and dependence of the magnetization on applied magnetic field; also, schematic representation of ferromagnetism, with examples of ferromagnetic materials. [Adapted from C. M. Hurd, *Contemp. Phys.*, **23**, 480 (1982), with permission from Taylor & Francis, Bristol, PA.]

5. *Ferrimagnetism.* Magnetic order in which two or more different magnetic species exist (atoms or ions) with collinear magnetic moments. In general, some moments couple in an antiparallel fashion. The resulting magnetization is nonzero (Fig. 1.11). Examples are magnetite, $FeO \cdot (Fe_2O_3)$ and $GdFe_2$.

6. *Metamagnetism.* This is a property of some substances in which the antiferromagnetic order is altered by the application of an external magnetic field, by virtue of its small anisotropy (Fig. 1.12); there is a type of itinerant metamagnetism in which the magnetic field that produces this alteration is the field around a magnetic impurity;.

7. *Enhanced Pauli Paramagnetism.* Also known as *incipient ferromagnetism*

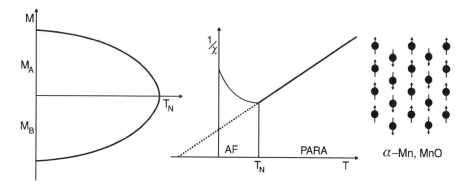

Figure 1.10 Schematic representation of the temperature dependence of the magnetization of the opposing sublattices in an antiferromagnetic material, with variation of the inverse susceptibility ($1/\chi$); schematic representation of antiferromagnetism, with examples of antiferromagnetic materials. [Adapted from C. M. Hurd, *Contemp. Phys.*, **23**, 482 (1982), with permission from Taylor & Francis, Bristol, PA.]

16 INTRODUCTION

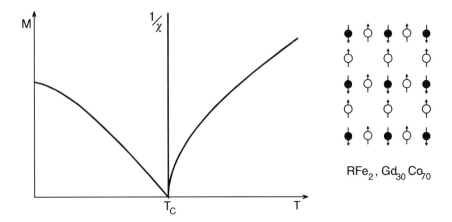

Figure 1.11 Temperature dependence of the magnetization M of a ferrimagnetic material, and of the inverse susceptibility $(1/\chi)$; also, schematic representation of ferrimagnetism, with examples of ferrimagnetic materials. [Adapted from C. M. Hurd, *Contemp. Phys.*, **23**, 483 (1982), with permission from Taylor & Francis, Bristol, PA.]

this type of itinerant paramagnetism is characterized by strong interactions between the electrons, but not sufficiently strong to produce spontaneous magnetic order—aligned moments may arise in limited regions, and are called *localized spin fluctuations*, or paramagnons.

8. *Superparamagnetism.* This is observed in small single-domain particles. In this type of magnetism the magnetic moments of the particles behave in a way analogous to a paramagnetic system, with total moment several orders of

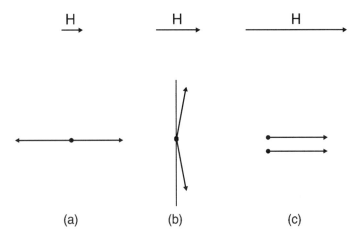

Figure 1.12 Schematic description of metamagnetism. The magnetic moments change from configurations *a* to *b* and finally *c*, aligning in parallel, as the external field is increased.

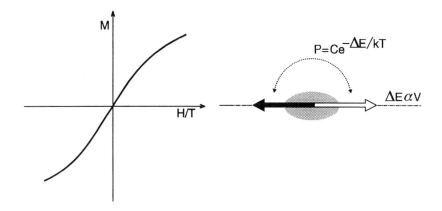

Figure 1.13 Curve of magnetization versus magnetic field of a superparamagnet, showing the absence of hysteresis; the probability of turning the magnetic moment is $C \times \exp(-\Delta E/kT)$, where ΔE is an activation energy. [Adapted from C. M. Hurd, *Contemp. Phys.*, **23**, 486 (1982), with permission from Taylor & Francis, Bristol, PA.]

magnitude larger than those of the individual atoms (therefore its name). The magnetic behavior is well described by the classical expression of Langevin (see Chapter 2); the curves of magnetization versus B/T are independent of temperature. The moments of each of these particles may point along different directions, defined by the crystal field (Fig. 1.13). Below a given temperature (called the *blocking temperature*), the changes in direction, which are due to thermal activation, occur in timescales longer than the observation time, causing the moments to appear frozen.

9. *Superferromagnetism.* A system of small particles that orders magnetically exhibits this type of magnetism (Mørup 1983).

10. *Canted Magnetism.* A type of magnetic order containing different and noncollinear magnetic moments

11. *Speromagnetism.* The ordered magnetic materials can also be speromagnets, in which the magnetic moments point along random directions (Fig. 1.14).

12. *Asperomagnetism.* In this type of magnetism the magnetic moments are distributed around a preferred direction.

13. *Sperimagnetism.* Magnetism in which there is more than one magnetic species, with the moments of at least one of the species pointing along a defined direction.

14. *Spin Glasses and Mictomagnetism.* Types of magnetism in which the magnetic moments "freeze" below a certain temperature T_f, pointing in random directions (as in a speromagnet). In spin glasses there is no correlation between neighbor moments. In mictomagnetic (micto = mixed) substances (or cluster glasses), there is short range correlation among the moments, with regions of resulting nonzero magnetization. One example of spin glass is given by a dilute solution of Mn in a Cu matrix, and the magnetic behavior is schematized

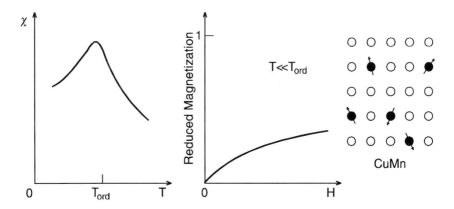

Figure 1.14 Dependence of the susceptibility χ on the temperature of a speromagnet, exhibiting a characteristic cusp, and the dependence of the reduced magnetization on magnetic field H; schematic representation and examples of a material that presents speromagnetism—many spin glasses order in this way at low temperature. [Adapted from C. M. Hurd, *Contemp. Phys.*, **23**, 487 (1982), with permission from Taylor & Francis, Bristol, PA.]

in Fig. 1.14. Spin glasses are formed when there is either spatial randomness, or randomness in the interaction between neighbors, combined with "frustration", which means the impossibility of satisfying the type of coupling "demanded" by each neighbor (e.g., parallel, or antiparallel) (e.g., Mydosh 1996).

There are several ways of dividing materials into classes, according to their magnetic properties: (1) in relation to the magnitude and orientation of the permanent magnetic moments—ferromagnets, ferrimagnets, and helimagnets (those in which the tip of the magnetization vector follows a helix); (2) according to the degree of mobility of the electrons responsible for the magnetism—localized and itinerant; and (3) according to the value of the coercive field (reversed magnetic field required to cancel the magnetization in the $M-H$ curve)—magnetically soft and hard (Fig. 1.15) (see Chapter 5).

This latter form of classification (3) is the most important in terms of the practical applications of magnetic materials. The hard magnetic materials (as NdFeB, e.g.) are employed in the fabrication of permanent magnets. The magnetically soft materials, like Permalloy, are used as magnetic shields, inductors, and transformer cores. The coercive field varies from about $1\,A\,m^{-1}$ in Permalloy, to $10^6\,A\,m^{-1}$ in NdFeB (see Tables 5.V and 5.VII). Materials used as magnetic recording media (e.g., Cr_2O_3 and Fe_2O_3) present intermediate values of magnetic hardness (see Table 5.VI).

The materials that present itinerant magnetism can still be subdivided according to the degree of filling of the conduction band, into strong itinerant magnets (only one magnetic subband partially filled, e.g., Ni) or weak (both sub-bands partially filled, e.g., Fe) (see Section 4.3 and Fig. 4.9).

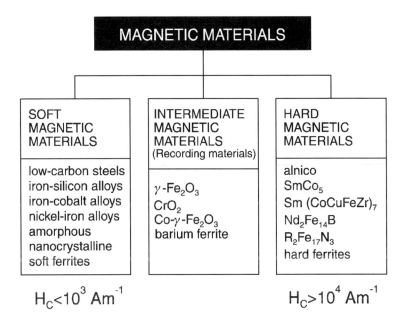

Figure 1.15 Types of magnetic materials with technological applications and some examples, with range of coercivities. The examples of intermediate magnetic materials are taken from materials used in magnetic recording (see Chapter 5).

Other systems present diamagnetic behavior at low temperature and behave as paramagnets at high temperature; they are said to present the induced paramagnetism of Van Vleck.

In order to classify a sample into one of these categories, the first properties usually studied are the shape of the magnetization curve, the dependence of the magnetization on an external applied magnetic field, the variation of the specific heat with temperature, and so on. In the last decades these studies have been supplemented with analysis employing local techniques, specifically, techniques using as probes atomic nuclei, muons and positrons. These experimental techniques include Mössbauer spectroscopy, nuclear magnetic resonance (NMR), angular correlations, muon spin rotation, and positron annihilation. The probes measure the magnetic and electrostatic interactions with the nuclei and with the electrons.

The technique of neutron diffraction allows the study of the spatial distribution, direction, and magnitude of the magnetic moments in condensed matter. In the inelastic scattering of neutrons, magnetic excitations (magnons) are created and annihilated, and using this technique, the spectrum of these excitations may be obtained.

The presence of magnetic order affects several properties of the materials; including electrical transport properties, elastic properties, and optical properties, among others.

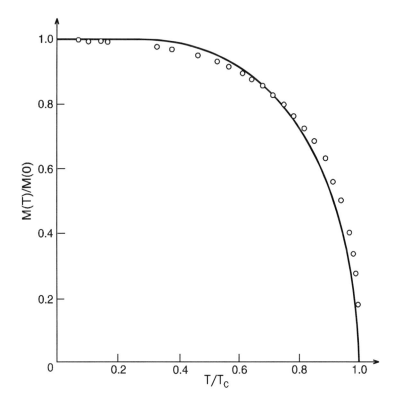

Figure 1.16 Temperature dependence of the spontaneous magnetization of metallic nickel (circles). The continuous curve is a Brillouin function for $J = \frac{1}{2}$ (see Sections 2.3 and 2.6) [Weiss and Forrer (1926)].

1.4 MAGNETIC PROPERTIES OF SOME MAGNETIC MATERIALS

The magnetic properties of matter are affected by variables that can be controlled experimentally, such as the temperature, pressure, and concentration of the different phases; the presence of defects; the intensity of applied magnetic fields; the degree of crystallinity; and the dimensionality. As an example of the temperature dependence of a magnetic property, we may show the variation of the magnetization of Ni with temperature (Fig. 1.16).

Pressure affects the temperature of magnetic order of magnetic materials, as can be seen, for example, in iron (Fig. 1.17). The variation of the magnetic moment per atom in alloys formed with the $3d$ transition elements exemplifies the importance of the concentration of components for the magnetic properties; the curve of magnetic moment versus number of conduction electrons per atom (a quantity related to the concentration) is known as the *Slater–Pauling* curve (Fig. 1.18).

The dimensionality of a sample, namely, its shape either as a solid body, a thin film, or a linear chain of atoms, affects its magnetic properties. This can be seen

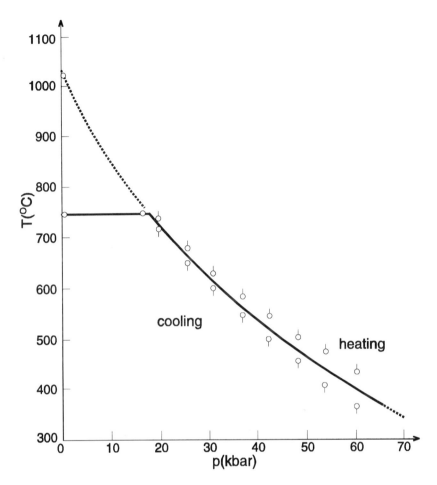

Figure 1.17 Variation of the magnetic ordering temperature [Curie temperature (T_C)] of iron, as a function of pressure. [Reprinted from Landolt-Börnstein, *Magnetic Properties of 3d Elements, New Series III/19a*, Springer-Verlag, New York, 1986, p. 39, with permission.]

from the variation of magnetic ordering temperature versus thickness in ultrathin metallic films (Fig. 1.19).

1.5 PERMANENT MAGNETS

One very important class of magnetic materials is formed of the materials employed in the fabrication of permanent magnets. In antiquity, the only known permanent magnets were naturally occurring fragments of magnetite (Fe_3O_4), but nowadays there exists a wide variety of permanent magnet materials. The first artificial magnets were made of iron alloys, such as iron carbon.

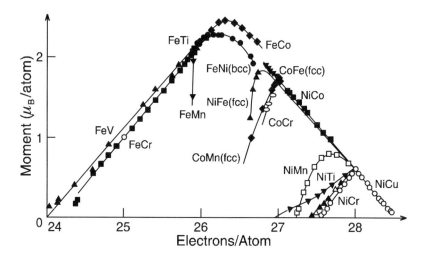

Figure 1.18 Dependence of the spontaneous magnetic moment of binary alloys of elements of the 3d series, as a function of electronic concentration, specifically of the number of 3d plus 4s electrons of the respective free atoms; this is known as the *Slater–Pauling* curve.

The utility of magnets derives from the possibility of maintaining a magnetic field in their vicinity, stable with time, and with no expense of energy. For economical and practical reasons, it is desirable to have magnets with the smallest possible dimensions that generate a given induction B. For a given

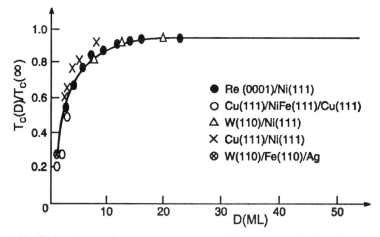

Figure 1.19 Ratios of magnetic ordering temperature (Curie temperature) of ultrathin metallic films to T_C of the corresponding bulk metals, as a function of the thickness, measured in numbers of atomic monolayers. [Reprinted from U. Gradman, in *Handbook of Magnetic Materials*, K. H. J. Buschow, Ed., North-Holland, Amsterdam, 1993, p. 36, with permission from Elsevier North-Holland.]

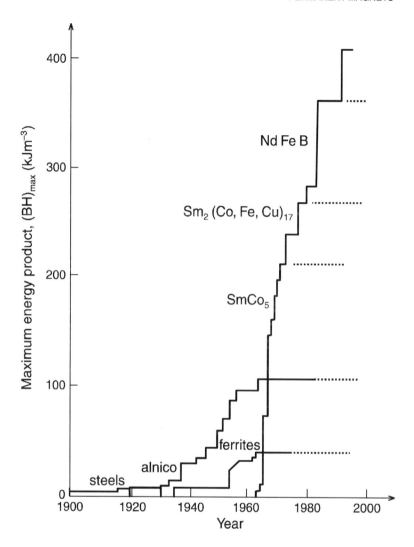

Figure 1.20 Evolution of the materials employed in the construction of permanent magnets; variation of the value of the energy product (*BH*) with time (see Chapter 5). [Reprinted from J. Evetts, Ed., *Concise Encyclopedia of Magnetic and Superconducting Materials*, Pergamon, London, 1992, p. xx, with permission from Elsevier Science.]

magnet size, the quantity that must be maximized is the energy product $(BH)_{max}$, the maximum product of the magnetic induction by the field H in the second quadrant of the $B-H$ curve (this will be discussed in more detail in Chapter 5) [see Eq. (1.17)]. Generally, the larger the energy product, the more appropriate is the material for use in permanent magnets.

Materials used in permanent magnets must in general possess: (1) high value of the magnetization M, (2) high uniaxial anisotropy, (3) high magnetic ordering temperature; these are usually intrinsic properties. Other necessary characteristics are a large value of the coercive force or of the coercivity, and a magnetization that is not affected by external magnetic fields. These properties depend on the microstructure, specifically, on the grain size, the presence of impurities, and other factors. Some of the most promising alloys for the use as permanent magnets (Fig. 1.20) are those associating rare earths—responsible for high anisotropy energies, to d transition metals—that give rise to elevated magnetic ordering temperatures.

GENERAL READING

Shadowitz, A., *The Electromagnetic Field*, McGraw-Hill, New York, 1975.
Wohlfarth, E. P., "Magnetism and Magnetic Materials: Historical Developments and Present Role in Industry and Technology, in Ferromagnetic Materials," in E. P. Wohlfarth, Ed., *Handbook of Magnetic Materials*, Vol. 3, North-Holland, Amsterdam, 1982, p 1.

REFERENCES

Cohen, E. R. and P. Giacomo, "Symbols, Units, Nomenclature and Fundamental Constants in Physics," *Physica A* **146**, 1 (1987).
Crooks, J. E., *J. Chem. Educ.* **56**, 301 (1979).
Evetts, J., Ed., *Concise Encyclopedia of Magnetic and Superconducting Materials*, Pergamon Press, London, 1992.
Gradman, U., in *Handbook of Magnetic Materials*, K. H. J. Buschow, Ed., North-Holland, Amsterdam, 1993, p. 1.
Grandjean, F. and G. J. Long, in *Supermagnets, Hard Magnetic Materials*, Kluwer Academic, Amsterdam, 1990, p. 27.
Grant, I. S. and W. R. Phillips, *Electromagnetism*, 2nd ed., Wiley, Chichester, 1990.
Hurd, C. M., *Contemp. Phys.* **23**, 469 (1982).
Landolt-Börnstein, *Tables of Magnetic Properties of 3d Elements*, Landolt-Börnstein, Springer-Verlag, Berlin, 1962.
Landolt-Börnstein, *Tables of Magnetic Properties of 3d Elements*, Landolt-Börnstein, New Series III/19a, Springer-Verlag, Berlin, 1986.
Mørup, S., M. B. Madsen, J. Franck, J. Villadsen, and C. J. W. Koch, *J. Mag. Mag. Mat.* **40**, 163 (1983).
Mydosh, J. A., *J. Mag. Mag. Mat.* **157/158**, 606 (1996).
Shadowitz, A., *The Electromagnetic Field*, McGraw-Hill, New York, 1975.
Taylor, B. N., *Guide for the Use of the International System of Units (SI)*. NIST Special Publication 811, 1995 edition, Gaithersburg, MD, 1995.
Weiss, P. and R. Forrer, *Ann. Phys.* **5**, 153 (1926).

EXERCISES

1.1 *Coercive Force of a Particle.* Consider a particle of a uniaxial ferromagnetic single domain. Let **H** be an applied field and $U_K = K \sin^2(\theta)$ is the anisotropy energy, with θ the angle between the direction of the applied field and the magnetization **M**. Write the total energy of the particle and show that the reverse field along the magnetization axis required to invert **M** is given by $H = 2K/M_s$.

1.2 *Magnetic Moment of a Sample.* The magnetic moment of a sample with magnetization **M** may be written as

$$\boldsymbol{\mu} = \int \mathbf{M}(\mathbf{r}) dv$$

We may define two quantities related to **M**, the *magnetic pole density*, $\rho_m = -\nabla \cdot \mathbf{M}(\mathbf{r})$ and the superficial density of magnetic poles, $\sigma_m = \mathbf{M}(\mathbf{r}) \cdot \mathbf{n}$, where **n** is the unit vector normal to the surface of the sample. From the vector expression

$$\nabla \cdot (f\mathbf{A}) = (\nabla f) \cdot \mathbf{A} + f\nabla \cdot \mathbf{A}$$

where f is a scalar function and **A** a vector function, show that

$$\boldsymbol{\mu} = \int_V \mathbf{r}\rho_m \, dv + \oint_S \mathbf{r}\sigma_m \, da$$

where S is the limiting surface of the sample of volume V.

1.3 *Energy of a Magnetized Sphere.* Compute the magnetic self-energy of a sphere with saturation magnetization M_s and radius R. Use $D = \frac{1}{3}$ for the demagnetizing factor of the sphere (SI).

1.4 *Magnetic Field inside a Sphere.* Compute the values of **H** and **B** inside a homogeneous sphere of permeability $\mu > 0$ placed in a uniform magnetic field of intensity $\mathbf{H} = H\mathbf{i}$ (Figs 1.1 and 1.2). Do **H** and **B** inside the sphere change in the same way, relative to their values in vacuum?

2
ATOMIC MAGNETIC MOMENTS

The magnetic moments carried by the atoms are related to the angular momenta of their unpaired electrons. There are two contributions to the electronic angular momenta: an orbital contribution and a spin contribution.

The orbital term of the atomic magnetic moment can be derived by making an analogy of the electronic orbit with an electrical circuit. An electric current I flowing through a circular coil of area A has an associated magnetic dipole moment $\mu = IA$ [from Eq. (1.2)], where $\mathbf{A} = A\hat{\mathbf{n}}$ and $\hat{\mathbf{n}}$ is the unit normal to the plane of the orbit. We may thus obtain the magnitude μ of the magnetic moment associated with the motion of one electron of charge $-e$, moving in a circular orbit with angular frequency ω:

$$\mu = I\pi r^2 = \frac{-e\omega\pi r^2}{2\pi} = \frac{-e\omega r^2}{2} \tag{2.1}$$

where r is the radius of the orbit.

The magnitude of the orbital angular momentum $\mathbf{J} = \mathbf{r} \times m_e\mathbf{v}$ of this electron is

$$J = m_e\omega r^2 \tag{2.2}$$

where m_e is the electron mass, and therefore the magnetic orbital moment of the electron is

$$\boldsymbol{\mu} = \frac{-e}{2m_e}\mathbf{J} \tag{2.3}$$

Since the component of the orbital angular momentum of the electron in a given direction (let us choose the z direction) is quantized, taking values $\hbar, 2\hbar, 3\hbar$, and so on, the smallest value of J_z is $J_z = \hbar$ ($\hbar$ is very small, equal to 1.0546×10^{-34} J s). The corresponding magnetic moment is $\mu_z = -e\hbar/2m_e$.

The quantity $e\hbar/2m_e$ is called the *Bohr magneton* (μ_B), and its value, in SI units, is 9.27×10^{-24} J T^{-1} (or 9.27×10^{-21} erg G^{-1} in the CGS system):

$$\mu_B = \frac{e\hbar}{2m_e} \tag{2.4}$$

We will, from now on, express J in units of $\hbar$; Eq. (2.3) is then written $\mu = (-e/2m_e)\hbar\mathbf{J}$.

Besides this orbital momentum, the electron has also an intrinsic angular momentum, or spin. The spin has an associated magnetic moment, in the same way as the orbital momentum, but with a proportionality constant twice as large. We may write

$$\mu = \gamma\hbar\mathbf{J} \begin{cases} \gamma = \dfrac{-e}{2m_e} & \text{for pure orbital angular momentum} \\ \gamma = \dfrac{-e}{m_e} & \text{for pure spin angular momentum} \end{cases} \tag{2.5}$$

where γ is the gyromagnetic ratio (or magnetogyric ratio), measured in s^{-1} T^{-1}; for electron orbital motion, $\gamma = -8.7941 \times 10^{10}$ s^{-1} T^{-1}.

This can also be written

$$\mu = -g\mu_B\mathbf{J} \begin{cases} g = 1 & \text{pure orbital momentum} \\ g = 2 & \text{pure spin momentum} \end{cases} \tag{2.6}$$

in terms of g, the electron g-factor (a more accurate value of g for spin angular momentum is $g = 2.0023$, but it is normally taken as equal to 2).

2.1 DIAMAGNETISM

The application of an external magnetic field to an electrical circuit induces in it a current that is opposed to the original current, an effect equivalent to Lenz's law; this is also observed for an electron that moves in an atomic orbit. The induced current decreases the orbital magnetic moment; since this moment decreases with increasing field, the differential susceptibility, or magnetic response (dM/dH), is negative. This phenomenon is called *diamagnetism*.

We can derive an approximate expression for the diamagnetic volume susceptibility by writing the induced magnetic moment of one electron (from Eq. (2.1)):

$$\Delta\mu = \frac{-e\Delta\omega\rho^2}{2} \tag{2.7}$$

where $\Delta\omega$ is the angular frequency of precession induced by the external field, and ρ is the mean square radius of the projection of the electron orbit onto the plane perpendicular to the field $B\mathbf{k}$:

$$\overline{\rho^2} = \overline{x^2} + \overline{y^2} \tag{2.8}$$

Assuming that with the applied field the radius of the electron orbit remains the same, the variation in the force acting on the electron is the Lorentz force $-e\omega\rho B$:

$$\Delta(m_e \omega^2 \rho) = m_e \Delta(\omega^2)\rho = -e\omega\rho \tag{2.9}$$

Assuming that the fractional change in frequency is small, the decrease in frequency is given by

$$\Delta\omega = \omega_L = \frac{eB}{m_e} \tag{2.10}$$

This frequency is the Larmor frequency; substituting ω_L into the expression of μ, we can obtain the volume diamagnetic susceptibility:

$$\chi = \frac{\partial M}{\partial H} = \mu_0 \frac{\partial M}{\partial B} = \mu_0 \frac{\partial(n\Delta\mu)}{\partial B} \tag{2.11}$$

where μ_0 is the vacuum magnetic permeability and n is the number of atoms per unit volume. Since $\overline{r^2} = \overline{x^2} + \overline{y^2} + \overline{z^2}$, and for a spherically symmetric charge distribution $\overline{x^2} = \overline{y^2} = \overline{z^2}$, it results that $\overline{\rho^2} = \frac{2}{3}\overline{r^2}$.

Therefore, the atomic diamagnetic susceptibility is obtained summing over Z electrons, where Z is the atomic number:

$$\chi = -\mu_0 n \frac{e^2}{6m_e} \sum_i^Z \overline{r_i^2} \tag{2.12}$$

Using average values of $\overline{r^2} \approx 10^{-20}$ m^2, $n \approx 5 \times 10^{28}$ m^{-3}, and taking an atomic number $Z = 10$, we obtain for the diamagnetic volume susceptibility $\chi \approx -10^{-5}$. Expression (2.12) is sometimes referred to as the *Larmor diamagnetic susceptibility*.

There is another contribution to diamagnetism, observed in the metals, that is associated with the orbit of the conduction electrons under the action of external magnetic fields – it is the Landau diamagnetic susceptibility, and its magnitude is one-third that of the Pauli susceptibility (defined in Chapter 4).

2.2 ELECTRONS IN ATOMS

The energy levels E_n of an atom are obtained from the Schrödinger equation:

$$\mathcal{H}\Psi = E_n \Psi \tag{2.13}$$

where $\mathcal{H}$ is the hamiltonian of the one particle system, built from the classical expression

$$\mathcal{H} = \frac{p^2}{2m_e} + V \tag{2.14}$$

substituting the components of the momentum p by $-i\hbar\partial/\partial x_i$ to obtain the corresponding operators; V is the potential in which the electron moves. The Schrödinger equation takes the form

$$\left[\frac{\hbar^2}{2m_e}\left(\frac{\partial^2}{\partial x^2} + \frac{\partial^2}{\partial y^2} + \frac{\partial^2}{\partial z^2}\right) + V\right]\Psi = E\Psi \tag{2.15}$$

The potential energy of an electron near a nucleus of charge e (case of the hydrogen atom) is given by

$$V = \frac{-e^2}{(x^2 + y^2 + z^2)^{1/2}} \tag{2.16}$$

If m_e is the electron mass and m_p the proton mass, the Schrödinger equation is written, in terms of the reduced mass $m_r = m_e m_p/(m_e + m_p)$, as follows:

$$\frac{\hbar^2}{2m_r}\nabla^2\Psi + (E - V)\Psi = 0 \tag{2.17}$$

Changing to spherical coordinates, adequate to the symmetry of the problem, and assuming that the wavefunction is the product of a radial function and two angular functions

$$\Psi = R(r)\Theta(\theta)\Phi(\phi) \tag{2.18}$$

we obtain three independent differential equations, with solutions of the angular parts that are the Legendre polynomials $\mathcal{P}_l^m(\cos\theta)$ and $\Phi = C\exp(im\phi)$, with l and m integers.

The solution of the radial part gives an exponential decay $\exp(-zr/a)$ modulated by a function that has zeros (for $l \neq 0$). From this it also results the expression for the energy

$$E_n = \frac{-e^2 m_r^4}{2\hbar^2}\frac{1}{n^2} \tag{2.19}$$

with n an integer. A set of electrons having the same n constitute a shell.
In conclusion, the solution involves the following numbers:

1. *Principal Quantum Number n.* This number essentially determines the energy of the shell. The shells are traditionally denoted K, L, M, N, and so on for $n = 1, 2, 3, 4 \ldots$.
2. *Orbital Quantum Number l.* Determines the orbital angular momentum of the electron, whose magnitude is given by

$$\sqrt{l(l+1)}\hbar \tag{2.20}$$

   The number l is an integer and may have values $0, 1, 2, \ldots, n-1$; the electrons are then called s, p, d, f, and so on. In the relativistic description of the atom, the energy of the electrons is also dependent on l.
3. *Magnetic Quantum Number m_l.* This gives the component of the orbital momentum along a specific direction. The number m_l may be equal to $l, l-1, l-2, \ldots, 0, \ldots, -(l-1), -l$; that is, it takes $2l+1$ values. In the spatial representation of the atomic quantities, known as the *vector model*, the orbital momentum can point only along certain directions (Fig. 2.1) and its projections are given by m_l; this is called *space quantization*.
4. *Spin Quantum Number m_s.* Dirac's theory introduces another number: the spin quantum number, which may take values $\frac{1}{2}$ and $-\frac{1}{2}$. Therefore, the state of the electron is characterized by four quantum numbers: n, l, m_l, and m_s.

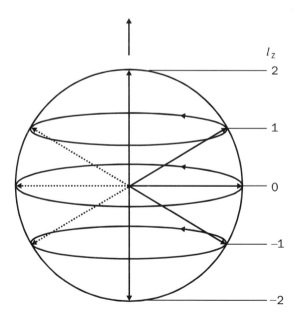

Figure 2.1 Orbital angular momentum of a 3d electron ($l = 2$), and the value of its projections along a direction z, showing space quantization.

The electrons have magnetic moments associated with their components of angular momentum. Thus, corresponding to their orbital momentum, there is a magnetic moment [from Eqs. (2.5) and (2.20)] of magnitude:

$$|\mu^l| = \frac{e}{2m_e}\sqrt{l(l+1)}\hbar \tag{2.21}$$

and the component of the magnetic orbital moment μ^l in a direction defined by an applied magnetic field (assumed to be parallel to the z axis) is

$$|\mu_z^l| = \frac{e}{2m_e}m_l\hbar \tag{2.22}$$

The magnetic moment corresponding to the spin angular momentum is

$$|\mu_z^s| = 2\frac{e}{2m_e}m_s\hbar \tag{2.23}$$

which differs from the value of the orbital magnetic moment only by the factor 2.

The orbital and spin momenta of an electron interact with each other, as well as with the momenta of different electrons of the same atom. The interactions can be described through the scalar products of the angular momentum operators (see Chapter 3). Considering the interactions between two electrons labeled i and j of the same atom described by

$$a_{ij}\boldsymbol{\ell}_i \cdot \mathbf{s}_j$$
$$b_{ij}\boldsymbol{\ell}_i \cdot \boldsymbol{\ell}_j$$
$$c_{ij}\mathbf{s}_i \cdot \mathbf{s}_j$$

a hierarchy is observed among the intensities of these interactions: the coupling parameters obey the relations

$$b_{ij} > a_{ii}, a_{ij}$$
$$c_{ij} > a_{ii}, a_{ij}$$

This leads to the coupling of spins and orbital momenta of different electrons, forming the total spin angular momentum (S) and the total orbital momentum (L). This is the most common angular momentum coupling, called LS coupling, or *Russell–Saunders coupling*. The spins couple to form the total spin **S** and the orbital momenta couple to form **L**:

$$\mathbf{s}_i \rightarrow \mathbf{S}$$
$$\boldsymbol{\ell}_i \rightarrow \mathbf{L}$$

In heavy atoms there is a strong coupling between the momenta l_i and s_i of each electron, leading to the total angular momentum per electron j_i. This type of coupling is called jj coupling. In LS coupling, a pair of values L, S characterize a term, denoted by ^{2S+1}X, where $X = S, P, D, \ldots$, depending on the value of L.

The total spin and orbital momenta interact through the atomic spin-orbit interaction; this is described by the equation

$$W_J = \lambda \mathbf{L} \cdot \mathbf{S} \tag{2.24}$$

$\mathbf{L}$ and $\mathbf{S}$ combine to form the total angular momentum $\mathbf{J}$

$$\mathbf{L} + \mathbf{S} \rightarrow \mathbf{J} \tag{2.25}$$

and the corresponding quantum number is J. In LS coupling an atomic level is characterized by a set L, S, and J. J may take the values

$$J = |L - S|, \; |L - S + 1| \cdots |L + S - 1|, \; |L + S| \tag{2.26}$$

and the levels defined by these values of J are called *multiplets*. The projection of $\mathbf{J}$ along an arbitrary direction is quantized, and the corresponding quantum number is M_J, which may take the values

$$M_J = J, J - 1, \cdots, -J + 1, -J \tag{2.27}$$

An atomic state is defined by a set of L, S, J, and M_J, or by L, S, J, M_S, and M_J. The maximum values of L and S are given by $\sum l_i$ and $\sum s_i$, but in each atom, the ground-state values of L and S follow empirical rules known as *Hund's rules*:

1. The combination of s_i that results in the smallest energy, and therefore is the most stable configuration, is that for which the quantity $2S + 1$ is maximum.
2. When the first rule is satisfied, there are several possible values of L (for the same value of $2S + 1$); the most stable is the one that makes L maximum.

These values define the ground-state atomic level.

Examples of the quantum numbers for the electrons in two transition ions are given below:

1. $Co^{2+}(3d^7)$ ion: $n = 3, l = 2$

m_s	$\frac{1}{2}$	$\frac{1}{2}$	$\frac{1}{2}$	$\frac{1}{2}$	$\frac{1}{2}$	$-\frac{1}{2}$	$-\frac{1}{2}$	$\rightarrow S = \frac{3}{2}$
m_l	2	1	0	-1	-2	2	1	$\rightarrow L = 3$

2. $Gd^{3+}(4f^7)$ ion: $n = 4, l = 3$

m_s	$\frac{1}{2}$	$\frac{1}{2}$	$\frac{1}{2}$	$\frac{1}{2}$	$\frac{1}{2}$	$\frac{1}{2}$	$\frac{1}{2}$	$\rightarrow S = 7/2$
m_l	3	2	1	0	-1	-2	-3	$\rightarrow L = 0$

The origin of Hund's rules is Pauli's principle, which forbids two electrons to have the same quantum numbers. The electrons of parallel spin avoid each other, and this reduces the Coulomb repulsion between them. This makes the spins couple in parallel, leading to a maximum value of S in the ground state.

When the spin–orbit coupling constant λ [Eq. (2.24)] is positive (which is the case when the shell is less than half full), the minimum energy configuration is obtained for L and S antiparallel, that is, for $J = L - S$. For a shell more than half full, the opposite is true and $J = L + S$.

The $3d$ subshell of Co^{2+} contains 7 of the 10 electrons that it may accommodate; the subshell is more than half full, and therefore the ground state is characterized by the quantum number $J = L + S = \frac{9}{2}$ (Fig. 2.2).

The coupling between the angular momenta $\mathbf{L}$ and $\mathbf{S}$ and between the associated magnetic moments μ_L and μ_S is represented in Fig. 2.3. The orbital and spin magnetic moments

$$\mu_L = -\mu_B \mathbf{L} \tag{2.28}$$

$$\mu_S = -2\mu_B \mathbf{S} \tag{2.29}$$

add vectorially to form the total magnetic moment μ. The total magnetic moment μ has a component μ_J along $\mathbf{J}$, and a component μ' that precesses around $\mathbf{J}$ and is not effective (Fig. 2.3):

$$\mu = \mu_J + \mu' \tag{2.30}$$

The magnitude of the part parallel to $\mathbf{J}$ may be obtained from Fig. 2.3:

$$|\mu_J| = -\mu_B \frac{\mathbf{L} \cdot \mathbf{J}}{|\mathbf{J}|} - 2\mu_B \frac{\mathbf{S} \cdot \mathbf{J}}{|\mathbf{J}|} \tag{2.31}$$

From $\mathbf{J} = \mathbf{L} + \mathbf{S}$ it follows

$$\mathbf{L} \cdot \mathbf{J} = \tfrac{1}{2}(J^2 + L^2 - S^2) \tag{2.32}$$

$$\mathbf{S} \cdot \mathbf{J} = \tfrac{1}{2}(J^2 + S^2 - L^2) \tag{2.33}$$

and therefore, substituting

$$|\mu_J| = -\frac{\mu_B}{2} \frac{3J^2 + S^2 - L^2}{|\mathbf{J}|} = -\frac{\mu_B}{2} \frac{3J(J+1) + S(S+1) - L(L+1)}{\sqrt{J(J+1)}} \tag{2.34}$$

writing

$$\mu_J = -g\mu_B \mathbf{J} \tag{2.35}$$

and thus

$$|\mu_J| = g\mu_B\sqrt{J(J+1)} \qquad (2.36)$$

it follows that g is given by

$$g = \frac{3J(J+1) + S(S+1) - L(L+1)}{2J(J+1)} \qquad (2.37)$$

or

$$g = 1 + \frac{J(J+1) + S(S+1) - L(L+1)}{2J(J+1)} \qquad (2.38)$$

This quantity is known as the Landé g-factor.

Although we have treated so far **L**, **S**, **J** and μ as vectors, they are in fact quantum mechanical operators. Therefore, for example, the measured total angular momentum is the expectation value $\langle \mathbf{J} \rangle = \langle J, M|\mathbf{J}|J, M\rangle = \int \phi_{JM} \mathbf{J} \phi_{JM}^* \, dv$. The expectation value of the magnetic moment $\langle \mu \rangle$ is parallel to $\langle \mathbf{J} \rangle$, as indicated above, since $\langle \mu \rangle = \langle \mu_J \rangle$. The expectation values of μ and **J** are also connected through the g-factor:

$$\langle \mu \rangle = -g\mu_B \langle \mathbf{J} \rangle \qquad (2.39)$$

The multiplets of an atom under the effect of a magnetic field are split into sublevels characterized by the projection M_J of the total angular momentum in the z direction. The magnetic energy is given by[1]

$$E_{M_J} = -\mu_J \cdot \mathbf{B} = g\mu_B M_J B \qquad (2.40)$$

In the presence of a magnetic induction of 1 tesla (T) (10,000 G) this energy is of the order of 10^{-23} J $\approx 10^{-4}$ eV ≈ 1 cm^{-1}. The thermal energy kT at room temperature is of the order of $\frac{1}{40} = 0.025$ eV, or 200 cm^{-1}.

It was shown in Fig. 2.2 how the atoms in the sublevels characterized by different M_J, previously degenerate, possess different energies in the magnetic field.

The probability of occupation of the sublevels, or the proportion of atoms of momentum M_J, depends on the temperature, and is given by a Boltzmann distribution:

$$P(M_J) = \frac{\exp(-E_{M_J}/kT)}{\sum_{M_J} \exp(-E_{M_J}/kT)} \qquad (2.41)$$

---

[1] The conditions of validity of this expression are discussed in Ashcroft and Mermin (1976).

36 ATOMIC MAGNETIC MOMENTS

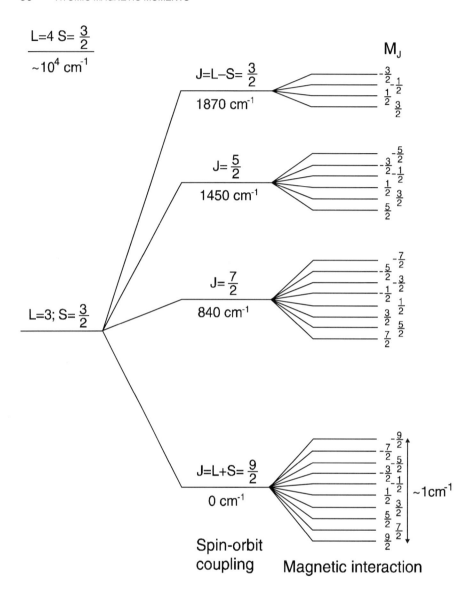

Figure 2.2 Energy levels of the free ion Co^{2+} (electron configuration $3d^7$). [Adapted from J. Crangle, *Solid State Magnetism*, Van Nostrand Rheinhold, New York, 1991, p. 21.]

We can in general consider the population of the sublevels M_J corresponding to the lowest J multiplet, since the next multiplet is usually too high in energy compared to the thermal energy kT; every atom is in the ground state characterized by the quantum number of the total angular momentum J. In the example of the Co^{2+} ion (Fig. 2.2) $J = L + S$ is the lowest multiplet, and the

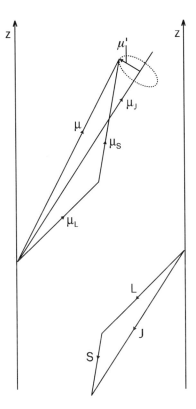

Figure 2.3 Vector representation of the angular momenta **L**, **S**, and **J**, and of their corresponding dipole magnetic moments, μ_L μ_S, and μ_J.

next one, $J = L + S - 1$, is at 1.7×10^{-20} J $= 1.0 \times 10^{-1}$ eV $= 840$ cm^{-1}, corresponding to a separation of 1200 K. In this case, only the lowest multiplet will be populated at room temperature.

The periodic table of elements (Table 2.I) brings to the light the periodicity of the physical and chemical properties of the elements as a function of the atomic number Z. This regularity arises from the way the electronic configurations vary with the atomic number, that is, the form in which the electronic states defined by the quantum numbers n and l are filled. Figure 2.4 shows a graph of the variation of the atomic radius as a function of the atomic number Z; this radius varies in a periodic way, showing minima at values of Z corresponding to the noble gases.

The electronic structure of each element, with very few exceptions, is identical to the structure of the preceding element, with the addition of one electron.

The energy of the electron, as a rule, increases with the quantum number n, but it is also dependent on the orbital quantum number l; the ground states are those states of minimum energy. The electrons with small l stay a longer time near the nucleus, and therefore their energy is lower. In this way, the filling order of the subshells [defined by the pair (n, l)], as Z increases, beginning with hydrogen,

Table 2.1 Periodic table of the elements*

Z	Element	Electronic Configuration	Ionization Potential (eV)	Z	Element	Electronic Configuration	Ionization Potential (eV)
1	H	$(1s)^1$	13.598	52	Te	$(Kr)(5s)^2(4d)^{10}(5p)^4$	9.009
2	He	$(1s)^2$	24.587	53	I	$(Kr)(5s)^2(4d)^{10}(5p)^5$	10.451
3	Li	$(He)(2s)^1$	5.392	54	Xe	$(Kr)(5s)^2(4d)^{10}(5p)^6$	12.130
4	Be	$(He)(2s)^2$	9.322	55	Cs	$(Xe)(6s)^1$	3.894
5	B	$(He)(2s)^2(2p)^1$	8.298	56	Ba	$(Xe)(6s)^2$	5.212
6	C	$(He)(2s)^2(2p)^2$	11.260	57	La	$(Xe)(6s)^2(5d)^1$	5.577
7	N	$(He)(2s)^2(2p)^3$	14.534	58	Ce	$(Xe)(6s)^2(4f)^1(5d)^1$	5.47
8	O	$(He)(2s)^2(2p)^4$	13.618	59	Pr	$(Xe)(6s)^2(4f)^3$	5.42
9	F	$(He)(2s)^2(2p)^5$	17.422	60	Nd	$(Xe)(6s)^2(4f)^4$	5.49
10	Ne	$(He)(2s)^2(2p)^6$	21.564	61	Pm	$(Xe)(6s)^2(4f)^5$	5.55
11	Na	$(Ne)(3s)^1$	5.139	62	Sm	$(Xe)(6s)^2(4f)^6$	5.63
12	Mg	$(Ne)(3s)^2$	7.646	63	Eu	$(Xe)(6s)^2(4f)^7$	5.67
13	Al	$(Ne)(3s)^2(3p)^1$	5.986	64	Gd	$(Xe)(6s)^2(4f)^7(5d)^1$	6.14
14	Si	$(Ne)(3s)^2(3p)^2$	8.151	65	Tb	$(Xe)(6s)^2(4f)^9$	5.85
15	P	$(Ne)(3s)^2(3p)^3$	10.486	66	Dy	$(Xe)(6s)^2(4f)^{10}$	5.93
16	S	$(Ne)(3s)^2(3p)^4$	10.360	67	Ho	$(Xe)(6s)^2(4f)^{11}$	6.02
17	Cl	$(Ne)(3s)^2(3p)^5$	12.967	68	Er	$(Xe)(6s)^2(4f)^{12}$	6.10
18	Ar	$(Ne)(3s)^2(3p)^6$	15.759	69	Tm	$(Xe)(6s)^2(4f)^{13}$	6.18
19	K	$(Ar)(4s)^1$	4.341	70	Yb	$(Xe)(6s)^2(4f)^{14}$	6.254
20	Ca	$(Ar)(4s)^2$	6.113	71	Lu	$(Xe)(6s)^2(4f)^{14}(5d)^1$	5.426
21	Sc	$(Ar)(4s)^2(3d)^1$	6.54	72	Hf	$(Xe)(6s)^2(4f)^{14}(5d)^2$	7.0
22	Ti	$(Ar)(4s)^2(3d)^2$	6.82	73	Ta	$(Xe)(6s)^2(4f)^{14}(5d)^3$	7.89
23	V	$(Ar)(4s)^2(3d)^3$	6.74	74	W	$(Xe)(6s)^2(4f)^{14}(5d)^4$	7.98
24	Cr	$(Ar)(4s)^1(3d)^5$	6.766	75	Re	$(Xe)(6s)^2(4f)^{14}(5d)^5$	7.88
25	Mn	$(Ar)(4s)^2(3d)^5$	7.435	76	Os	$(Xe)(6s)^2(4f)^{14}(5d)^6$	8.7

Z	Element	Configuration	Density	Value
26	Fe	$(Ar)(4s)^2(3d)^6$	7.870	
27	Co	$(Ar)(4s)^2(3d)^7$	7.86	
28	Ni	$(Ar)(4s)^2(3d)^8$	7.635	
29	Cu	$(Ar)(4s)^1(3d)^{10}$	7.726	
30	Zn	$(Ar)(4s)^2(3d)^{10}$	9.394	
31	Ga	$(Ar)(4s)^2(3d)^{10}(4p)^1$	5.999	
32	Ge	$(Ar)(4s)^2(3d)^{10}(4p)^2$	7.899	
33	As	$(Ar)(4s)^2(3d)^{10}(4p)^3$	9.81	
34	Se	$(Ar)(4s)^2(3d)^{10}(4p)^4$	9.752	
35	Br	$(Ar)(4s)^2(3d)^{10}(4p)^5$	11.814	
36	Kr	$(Ar)(4s)^2(3d)^{10}(4p)^6$	13.999	
37	Rb	$(Kr)(5s)^1$	4.177	
38	Sr	$(Kr)(5s)^2$	5.695	
39	Y	$(Kr)(5s)^2(4d)^1$	6.38	
40	Zr	$(Kr)(5s)^2(4d)^2$	6.84	
41	Nb	$(Kr)(5s)^1(4d)^4$	6.88	
42	Mo	$(Kr)(5s)^1(4d)^5$	7.099	
43	Tc	$(Kr)(5s)^2(4d)^5$	7.28	
44	Ru	$(Kr)(5s)^1(4d)^7$	7.37	
45	Rh	$(Kr)(5s)^1(4d)^8$	7.46	
46	Pd	$(Kr)(4d)^{10}$	8.34	
47	Ag	$(Kr)(5s)^1(4d)^{10}$	7.576	
48	Cd	$(Kr)(5s)^2(4d)^{10}$	8.993	
49	In	$(Kr)(5s)^2(4d)^{10}(5p)^1$	5.786	
50	Sn	$(Kr)(5s)^2(4d)^{10}(5p)^2$	7.344	
51	Sb	$(Kr)(5s)^2(4d)^{10}(5p)^3$	8.641	
77	Ir	$(Xe)(6s)^2(4f)^{14}(5d)^7$		9.1
78	Pt	$(Xe)(6s)^1(4f)^{14}(5d)^9$		9.0
79	Au	$(Xe)(6s)^1(4f)^{14}(5d)^{10}$		9.225
80	Hg	$(Xe)(6s)^2(4f)^{14}(5d)^{10}$		10.437
81	Tl	$(Xe)(6s)^2(4f)^{14}(5d)^{10}(6p)^1$		6.108
82	Pb	$(Xe)(6s)^2(4f)^{14}(5d)^{10}(6p)^2$		7.416
83	Bi	$(Xe)(6s)^2(4f)^{14}(5d)^{10}(6p)^3$		7.289
84	Po	$(Xe)(6s)^2(4f)^{14}(5d)^{10}(6p)^4$		8.42
85	At	$(Xe)(6s)^2(4f)^{14}(5d)^{10}(6p)^5$		—
86	Rn	$(Xe)(6s)^2(4f)^{14}(5d)^{10}(6p)^6$		10.748
87	Fr	$(Rn)(7s)^1$		—
88	Ra	$(Rn)(7s)^2$		5.279
89	Ac	$(Rn)(7s)^2(6d)^1$		6.9
90	Th	$(Rn)(7s)^2(6d)^2$		—
91	Pa	$(Rn)(7s)^2(5f)^2(6d)^1$		—
92	U	$(Rn)(7s)^2(5f)^3(6d)^1$		—
93	Np	$(Rn)(7s)^2(5f)^4(6d)^1$		—
94	Pu	$(Rn)(7s)^2(5f)^6$		5.8
95	Am	$(Rn)(7s)^2(5f)^7$		6.0
96	Cm	$(Rn)(7s)^2(5f)^7(6d)^1$		—
97	Bk	$(Rn)(7s)^2(5f)^8(6d)^1$		—
98	Cf	$(Rn)(7s)^2(5f)^9(6d)^1$		—
99	Es	$(Rn)(7s)^2(5f)^{11}$		—
100	Fm	$(Rn)(7s)^2(5f)^{12}$		—
101	Md	$(Rn)(7s)^2(5f)^{13}$		—
102	No	$(Rn)(7s)^2(5f)^{14}$		—
103	Lr	$(Rn)(7s)^2(5f)^{14}(6d)^1$		—

Source: Adapted from R. A. Meyers, Ed. *Encyclopedia of Physical Science and Technology*, Vol. 2, Academic Press, Orlando, FL (1987). The electronic configuration for Bk and Cf taken from Landolt-Börnstein, *Magnetic Properties of Metals, New Series III/19f*, Springer-Verlag, New York (1991).

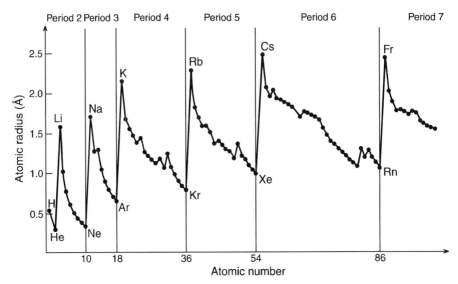

Figure 2.4 Variation of the atomic radius r of the elements versus atomic number Z, showing the periodicity of $r(Z)$. Note the minima at the radii of the noble gases. [Reprinted from R. A. Meyers, Ed., *Encyclopedia of Physical Science and Technology*, Vol. 10, Academic Press, Orlando, FL, 1987, p. 265.]

is: 1s, 2s, 2p, 3s, 3p, 4s, 3d, 4p, 5s, 4d, 5p, 6s, 4f, 5d, and 6p. Therefore, the energy of an electron may be lower in the next $(n + 1)$ orbit, with an orbital number $l - 1$ lower than that it would have by entering the n orbit. The (n, l) subshell thus remains incomplete.

The elements that present these incomplete shells are called *transition elements* and belong to the groups: 3d (iron group), 4d (palladium group), 5d (platinum group), 4f (lanthanides), and 5f (actinides). As opposed to what occurs with the closed shells, where the sum of the projections of angular momenta m_l and m_s is zero, the incomplete shells have nonzero angular momentum, and as a consequence, nonzero magnetic moment. For this reason, the elements important for magnetism are the transition elements. An incomplete outer subshell (e.g., 4s), however, does not lead to magnetic effects, since the unpaired electron participates in the chemical bond.

The metallic elements of the Periodic Table may be classified as transition metals (already mentioned), as noble metals [those that have just filled the d subshell (copper, silver, and gold)], and as normal metals, like aluminum, that are formed by adding one electron to the outer shell (4s, 5s, etc).

The rare earths are defined as the set of elements of atomic number between 57 (La) and 71 (Lu), (i.e., the lanthanides) plus the elements Sc and Y.

2.3 MAGNETIC MOMENT OF AN ASSEMBLY OF ATOMS

The projection of the magnetic moment of each atom, in the direction defined by the magnetic field **B**, will be (we assume $\mathbf{B} = B\mathbf{k}$; the z axis is parallel to $\langle \boldsymbol{\mu} \rangle$, i.e., antiparallel to $\langle \mathbf{J} \rangle$):

$$\mu_J^z = g\mu_B M_J \tag{2.42}$$

where M_J may be $J, J-1, \ldots, -(J-1), -J$.

The average magnetic moment per atom will be, at a temperature T, a sum over the sublevels M_J:

$$\langle \mu_J^z \rangle_T = g\mu_B \sum_{M_J} M_J P(M_J) \tag{2.43}$$

where $P(M_J)$, the probability of occupation of a sublevel characterized by M_J, is as given by Eq. (2.41) (Boltzmann distribution). This is a thermal average, and we use the notation $\langle \cdots \rangle_T$.

The preceding expression is equal to

$$\langle \mu_J^z \rangle_T = g\mu_B \frac{\sum_{M_J} M_J \exp(g\mu_B B M_J / kT)}{\sum_{M_J} \exp(g\mu_B B M_J / kT)} \tag{2.44}$$

This function can be put into a more compact form. Making $x = g\mu_B JB/kT$ and $v = \sum_{M_J} \exp(xM_J/J)$ we see that

$$\langle \mu_J^z \rangle_T = g\mu_B J \frac{\sum_{M_J}(M_J/J)\exp(xM_J/J)}{\sum_{M_J}\exp(xM_J/J)} = g\mu_B J \frac{dv/dx}{v} \tag{2.45}$$

We may easily compute v, since v is the sum of the terms of a geometric progression. Making $z = \exp(x/J)$, we obtain

$$v = \sum_{M_J=-J}^{J} z^{M_J} = z^{-J}(1 + z + z^2 + \cdots + z^{2J}) \tag{2.46}$$

(since $M_J = -J, -J+1, \cdots, +J$).

Recalling that

$$S_n = a_0 + a_0 x + a_0 x^2 + \cdots + a_0 x^{n-1} = \frac{a_0(x^n - 1)}{x - 1} \tag{2.47}$$

we have

$$v = z^{-J} \frac{z^{2J+1} - 1}{z - 1} = \frac{z^{J+1/2} - z^{-(J+1/2)}}{z^{1/2} - z^{-1/2}} \tag{2.48}$$

ATOMIC MAGNETIC MOMENTS

$$z^{J+1/2} = \left(\exp\left(\frac{x}{J}\right)\right)^{J+1/2} = \exp\left(x + \frac{x}{2J}\right) \tag{2.49}$$

$$v = \frac{\exp[(1 + 1/2J)x] - \exp[-(1 + 1/2J)x]}{\exp(x/2J) - \exp(-x/2J)} \tag{2.50}$$

but $\sinh(x) = [\exp(x) - \exp(-x)]/2$; therefore,

$$v = \frac{\sinh(1 + 1/2J)x}{\sinh(x/2J)} \tag{2.51}$$

Computing the derivative of v, we obtain

$$\frac{dv}{dx} = \frac{\sinh(x/2J)(1 + 1/2J)\cosh[(1 + 1/2J)x]}{[\sinh(x/2J)]^2}$$

$$- \frac{\sinh[(1 + 1/2J)x](1/2J)\cosh(x/2J)}{[\sinh(x/2J)]^2} \tag{2.52}$$

and from Eq. (2.45):

$$\langle \mu_J^z \rangle_T = g\mu_B J \frac{dv/dx}{v} = g\mu_B J \left[\frac{(1 + 1/2J)\cosh[(1 + 1/2J)x]}{\sinh[(1 + 1/2J)x]} \right.$$

$$\left. - \frac{(1/2J)\cosh(x/2J)}{\sinh(x/2J)} \right] \tag{2.53}$$

or

$$\langle \mu_J^z \rangle_T = g\mu_B J \left[\left(1 + \frac{1}{2J}\right)\coth\left[\left(1 + \frac{1}{2J}\right)x\right] - \frac{1}{2J}\coth\left(\frac{x}{2J}\right) \right] \tag{2.54}$$

Finally, we obtain for the projection of the average magnetic moment in the z direction, as a function of the parameter x:

$$\langle \mu_J^z \rangle_T = g\mu_B J B_J(x) \tag{2.55}$$

where

$$x = \frac{g\mu_B J B}{kT} \tag{2.56}$$

and $B_J(x)$ is the Brillouin function, defined by

$$B_J(x) = \left(1 + \frac{1}{2J}\right)\left[\coth\left(1 + \frac{1}{2J}\right)x\right] - \frac{1}{2J}\coth\left(\frac{x}{2J}\right) \qquad (2.57)$$

In the special case $J = \frac{1}{2}$,

$$B_{\frac{1}{2}}(x) = 2\coth(2x) - \coth(x) = \tanh(x) \qquad (2.58)$$

We can see in Fig. 2.5 the dependence of $B_J(x)$ on x and J. Experimental

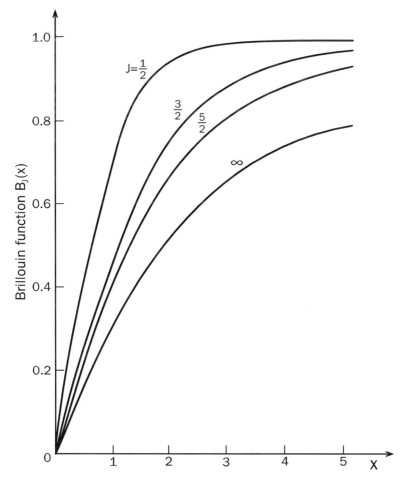

Figure 2.5 Plot of the Brillouin function $B_J(x)$ as a function of $x = g\mu_B JB/kT$, for $J = \frac{1}{2}, \frac{3}{2}, \frac{5}{2}$, and $J = \infty$.

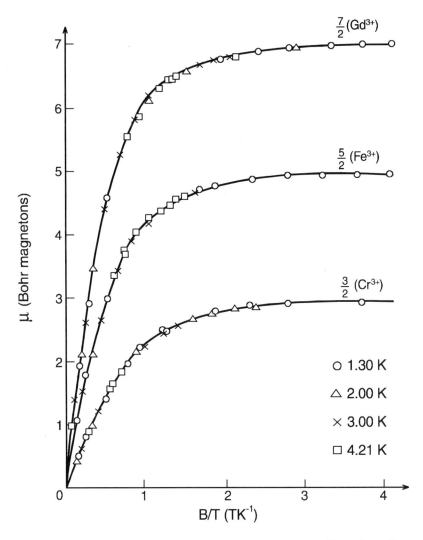

Figure 2.6 Magnetic moment per ion in salts containing the ions $Gd^{3+}(J=\frac{7}{2})$, $Fe^{3+}(J=\frac{5}{2})$ and $Cr^{3+}(J=\frac{3}{2})$ measured at different temperatures, versus $B/T = kx/g\mu_B J$. The curves are the Brillouin functions $B_J(x)$ for the corresponding values of $J \equiv S$. [Reprinted from W. E. Henry, *Phys. Rev.* **88**, 559 (1952).]

results for the magnetization of paramagnetic salts are presented in Fig. 2.6 (Henry 1952). We note the good agreement between the measurements with salts of Gd, Cr, and Fe, and the Brillouin functions for the corresponding values of J. The figure shows the magnetic moments described by $|\langle\mu_J^z\rangle_T| = g\mu_B J B_J$, and therefore, the curves tend to the saturation value $g\mu_B J$. Note also that the experimental points fall on the curves independently of the temperature of the measurements, depending only on the ratio B/T. The measurements were made at $T = 1.3$ K, $T = 2.0$ K, $T = 3.0$ K, and $T = 4.2$ K.

In the experiments with paramagnetic samples, with values of B and T attained in the most common experimental conditions, the values of $x = g\mu_B JB/kT$ are small. For small argument, $\coth(x)$ is equal to

$$\coth(x) = \frac{1}{x} + \frac{x}{3} + \cdots \tag{2.59}$$

Substituting into $B_J(x)$ [Eq. (2.57)]:

$$B_J(x) \approx \left(1 + \frac{1}{2J}\right)\left\{\left[\frac{1}{(1+1/2J)x} + \frac{(1+1/2J)}{3}x\right]\right\} - \frac{1}{2J}\left[\frac{2J}{x} + \frac{x}{6J}\right] = \frac{J+1}{3J}x \tag{2.60}$$

Therefore, in this limit (small x), the magnetization is proportional to x: this is visible through the initial linearity in the graph of $B_J(x)$ (Fig. 2.5).

From this result we may determine the susceptibility, that is, the rate of change $\partial M/\partial H$ in this region (small x). The volume susceptibility, χ, is obtained from the knowledge that in a unit volume we have n atoms; the total magnetic moment per unit volume ($\equiv M$) is therefore

$$M = n\langle \mu_J^z \rangle_T = ng\mu_B J B_J(x) \approx ng\mu_B J \frac{g\mu_B JB(J+1)}{kT \quad 3J} = \frac{ng^2\mu_B^2 J(J+1)B}{3kT} \tag{2.61}$$

The susceptibility is

$$\chi = \frac{\partial M}{\partial H} = \mu_0 \frac{\partial M}{\partial B} \tag{2.62}$$

or

$$\chi = \frac{\mu_0 ng^2 \mu_B^2 J(J+1)}{3kT} = \frac{C}{T} \tag{2.63}$$

a relation known as Curie law, obeyed by the susceptibility of many substances; C, the Curie constant, is given by

$$C = \frac{\mu_0 ng^2 \mu_B^2 J(J+1)}{3k} \tag{2.64}$$

When the separation between the multiplets is not much larger than kT, deviations from the Curie law are observed. The constant C contains $g^2 J(J+1)$, which is the square of the effective paramagnetic moment p_{eff}:

$$p_{\text{eff}} = g\sqrt{J(J+1)} \tag{2.65}$$

46 ATOMIC MAGNETIC MOMENTS

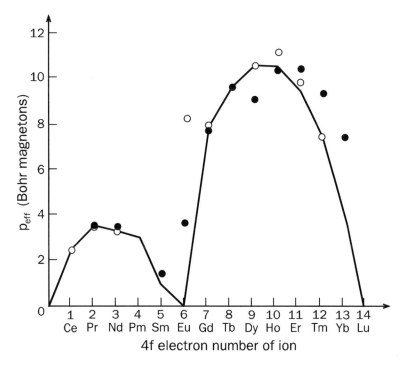

Figure 2.7 Experimental values of the effective paramagnetic moment p_{eff} for rare earths, in the oxides of formula R_2O_3 (open circles) and in the metals (full circles), as a function of the number of 4f electrons. The curve corresponds to the values computed with Eq. (2.65).

The values of the moments p_{eff} of the rare earths determined experimentally are in good agreement with the p_{eff} computed with Eq. (2.65). Figure 2.7 shows that the moments given by $g\sqrt{J(J+1)}$ coincide with those obtained experimentally, for the metallic rare earths and for the R_2O_3 oxides. The fractional deviations are larger in the cases of Eu and Sm; the separation between the lowest states and the states immediately above is smaller among all the rare earths in the case of Eu^{3+} (350 cm^{-1}) and Sm^{3+} (1000 cm^{-1}) (Table 2.II). This is the explanation for the observed disagreements. To compute the magnetic moment of the ions that have a separation between the multiplets comparable to kT, we have to take into account the occupation of the higher multiplets, and of their magnetic sublevels.

If we compare the effective paramagnetic moments of the transition elements of the d series, we will find a large disagreement between the computed effective moments p_{eff} and the measured moments. The agreement may be recovered if we write S instead of J in the expression of p_{eff}. This is an evidence of the importance of the interaction of these ions with the electrostatic crystalline field (see Section 2.7). This interaction, in these ions, is larger than the interaction $\lambda \mathbf{S} \cdot \mathbf{L}$ (spin–orbit). The smaller extension of the 4f shell leads to a partial shielding of

Table 2.II Properties of the 3^+ rare earth ions[a]

Z	$4f^n$	3^+ Ion		L	S	J	g	$(g-1)^2 J(J+1)$	$\Delta_0 (\text{cm}^{-1})$
57	0	La	Ce^{4+}	0	0	0	0	0	
58	1	Ce		3	$\frac{1}{2}$	$\frac{5}{2}$	$\frac{6}{7}$	0.18	2200
59	2	Pr		5	1	4	$\frac{4}{5}$	0.80	2150
60	3	Nd		6	$\frac{3}{2}$	$\frac{9}{2}$	$\frac{8}{11}$	1.84	1900
61	4	Pm		6	2	4	$\frac{3}{5}$	3.20	1600
62	5	Sm		5	$\frac{5}{2}$	$\frac{5}{2}$	$\frac{2}{7}$	4.46	1000
63	6	Eu		3	3	0	—	0	350
64	7	Gd	Eu^{2+}	0	$\frac{7}{2}$	$\frac{7}{2}$	2	15.75	—
65	8	Tb		3	3	6	$\frac{3}{2}$	10.50	2000
66	9	Dy		5	$\frac{5}{2}$	$\frac{15}{2}$	$\frac{4}{3}$	7.08	3300
67	10	Ho		6	2	8	$\frac{5}{4}$	4.50	5200
68	11	Er		6	$\frac{3}{2}$	$\frac{15}{2}$	$\frac{6}{5}$	2.55	6500
69	12	Tm		5	1	6	$\frac{7}{6}$	1.17	8300*

* $^3H_4$ is lower at $5900\,\text{cm}^{-1}$.
[a] In this table g is Lande's factor and Δ_0 is the spin–orbit splitting to the next J level.
Source: Reprinted from R. J. Elliott, in *Magnetic Properties of Rare Earth Metals*, R. J. Elliott, Ed., Plenum Press, London, 1972, p. 2.

this shell to the effects of the crystalline field, making the 4f electrons relatively insensitive to the chemical bonds.

2.4 LANGEVIN PARAMAGNETISM

In the derivation of the expression of the magnetic moment of an assembly of atoms (Section 2.3), the quantization of the angular momentum was taken into account. If the angular momentum were not quantized, as in the classical case, any value of μ^z would be allowed, and the magnetic moments could point along any direction in relation to the direction of the external field **B**.

The projection of the magnetic moment along the z direction in the classical case is given by:

$$\mu^z = \boldsymbol{\mu} \cdot \mathbf{k} = \mu \cos(\theta) \qquad (2.66)$$

where θ can take any value between 0 and π.

Making the average over θ, one arrives, after some algebra (Exercise 2.5) at the expression for the z projection of the magnetic moment

$$\langle \mu^z \rangle_T = \mu L(x) \qquad (2.67)$$

where $x = \mu B/kT$ and $L(x)$ is the Langevin function given by

$$L(x) = \coth\ x - \frac{1}{x} \qquad (2.68)$$

The Langevin function $L(x)$ is therefore the classical analog of the Brillouin function. This function describes well the magnetization of small particles formed of large clusters of atoms, in systems known as *superparamagnetic* (Chapter 1). In superparamagnets the effective moments are very large, reaching 10^5 Bohr magnetons, for instance, and for this reason their magnetization is well described by a classical model like that of Langevin (Fig. 2.8).

2.5 NUCLEAR MAGNETISM

The atomic nuclei may also have an angular momentum (I), and therefore a magnetic moment, given, in analogy with the electronic case, by

$$\mathbf{m} = g\mu_N \mathbf{I} \qquad (2.69)$$

where g is the nuclear g-factor and μ_N is the nuclear magneton, equivalent to the Bohr magneton, but involving the proton mass m_p:

$$\mu_N = \frac{e\hbar}{2m_p} \qquad (2.70)$$

Since this mass is 1836 times larger than the electron mass, the nuclear magneton is smaller than the Bohr magneton, in the same ratio. This fact explains why the magnetic effects associated with the nuclear magnetism are much weaker than those due to the magnetism of the electrons. The nuclear magnetic moment is also written in terms of the gyromagnetic ratio $\gamma = g\mu_N/\hbar$ as $\mathbf{m} = \gamma\hbar\mathbf{I}$.

The vectors $\mathbf{I}$ and $\mathbf{J}$ combine to form the total momentum $\mathbf{F}$, and the interaction between $\mathbf{I}$ and $\mathbf{J}$ is the hyperfine interaction (see Chapter 6).

The nuclear magnetic susceptibility is given by the Curie law:

$$\chi_n = \frac{\mu_0 n g^2 \mu_N^2 I(I+1)}{3kT} = \frac{C}{T} \qquad (2.71)$$

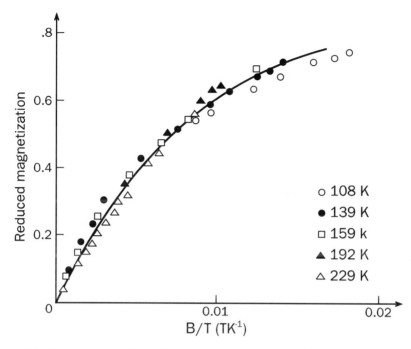

Figure 2.8 Experimental variation of the reduced magnetization (M_T/M_0) of superparamagnetic ferrite particles obtained at different temperatures, versus B/T; the continuous line is the Langevin function. [Reprinted from J. Crangle, *Solid State Magnetism*, Van Nostrand Rheinhold, New York, 1991, p. 172.]

where n is the number of nuclei per unit volume. Because the ratio of the magnitudes of the electronic and nuclear moments is so large, this susceptibility is negligible in comparison to the electronic susceptibility. Usually, it is even smaller than the diamagnetic susceptibility. However, at very low temperatures, the nuclear susceptibility may be comparable to the latter.

2.6 FERROMAGNETISM

We have so far studied an assembly of atoms whose unpaired electrons, under the action of an external field B, occupy nondegenerate energy levels; from the unequal occupation of these states arises the existence of magnetic moments $\langle \mu_J \rangle_T$.

In ferromagnetic materials there is a non-zero magnetic moment (inside a domain), even in the absence of an external field. The first explanation for this fact, proposed by P. Weiss in 1907, is that each individual atomic moment is oriented under the influence of all the other magnetic moments, which act through an effective magnetic field.

To obtain the magnetization using this hypothesis, we may follow the same steps of the reasoning of Section 2.3, this time assuming that each ion feels, instead of **B**, a field $\mathbf{B} + \mathbf{B}_m$, where $\mathbf{B}_m$ is this effective field.

The magnetization under a field **B**, at temperature T, assuming n atoms per unit volume (number per cubic meter in the SI) is

$$M_{BT} = n\langle \mu_J^z \rangle_T \tag{2.72}$$

The effective magnetic field due to the other ions, or mean field, in the simplest hypothesis, due to Weiss, is called the *molecular field*, and is proportional to the magnetization:

$$\mathbf{B}_m = \lambda_m \mathbf{M}_{BT} = \lambda_m n \langle \mu_J^z \rangle_T \mathbf{k} \tag{2.73}$$

where λ_m is the molecular field constant, or molecular field coefficient.

The magnetization may be computed as in the preceding case; the moment per atom as a function of T is

$$\langle \mu_J^z \rangle_T = g\mu_B J B_J(x') \tag{2.74}$$

where x' is the equivalent to $x = g\mu_B JB/kT$ of the paramagnetic case, with the addition of the molecular field $\mathbf{B}_m$:

$$x' = g\mu_B J \frac{(B + \lambda_m n \langle \mu_J^z \rangle_T)}{kT} \tag{2.75}$$

Therefore

$$\langle \mu_J^z \rangle_T = g\mu_B J B_J \left(g\mu_B J \frac{B + \lambda_m n \langle \mu_J^z \rangle_T}{kT} \right) \tag{2.76}$$

This expression, which gives the magnetic moment (per atom) as a function of the temperature, is more complex than in the preceding case (the paramagnetic case), since now $\langle \mu_J^z \rangle_T$ is present on both sides of the equation.

Assuming initially $B = 0$, we have

$$x' = \frac{g\mu_B J \lambda_m n \langle \mu_J^z \rangle_T}{kT} \tag{2.77}$$

and

$$\langle \mu_J^z \rangle_T = g\mu_B J B_J \left(g\mu_B J \frac{\lambda_m n \langle \mu_J^z \rangle_T}{kT} \right) \tag{2.78}$$

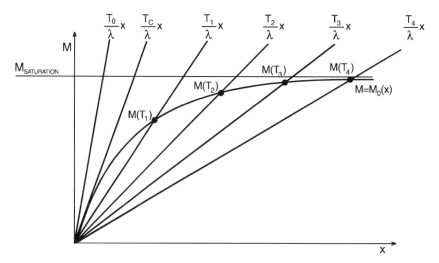

Figure 2.9 Graphical solution of the system of Eqs. (2.79). The straight lines are representations of Eq. (2.79a) for the temperatures T_0, T_1, T_2, T_3, and T_4. The temperatures T_1, T_2, T_3, and T_4 are below T_C, and in these cases the system has two solutions; for temperatures above T_C, as T_0, there is only the trivial solution ($x' = 0$, $\langle \mu_J^z \rangle_T = 0$).

From (2.77), it follows that

$$\langle \mu_J^z \rangle_T = \frac{x'}{g\mu_B J \lambda_m n / kT} \tag{2.79a}$$

and from (2.74)

$$\langle \mu_J^z \rangle_T = g\mu_B J B_J(x') \tag{2.79b}$$

We may find the values of x' and $\langle \mu_J^z \rangle_T$ that solve the system of equations above (2.79) making a graph of $\langle \mu_J^z \rangle_T$ [using (2.79b)] and finding the intersections with the straight lines that describe (2.79a), for different values of T. This graphical method was used by Weiss. Alternatively, we may compute $\langle \mu_J^z \rangle_T$ by solving self-consistently these equations using a computer.

The graph of the two functions (Fig. 2.9) shows two intersections in the plane $(x', \langle \mu_J^z \rangle_T)$; the solution $x' = 0$, $\langle \mu_J^z \rangle_T = 0$ always exists, but of course it is of no interest. We may note also that as we approach the solution with $\langle \mu_J^z \rangle_T$ tending to zero, $|d\langle \mu_J^z \rangle_T / dx'|$ increases, that is, $\langle \mu_J^z \rangle_T$ falls more rapidly. The spontaneous magnetizations computed this way are shown in Fig. 2.10, under the form of reduced magnetization M_{0T}/M_{00} versus reduced temperature T/T_C (T_C is the Curie temperature). The reduced magnetization is

$$\frac{M_{0T}}{M_{00}} = B_J(x') = \frac{\langle \mu_J^z \rangle_T}{g\mu_B J} = \frac{x' kT}{(g\mu_B J)^2 n \lambda_m} \tag{2.80}$$

52 ATOMIC MAGNETIC MOMENTS

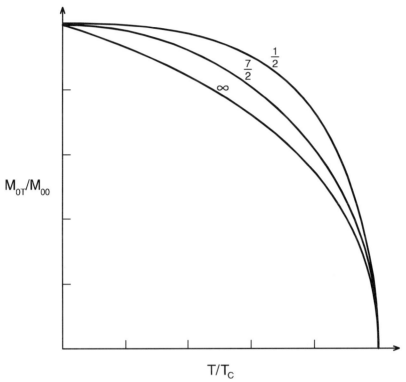

Figure 2.10 Reduced magnetization (M_{0T}/M_{00}) versus reduced temperature (T/T_C) for different values of the angular momentum J, in the Weiss mean field model.

If we make $B \neq 0$, the magnetizations may still be computed; now the magnetization curves change shape slightly, with a tail that extends beyond the ferromagnetic Curie temperature. One can see from Fig. 2.11 that there is a finite magnetization above T_C, with $B \neq 0$.

Just below T_C, $B_J(x')$ is small, and may be approximated by Eq. (2.60):

$$B_J(x') \approx \frac{J+1}{3J} x' \qquad (2.81)$$

Using Eq. (2.80), we have

$$\frac{J+1}{3J} x' = \frac{x'kT}{(g\mu_B J)^2 n\lambda_m} \qquad (2.82)$$

which is valid for T tending to T_C (i.e., for a magnetization tending to zero). Thus the Curie temperature results:

$$T_C = \frac{g^2 \mu_B^2 n \lambda_m J(J+1)}{3k} \qquad (2.83)$$

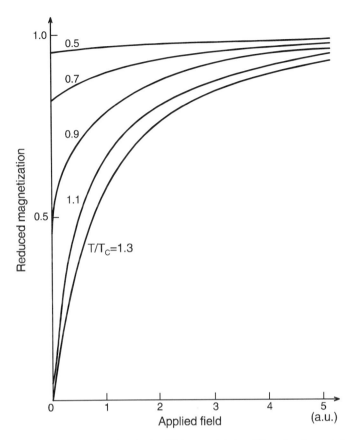

Figure 2.11 Reduced magnetization in the Weiss model, for different values of the applied magnetic field.

From this relation we may compute, for a value of T_C determined experimentally, the value of the molecular field parameter λ_m. For example, in metallic gadolinium (Fig. 2.12), $T_C = 293.4$ K, $S = 7/2, L = 0, g = 2$, atomic mass $M = 157.3$, and density 7.9 g/cm^3. With $N = 6.023 \times 10^{23}$ mol^{-1}, $\mu_B = 9.27 \times 10^{-24}$ J T^{-1}, $k = 1.381 \times 10^{-23}$ J K^{-1}, it follows that $\lambda_m = 0.742 \times 10^{-4}$ J^{-1} T^{2}m^3. The molecular field is obtained from the saturation magnetization $M_{00} = 2.12 \times 10^6$ A m^{-1}; the molecular field at $T = 0$ K is $B_m = \lambda_m M_{00} = 157$ T$(= 1.57 \times 10^6$ G$)$.

Table 2.III gives Curie temperatures and magnetic moments per atom for some ferromagnetic elements.

A ferromagnet above T_C presents no spontaneous magnetization; in other words, it has no magnetization with $B = 0$. However, under the influence of an external field, a nonzero magnetization appears, as in the case of a paramagnet. We may measure this magnetic response through the susceptibility $\chi = \partial M/\partial H$; this quantity may be computed within the Weiss model.

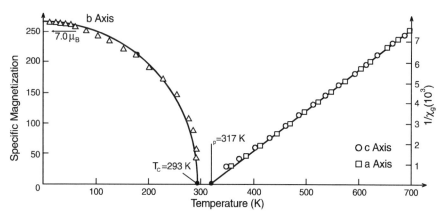

Figure 2.12 Values of the magnetic moment per unit mass of metallic Gd, as a function of temperature. The continuous curve is the magnetic moment given by the Weiss mean field model, for $J = \frac{7}{2}$. On the right side of the graph, measurements of the inverse of the susceptibility $(1/\chi)$. [Reprinted from J. J. Rhyne, in *Magnetic Properties of Rare Earth Metals*, R. J. Elliott, Ed., Plenum Press, London, 1972, p. 132.]

The magnetization in this temperature region is small, and therefore we may use Eq. (2.81):

$$B_J(x') \approx \frac{J+1}{3J} x' \tag{2.84}$$

and the magnetic moment with applied field is

$$\langle \mu_J^z \rangle_T = g\mu_B J B_J(x') \approx -\tfrac{1}{3} g\mu_B (J+1) x' \tag{2.85}$$

substituting

$$x' = g\mu_B J \frac{B + \lambda_m n \langle \mu_J^z \rangle_T}{kT} \tag{2.86}$$

we obtain

$$\langle \mu_J^z \rangle_T \approx g^2 \mu_B^2 J(J+1) \frac{B + \lambda_m n \langle \mu_J^z \rangle_T}{3kT} \tag{2.87}$$

Table 2.III Curie temperature (T_C) and magnetic moment per atom (μ_{at}) of some ferromagnetic elements, for different crystalline structures

Element	Fe(bcc)	Co(fcc)	Co(hcp)	Ni(fcc)	Gd(hcp)
T_C (K)	1044	1388	1360	627.4	293.4
μ_{at} at 0 K (μ_B)	2.217	1.753	1.721	0.6157	7.56

Source: Reprinted from L. J. Swartzendruber, *J. Mag. Magn. Mat.* **100**, 573 (1991), with permission from Elsevier North-Holland, NY.

With $C = \mu_0 g^2 \mu_B^2 n J(J+1)/3k$ [Eq. (2.64)] it becomes

$$n\langle\mu_J^z\rangle_T T = \frac{C}{\mu_0}(B + \lambda_m n\langle\mu_J^z\rangle_T) \tag{2.88}$$

and it follows that

$$M_{BT} = n\langle\mu_J^z\rangle_T = \frac{CB/\mu_0}{T - C\lambda_m/\mu_0} \tag{2.89}$$

Using $H = B/\mu_0$ we obtain the susceptibility per unit volume $\chi = \partial M_{BT}/\partial H$:

$$\chi = \frac{\partial(n\langle\mu_J^z\rangle_T)}{\partial H} = \frac{C}{T - C\lambda_m/\mu_0} = \frac{C}{T - \theta_p} \tag{2.90}$$

with

$$\theta_p = \frac{C\lambda_m}{\mu_0} = \frac{g^2 \mu_B^2 n \lambda_m J(J+1)}{3k} \tag{2.91}$$

Equation (2.90) expresses the Curie–Weiss law, and θ_p is the paramagnetic Curie temperature. It should be noted that the paramagnetic Curie temperature (θ_p) is given, in the Weiss model, by the same expression describing the Curie temperature (T_C) [Eq. (2.83)]. However, the values for θ_p and T_C observed experimentally do not show, in general, this coincidence.

Therefore, the behavior of the susceptibility of a ferromagnet above the temperature of magnetic order T_C is analogous to that of a paramagnetic material, with the difference that θ_p is not zero for a ferromagnet.

2.7 CRYSTAL FIELDS

The incomplete electronic shells of the transition elements with nonzero orbital momentum ($L \neq 0$) do not have spherical symmetry. When a transition metal atom is located in a crystal, the charges of the electrons in these shells interact with the charges of the crystalline lattice—this is the crystal field interaction. The crystal field (CF) interaction depends on the orientation of the charge cloud relative to the crystal axes. The closed shells practically do not contribute to this interaction.

The crystal field interaction represents another term that must be added to the hamiltonian of the free ion (or atom), which already contains the electron–nucleus Coulomb interaction ($\mathcal{H}_{coul}$), the electron–electron interaction, and the spin–orbit interaction ($\mathcal{H}_{LS}$).

There are three regimes for the crystal field interaction, defined according to its relative intensity:

1. *Strong crystal field interaction*—observed in the elements of the 4d and 5d transition series. In this case

$$\mathcal{H}_{cf} > \mathcal{H}_{coul} \gg \mathcal{H}_{LS} \tag{2.92}$$

2. *Medium crystal field interaction*—observed in the 3d series. We have here

$$\mathcal{H}_{cf} \approx \mathcal{H}_{coul} \gg \mathcal{H}_{LS} \tag{2.93}$$

3. *Weak crystal field interaction*—in the 4f series (rare earths). The interactions follow the relation

$$\mathcal{H}_{coul} \gg \mathcal{H}_{LS} \gg \mathcal{H}_{cf} \tag{2.94}$$

We now discuss how the crystal field problem is formulated for the rare earths, therefore, in the limit of weak crystal field interactions.

The most immediate form of obtaining the interaction with the crystal field is to start from the computation of the potential energy of the electronic charges q_i in the potential V due to the point charges of the lattice:

$$W_c = \sum_i^N q_i V(x_i, y_i, z_i) \tag{2.95}$$

The hamiltonian of the magnetic ion, in the presence of the magnetic interaction (with the exchange field) and of the interaction with the crystal field, is given by

$$\mathcal{H} = \mathcal{H}_{mag} + \mathcal{H}_{cf} \tag{2.96}$$

The matrix elements of $\mathcal{H}_{cf}$ may be derived from the classical potential energy of the charges [Eq. (2.95)], through the substitution $x \to x_{op}, y \to y_{op}$ (the position operators) and so on, and summing over all the N magnetic electrons.

A more practical method, the method of the operator equivalents (or Stevens's operators) (Stevens 1952), consists in substituting $x \to J_x, y \to J_y$, and so forth, observing the appropriate commutation rules. For this purpose, the products of x, y, and z are substituted by all possible combinations of J_x, J_y and J_z, divided by the total number of permutations.

We may give the following as examples:

$$\sum_i (3z_i^2 - r_i^2) \equiv \alpha_J \langle r^2 \rangle [3J_z^2 - J(J+1)] = \alpha_J \langle r^2 \rangle O_2^0 \tag{2.97}$$

$$\sum_i x_i y_i \equiv \alpha_J \langle r^2 \rangle \tfrac{1}{2}[J_x J_y + J_y J_x] \tag{2.98}$$

where α_J is a numerical constant for the second-order term that depends on l; for the fourth-order term it is β_J and for the sixth-order term, γ_J. These constants are determined by direct integration, and are tabulated.

The interaction is usually written in terms of the operators O_n^m of order n in the components of **J**; the B_n^m are numerical coefficients:

$$\mathcal{H}_{cf} = \sum_{n,m} B_n^m O_n^m \qquad (2.99)$$

The operators O_n^m are polynomials that involve the angular momentum operators J_z, J^2, J_+ and J_-. The maximum value of n in the hamiltonian is 6 for f electrons and 4 for d electrons. The presence of the different operators in this expression depends on the point symmetry of the sites where the ion is located, and on the choice of the crystal axes.

In magnetic samples, the main interaction in the hamiltonian is the magnetic interaction. The effect of the crystal field interaction is to admix excited states to the ground state defined by the magnetic interaction $|J, M = J\rangle$, leading to a reduction in $\langle J_z \rangle$. This effect is known as "quenching"; it causes a reduction in the magnetization and in the hyperfine field acting on the nucleus of the respective ion. One image to describe this effect is that, under the influence of the crystal field, the orientation of the electronic orbits varies continuously with time, and this leads, in the limit, to a null projection of the orbital moment along any direction.

In the 3d series the attenuation takes a different form: since $\mathcal{H}_{cf}$ is a strong perturbation in relation to the spin–orbit interaction, L and S decouple, and the mean value $\langle L_z \rangle$ is reduced. This explains, for example, why the magnetic moments found in the 3d series are nearer to $g\mu_B \langle S \rangle$ than to $g\mu_B \langle J \rangle$; in other words, the measured moments relate only to the spin angular momentum.

The values of the constants α_J, β_J, and γ_J, and the expressions of the operators O_n^m may be found in Hutchings (1966).

The higher the symmetry, the smaller the number of operators necessary to write the crystal field hamiltonian. For a crystal field of cubic symmetry, only four terms suffice:

$$\mathcal{H}_{cf} = B_4^0(O_4^0 + 5O_4^4) + B_6^0(O_6^0 - 21O_6^4) \qquad (2.100)$$

For hexagonal symmetry in the case of ideal c/a ratio, the hamiltonian is written

$$\mathcal{H}_{cf} = B_4^0 O_4^0 + B_6^0(O_6^0 + \tfrac{77}{8} O_6^6) \qquad (2.101)$$

The form of the hamiltonian, the operators that appear in its expression, vary depending on the choice of axes. For example, expression (2.100) was obtained for the z axis coinciding with the (100) direction. For $z \parallel$ to the (111) direction,

Table 2.IV Crystal field parameters for some rare-earth metals (in meV)

Rare earth	B_2^0	B_4^0	B_6^0	B_6^6
Ho	0.024	0.0	-9.6×10^{-7}	9.2×10^{-6}
Er	-0.027	-0.7×10^{-5}	8.0×10^{-7}	-6.9×10^{-6}
Tm	-0.096	0.0	-9.2×10^{-6}	8.9×10^{-5}

Source: Reprinted from J. Jensen and A. R. Mackintosh, *Rare Earth Magnetism: Structures and Excitations*, 1991, p. 114. By permission of Oxford University Press, Oxford, UK.

the cubic hamiltonian is

$$\mathcal{H}_{cf} = -\frac{2}{3}B_4^0(O_4^0 - 20\sqrt{2}O_4^3) + \frac{16}{9}B_6^0\left(O_6^0 + \frac{35\sqrt{2}}{4}O_6^3 + \frac{77}{8}O_6^6\right) \quad (2.102)$$

The attenuation of the magnetic moment due to the crystal field may be computed from the complete hamiltonian of the ion, containing the magnetic term and the crystal field term [Eq. (2.96)]. The eigenvectors are obtained by diagonalizing $\mathcal{H}$, and then computing $\langle \mu \rangle$. For the rare earths, the computed angular moments at $T = 0$ (McCausland and Mackenzie 1980) show the effect of the crystal field. Examples are (1) Tb: $\langle J_z \rangle = 0.9923\,J$ and (2) Dy: $\langle J_z \rangle = 0.988\,J$.

The parameters B_n^m are usually determined experimentally (see Table 2.IV). They may also be computed, but this involves a considerable degree of uncertainty, mainly because the B parameters contain the terms $\langle r^n \rangle$, and the shielding factors. The computation of the ratio between the B_n^m, however, does not present these difficulties.

An alternative notation for the crystal field coefficients (Lea 1962) uses a parameter x to measure the ratio between the terms of fourth and sixth orders:

$$\frac{x}{1-|x|} = \frac{F(4)B_4^0}{F(6)B_6^0} \quad (2.103)$$

where $F(4)$ and $F(6)$ are tabulated factors for the different $4f$ ions. With O_4 and O_6 the expressions of the operators of fourth and sixth orders of the cubic hamiltonian [in parentheses in Eq. (2.100)], and introducing a scaling parameter W (with dimension of energy), we have, for cubic symmetry:

$$\mathcal{H}_{cf} = W\left(\frac{xO_4}{F(4)} + \frac{(1-|x|)O_6}{F(6)}\right) \quad (2.104)$$

GENERAL READING

Craik, D., *Magnetism, Principles and Applications*, Wiley, Chichester, 1995.
Crangle, J., *Solid State Magnetism*, Van Nostrand Rheinhold, New York, 1991.
Hutchings, M. T., *Solid State Phys.* **16**, 227 (1966).
Morrish, A. H., *Physical Principles of Magnetism*, Wiley, New York, 1965.
Taylor, K. N. R. and M. I. Darby, *Physics of Rare Earth Solids*, Chapman & Hall, London, 1972.
Wallace, W. E., *Rare Earth Intermetallics*, Academic Press, New York, 1973.

REFERENCES

Ashcroft, N. W. and N. D. Mermin, *Solid State Physics*, Holt Rinehart and Winston, New York, 1976.
Crangle, J., *Solid State Magnetism*, Van Nostrand Rheinhold, New York, 1991.
Elliott, R. J. in R. J. Elliott, Ed., *Magnetic Properties of Rare Earth Metals*, Plenum Press, London, 1972, p. 1.
Henry, W. E., *Phys. Rev.* **88**, 559 (1952).
Hutchings, M. T., *Solid State Phys.* **16**, 227 (1966).
Jensen, J. and A. R. Mackintosh, *Rare Earth Magnetism: Structures and Excitations*, Clarendon Press, Oxford, 1991.
Landolt-Börnstein, *Magnetic Properties of Metals*, Landolt-Börnstein Tables, New Series III/19f1, Springer-Verlag, New York, 1991.
Lea, R. R., M. J. M. Leask, and W. P. Wolf, *J. Phys. Chem. Solids* **23**, 1381 (1962).
McCausland, M. A. H. and I. S. Mackenzie, *Adv. Phys.* **28**, 305 (1979); also *Nuclear Magnetic Resonance in Rare Earth Metals*, Taylor & Francis, London, 1980.
Meyers, R. A., Ed., *Encyclopedia of Physical Science and Technology*, Academic Press, Orlando, 1987.
Rhyne, J. J. in R. J. Elliott, Ed., *Magnetic Properties of Rare Earth Metals*, Plenum Press, London, 1972, p. 129.
Stevens, K. W. H., *Proc. Phys. Soc.* (London) **A65**, 209 (1952).
Swartzendruber, L. J., *J. Mag. Mag. Mat.* **100**, 573 (1991).

EXERCISES

2.1 *Larmor Frequency of one Electron.* Consider one electron subject to a Coulomb force, moving in a circular orbit around a nucleus of charge e. Write the expression of the total force acting on the electron assuming that a magnetic field **B** is applied, and show that the frequency of the electron motion around the direction of the field is given by

$$\omega = \pm \sqrt{\left(\frac{eB}{2m_e}\right)^2 + \left(\frac{e^2}{m_e r^3}\right)} + \frac{eB}{2m_e}$$

Evaluate the magnitudes of the different terms and make an approximation to obtain the Larmor frequency.

2.2 *Diamagnetic Susceptibility of Atomic Hydrogen.* The ground state of the hydrogen atom ($1s$) is described by the wavefunction $\psi = (\pi a_0^3)^{-1/2} \exp(-r/a_0)$ where $a_0 = \hbar^2/m_e e^2 = 0.529 \times 10^{-8}$ cm. Obtain the expectation value of r and r^2 for this state, and compute the diamagnetic susceptibility of hydrogen.

2.3 *Magnetic Moment of Iron.* The saturation magnetization of iron is 1.7×10^6 A m^{-1}. Assuming the density of iron is 7970 kg m^{-3} and Avogadro constant is 6.025×10^{26} kg^{-1}, compute the magnetic moment per iron atom in units of Bohr magnetons (atomic mass of iron = 56).

2.4 *Néel Temperature.* Consider an antiferromagnet formed of two sublattices A and B. Let $\lambda_{AB} = \lambda_{BA} = -\lambda$ be the molecular field coefficients of the two sublattices and $\lambda_{AA} = \lambda_{BB} = \lambda'$ in each sublattice. Let **B** be an external applied magnetic field.

(a) Write the expression of the total field acting on each sublattice, $\mathbf{B}_A$ and $\mathbf{B}_B$.

(b) Substituting the expressions obtained in the Brillouin function, make an expansion for high temperatures and show that the magnetization in each sublattice is given by

$$\mathbf{M}_A = \frac{C_A}{T\mu_0}(\mathbf{B} - \lambda\mathbf{M}_B + \lambda'\mathbf{M}_A)$$

and

$$\mathbf{M}_B = \frac{C_B}{T\mu_0}(\mathbf{B} - \lambda\mathbf{M}_A + \lambda'\mathbf{M}_B)$$

(c) Making $C_A = C_B = C$, show that the Néel temperature is given by $T_N = C(\lambda \pm \lambda')$ (*Suggestion:* The Néel temperature is the temperature for which $\mathbf{M}_A$ and $\mathbf{M}_B \neq 0$ for $\mathbf{B} = 0$.)

2.5 *Langevin Magnetism.* Derive the expression for the magnetization of an ensemble of classical magnetic moments (Langevin function).

2.6 *Relativistic Spin–Orbit Interaction.* One electron with velocity $\mathbf{v} = \mathbf{p}/m$, moving in a central force potential $-V/e$, feels a magnetic field equal to $\mathbf{B} = -(\mu_0/4\pi)\mathbf{v} \times \mathbf{E}$, where $\mathbf{E} = -\nabla V$. Show that the interaction energy between the electron spin and the field **B** may be written

$$\mathcal{H} = \frac{\mu_0}{4\pi}\frac{\hbar^2}{m_e^2}\frac{1}{r}\frac{dV}{dr}\mathbf{l}\cdot\mathbf{s} = \xi(r)\mathbf{l}\cdot\mathbf{s}$$

where $\mathbf{l} = \hbar \mathbf{r} \times \mathbf{p}$. The result obtained from Dirac's relativistic equation for the coupling constant is two times smaller than the result obtained above.

2.7 *Crystal Field and Direction of Magnetization* Let the crystal field hamiltonian be given by

$$\mathcal{H}_{cf}^q = B_2^0 O_2^0 = B_2^0 [3J_c - J^2]$$

where J_c is the component of $\mathbf{J}$ along the $\mathbf{c}$ direction of the crystal. Let z be the direction of magnetization of the crystal (direction of $\langle \mathbf{J} \rangle$). Considering the electrostatic interaction as a perturbation on the magnetic one, show that the expectation value of $\mathcal{H}_{cf}^q$ in the state $|J;J\rangle$ of J_z is given by

$$\langle \mathcal{H}_{cf}^q \rangle = \langle J;J | \mathcal{H}_{cf}^q | J;J \rangle = B_2^0 J(2J-1) \mathcal{P}_2(\cos\theta)$$

where θ is the angle between J_c and $\langle \mathbf{J} \rangle$ and $\mathcal{P}_2(\cos\theta)$ is the Legendre polynomial of order 2. Show that if $B_2^0 > 0$, the direction of magnetization will be perpendicular to the $\mathbf{c}$ axis and if $B_2^0 < 0$, $\mathbf{M}$ will be parallel to $\mathbf{c}$.

2.8 *Quenching of the Angular Momentum of a p Electron.* An atom containing a single electron in a p orbital is affected by a crystal field with octahedral symmetry, due to six equal charges Q located along the axes x, y, and z. The charges on the axes x and y are at the same distance r_0 from the center of the atom (origin of the coordinate system) and those on the z axis are at a distance r_1.

(a) Show that the dominant term in the crystal field is given by $\mathcal{H}_{cf} = A(3z^2 - r^2)$ and discuss the sign of A.

(b) Writing the p wavefunctions as $p_x = xf(r)$, $p_y = yf(r)$ and $p_z = zf(r)$, find the eigenenergies of the states in terms of A and $\langle r^2 \rangle$, the root mean square radius of the p orbital.

(c) Assume that a magnetic field is applied along the z direction. Compute the 3×3 matrix of the total hamiltonian $\mathcal{H}$.

(d) Evaluate the eigenstates of $\mathcal{H}$. In which states is the degeneracy lifted? Which state has the angular momentum quenched by the field?

3

INTERACTION BETWEEN TWO SPINS

3.1 EXCHANGE INTERACTION

In Chapter 2 we discussed the phenomenon of ferromagnetism, and its description within the Weiss theory, or mean field approximation. In this chapter we discuss the interaction between two electron spins that provides the physical basis for the onset of ferromagnetic order.

The molecular field postulated by Weiss to describe ferromagnetism remained without physical explanation until the birth of quantum mechanics. The magnetic fields required by the Weiss model were much larger than those associated with the magnetic dipolar interaction, and therefore this interaction could not explain ferromagnetic order. The physical phenomenon that is at the origin of the ordering of the magnetic ions is the exchange interaction, an interaction of electrostatic origin that results from the indistinguishability of the electrons. We shall discuss the formulation of the exchange interaction, arriving at the Heisenberg hamiltonian, and its connection with the molecular field concept (e.g., Patterson 1971).

Schrödinger's equation is written

$$\mathcal{H}\Psi = i\hbar \left(\frac{\partial \Psi}{\partial t} \right) \quad (3.1)$$

Assuming that the wavefunction Ψ can be separated into a spatial and a temporal part:

$$\Psi(r, t) = \Phi(\mathbf{r}) T(t) \quad (3.2)$$

it follows that the spatial part Φ obeys

$$\mathcal{H}\Phi(\mathbf{r}) = E\Phi(\mathbf{r}) \tag{3.3}$$

where E is the energy.

Let us consider a system formed of two electrons, spatial coordinates $\mathbf{r}_1$, $\mathbf{r}_2$, spin coordinates σ_1 and σ_2, and nondegenerate energy states. Conventionally, σ can be $+1$ or -1, corresponding to the z projection of the spin equal to $+\frac{1}{2}$ and $-\frac{1}{2}$, respectively.

The individual wavefunctions satisfy

$$\mathcal{H}_0^1 \varphi_m(\mathbf{r}_1) = E_1 \varphi_m(\mathbf{r}_1) \tag{3.4a}$$

and

$$\mathcal{H}_0^2 \varphi_n(\mathbf{r}_2) = E_2 \varphi_n(\mathbf{r}_2) \tag{3.4b}$$

where m and n are quantum numbers labeling the states of the electrons and E_1 and E_2 are the corresponding energies. The hamiltonian for the pair of electrons, assuming for the moment that they do not interact, is the sum of the partial hamiltonians

$$\mathcal{H}_0 = \mathcal{H}_0^1 + \mathcal{H}_0^2 \tag{3.5}$$

From the one-electron wavefunctions one can form

$$\Phi_1 = \varphi_m(\mathbf{r}_1)\varphi_n(\mathbf{r}_2) \tag{3.6a}$$

and

$$\Phi_2 = \varphi_n(\mathbf{r}_1)\varphi_m(\mathbf{r}_2) \tag{3.6b}$$

which are eigenfunctions of the total hamiltonian [Eq. (3.3)], with eigenvalue $E_0 = E_m + E_n$.

We will now assume that there is an interaction between the electrons. This is accounted for by introducing into the hamiltonian [Eq. (3.3)] a Coulomb potential term $V_{12}(\mathbf{r}_1, \mathbf{r}_2) = e^2/r_{12}$ to describe it:

$$\mathcal{H} = \mathcal{H}_0 + V_{12}(\mathbf{r}_1, \mathbf{r}_2) \tag{3.7}$$

The energy states of the system in the presence of this perturbation are

$$E = E_0 + E_{12} \tag{3.8}$$

obtained using the unperturbed wavefunctions Φ_1 and Φ_2 and solving

$$\begin{vmatrix} \langle 1|\mathcal{H}_0 + V_{12}|1\rangle - E & \langle 1|V_{12}|2\rangle \\ \langle 2|V_{12}|1\rangle & \langle 2|\mathcal{H}_0 + V_{12}|2\rangle - E \end{vmatrix} = 0 \quad (3.9)$$

where $|1\rangle$ and $|2\rangle$ refer to Φ_1 and Φ_2, respectively and

$$\begin{aligned} \langle 1|V_{12}|2\rangle &\equiv \int \varphi_m^*(\mathbf{r}_1)\varphi_n^*(\mathbf{r}_2) V_{12}\varphi_n(\mathbf{r}_1)\varphi_m(\mathbf{r}_2) d\tau \\ &\equiv \int \varphi_m^*(\mathbf{r}_2)\varphi_n^*(\mathbf{r}_1) V_{21}\varphi_n(\mathbf{r}_2)\varphi_m(\mathbf{r}_1) d\tau \\ &\equiv \int \varphi_n^*(\mathbf{r}_1)\varphi_m^*(\mathbf{r}_2) V_{12}\varphi_m(\mathbf{r}_1)\varphi_n(\mathbf{r}_2) d\tau \\ &\equiv \langle 2|V_{12}|1\rangle \end{aligned} \quad (3.10)$$

and $d\tau$ is a volume element; we have used $V_{21} \equiv V_{12}$. We can also show that $\langle 1|V_{12}|1\rangle = \langle 2|V_{12}|2\rangle$.

The eigenvalues obtained from Eq. (3.9) are

$$E_\pm = E_0 + K_{12} \pm \mathcal{J}_{12} \quad (3.11)$$

with

$$K_{12} \equiv \langle 1|V_{12}|1\rangle = \langle 2|V_{12}|2\rangle \quad (3.12a)$$

$$\mathcal{J}_{12} \equiv \langle 1|V_{12}|2\rangle = \langle 2|V_{12}|1\rangle \quad (3.12b)$$

where K_{12} is the Coulomb energy, specifically, the electrostatic energy of electrons in the unperturbed states, and $\mathcal{J}_{12}$ is called the *exchange integral*, also measured in energy units. The eigenvectors are

$$\Phi_\pm = \frac{1}{\sqrt{2}} (\Phi_1 \pm \Phi_2) \quad (3.13)$$

Experimentally it is observed that the total wavefunctions of the electrons and of the totality of the particles with half-integer spin (called *fermions*, since they follow Fermi–Dirac statistics) are antisymmetric; that is, they change sign when two particles are interchanged. The particles with integer spin (*bosons*, from the Bose statistics) have symmetric wavefunctions.

We may obtain this antisymmetry by combining a spatial function ϕ with a spin function χ in two different ways (using subscripts S and A for the symmetric and antisymmetric functions, respectively):

$$\phi_A(\mathbf{r}_1, \mathbf{r}_2)\chi_S(\sigma_1, \sigma_2) \quad (3.14a)$$

$$\phi_S(\mathbf{r}_1,\mathbf{r}_2)\chi_A(\sigma_1,\sigma_2) \tag{3.14b}$$

Thus an antisymmetric χ must multiply ϕ_S and a symmetric χ must multiply ϕ_A.

From the "spin up" wave function of the i electron $\alpha(i)$ and the "spin down" wave function of the j electron $\beta(j)$, we can construct the antisymmetric χ_A:

$$\chi_A = \frac{1}{\sqrt{2}}[\alpha(1)\beta(2) - \alpha(2)\beta(1)] \tag{3.15}$$

and the symmetric χ_S, that can take the forms:

$$\chi_S = \begin{cases} \alpha(1)\alpha(2) \\ \frac{1}{\sqrt{2}}[\alpha(1)\beta(2) + \alpha(2)\beta(1)] \\ \beta(1)\beta(2) \end{cases} \tag{3.16}$$

Thus, there exist, for two spins $\frac{1}{2}$, three symmetric spin functions χ_S, corresponding to a total spin $S = 1$ ("parallel spins"), and one single antisymmetric function χ_A, corresponding to $S = 0$ ("antiparallel spins").

We therefore have two cases:

$$\begin{cases} \phi_S \text{ and } \chi_A & \text{giving} \quad S = 0 \text{ (singlet)} \\ \phi_A \text{ and } \chi_S & \text{giving} \quad S = 1 \text{ (triplet)} \end{cases} \tag{3.17}$$

The sign in Eq. (3.11) is the same as that in Eq. (3.13); if it is positive, the spatial part of the wavefunction is symmetric [from (3.13)], and therefore the spin function is antisymmetric (↑↓). The state of minimum energy, or ground state, will in this case correspond to $\mathcal{J}_{12}\langle 0$ [from (3.11)].

The negative sign in (3.11) corresponds to the symmetric spin function (↑↑); the ground state is obtained in this case for $\mathcal{J}_{12}\rangle 0$. The two situations are then

$$\mathcal{J}_{12} < 0 : \text{ground state is } \uparrow\downarrow \text{ (singlet)}$$
$$\mathcal{J}_{12} > 0 : \text{ground state is } \uparrow\uparrow \text{ (triplet)}$$

Consequently, the energy E in these two cases depends on the relative orientation of the electronic spins; thus, to represent the interaction between the electrons, it suffices to introduce a term in the hamiltonian containing a factor

$$\mathbf{s}_1 \cdot \mathbf{s}_2 \tag{3.18}$$

Thus, the connection between the spin and spatial parts is indirect, although necessary, imposed by the antisymmetry of the total wavefunction. Because of this connection, the effect of the electrostatic interaction between the electronic charges may be described as an interaction between spins. Also, the motions of the electrons with parallel or antiparallel spins are correlated; for example, electrons of parallel spins tend to avoid each other.

Expanding the spin product, we obtain

$$\mathbf{s}_1 \cdot \mathbf{s}_2 = \tfrac{1}{2}[(\mathbf{s}_1 + \mathbf{s}_2)^2 - \mathbf{s}_1^2 - \mathbf{s}_2^2] \tag{3.19}$$

For electrons, $s = \tfrac{1}{2}$ and

$$\langle \mathbf{s}_1^2 \rangle = \langle \mathbf{s}_2^2 \rangle = \tfrac{3}{4} \tag{3.20}$$

in units of $\hbar$. The brackets $\langle \cdots \rangle$ indicate the expectation value, or the quantum average of the operator. The expectation value of the operator total spin squared is given by

$$\langle (\mathbf{s}_1 + \mathbf{s}_2)^2 \rangle = S(S+1) \tag{3.21}$$

This mean value will be equal to 0, for antiparallel spins ($S = 0$), or equal to 2, in the parallel case ($S = 1$).

The corresponding energies become

$$E_+ = E_0 + K_{12} + \mathcal{J}_{12} \quad \text{for} \quad \langle \mathbf{s}_1 \cdot \mathbf{s}_2 \rangle = -\tfrac{3}{4} \quad (S = 0) \tag{3.22a}$$

$$E_- = E_0 + K_{12} - \mathcal{J}_{12} \quad \text{for} \quad \langle \mathbf{s}_1 \cdot \mathbf{s}_2 \rangle = +\tfrac{1}{4} \quad (S = 1) \tag{3.22b}$$

Adding $(-2\mathcal{J}_{12}\langle \mathbf{s}_1 \cdot \mathbf{s}_2 \rangle - 3\mathcal{J}_{12}/2)$ to the first equation and $(-2\mathcal{J}_{12}\langle \mathbf{s}_1 \cdot \mathbf{s}_2 \rangle + \mathcal{J}_{12}/2)$ to the second equation (which does not alter them), we obtain the equation

$$E_\pm = E_0 + K_{12} - \tfrac{1}{2}\mathcal{J}_{12} - 2\mathcal{J}_{12}\langle \mathbf{s}_1 \cdot \mathbf{s}_2 \rangle \tag{3.23}$$

The conclusion is that the introduction of the interaction term V_{12} between the spins leads to the appearance of a new energy term; this result can be accounted for by including in the energy a term dependent on the relative orientation of these spins:

$$-2\mathcal{J}_{12}\langle \mathbf{s}_1 \cdot \mathbf{s}_2 \rangle \tag{3.24}$$

which can be used to express the two energy states of Eq. (3.11) (Fig. 3.1).

In a solid, the hamiltonian describing the interaction is:

$$\mathcal{H} = -2\mathcal{J} \sum_{i<j} \mathbf{S}_i \cdot \mathbf{S}_j \tag{3.25}$$

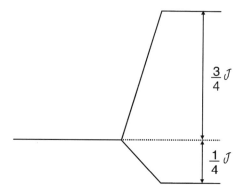

Figure 3.1 Energy levels of a system of two spins $\frac{1}{2}$ for $\mathcal{J}_{12} > 0$. The highest energy state corresponds in this case to one spin function with $S = 0$; the lowest state corresponds to three spin functions, with $S = 1$ (threefold degenerate).

where now the sum is performed on each pair of atoms (i,j) and $\mathcal{J}$ is an effective exchange parameter. This is known as the *Heisenberg hamiltonian*, and it is widely used for the description of many magnetic properties of materials, particularly insulators.

In the formulas in this chapter, S is measured in units of $\hbar$; otherwise, this expression would appear divided by $\hbar^2$.

In the study of the magnetism of the rare earths, since J is a good quantum number, the preceding interaction [Eq. (3.25)] is written, using the projection of $\mathbf{S}$ on the direction of $\mathbf{J}$, $\boldsymbol{\sigma} = (g-1)\mathbf{J}$, where g is the Landé g-factor:

$$\mathcal{H} = -2\mathcal{J}\sum_{i<j}\boldsymbol{\sigma}_i \cdot \boldsymbol{\sigma}_j \tag{3.26}$$

Expanding the scalar product, the Heisenberg hamiltonian is written

$$\mathcal{H} = -2\mathcal{J}\sum_{i<j}(S_i^x S_j^x + S_i^y S_j^y + S_i^z S_j^z) \tag{3.27}$$

A spin system with a privileged direction, defined, for example, by an external magnetic field, or by an axial crystalline anisotropy, may be described by a simplified hamiltonian. Its expression is

$$\mathcal{H} = -2\mathcal{J}\sum_{i<j}S_i^z S_j^z \tag{3.28}$$

known as the *Ising hamiltonian*. This hamiltonian accurately describes the magnetism of a large number of physical systems.

3.2 THE MEAN FIELD

We will show the relation between the Weiss molecular field and the Heisenberg hamiltonian. We start by describing, with the Heisenberg hamiltonian, the interaction of an atom of spin $\mathbf{S}_i$ with its z near neighbors:

$$\mathcal{H}_i = -2J \sum_j^z \mathbf{S}_i \cdot \mathbf{S}_j \qquad (3.29)$$

Expressing this in terms of the projection σ of $\mathbf{S}$ on the direction of the total angular momentum $\mathbf{J}$, we have

$$\mathcal{H}_i = -2J \sum_j^z \sigma_i \cdot \sigma_j \qquad (3.30)$$

If there exists spontaneous magnetic order, with magnetization $\mathbf{M}$, we can assume that the individual magnetic moments feel a mean field; in the molecular field approximation, this is given by $\lambda_m \mathbf{M}$, which is proportional to the average magnetic moment $\overline{\mu}$:

$$\mathbf{B}_m = \lambda_m \mathbf{M} = \lambda_m n \overline{\mu} = -\lambda_m n g \mu_B \langle \mathbf{J} \rangle_T \qquad (3.31)$$

where n is the number of magnetic moments per unit volume and $\langle \mathbf{J} \rangle_T$ is the thermal average of $\mathbf{J}$. The concept of a mean field (or of a molecular field) is applicable if the amplitude of the fluctuations in the magnetic field acting on the atomic moments is not very large on a given site, and if it is small from point to point.

One can approximate the interaction of the ion i:

$$\mathcal{H}_i = -2J(g-1)^2 \left(\sum_j^z \mathbf{J}_j \right) \cdot \mathbf{J}_i \approx -2J(g-1)^2 z \langle \mathbf{J} \rangle_T \cdot \mathbf{J}_i \qquad (3.32)$$

Equating the exchange interaction of the spin i Eq. [(3.29)] to the interaction of the moment μ that is acted on by the molecular field

$$\lambda_m n g \mu_B \langle \mathbf{J} \rangle_T \cdot \mu = -2J(g-1)^2 z \langle \mathbf{J} \rangle_T \cdot \mathbf{J}_i \qquad (3.33)$$

we finally obtain, using $\mu = -g\mu_B \mathbf{J}$ and assuming that the sum is done on the z near neighbors

$$J = \left(\frac{n g^2 \mu_B^2}{2z(g-1)^2} \right) \lambda_m \qquad (3.34)$$

From this expression one sees that the exchange integral $\mathcal{J}$ is proportional to the molecular field constant λ_m.

As examples of the magnitudes of $\mathcal{J}$, one can quote $\mathcal{J}(\text{Fe}) = 0.015$ meV, $\mathcal{J}(\text{Ni}) = 0.020$ meV.

3.3 INDIRECT INTERACTIONS IN METALS

The values of the magnetic moments of the pure rare earths are approximately the same as the values corresponding to the free ions. This happens since the 4f electrons are localized; that is, they have a mean radius $\langle r \rangle$ much smaller than the interionic distance d_{RR} (Fig. 3.2), and therefore are not much affected by the chemical bonds. One consequence of this localization is that the mechanism that gives rise to rare earth magnetic order is not the superposition of the 4f orbitals in neighbor atoms; other electrons besides the 4f electrons must be responsible for this order. It turns out that the conduction electrons, which have an itinerant character, play a decisive role in the ordering mechanism.

The first theoretical treatment of this coupling between the atomic spins through the conduction electrons is due to Zener (Zener 1951), who assumed three effective exchange constants; one between each atomic spin and its first neighbors, another between the atomic spins and every conduction electron, and a third connecting each conduction electron to all the others. This can be simplified to an effective hamiltonian containing interactions between the

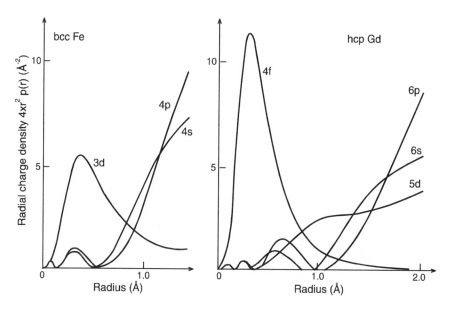

Figure 3.2 Normalized charge density of the electrons in bcc iron and hcp gadolinium, as a function of the radius (in atomic units). [Reprinted from R. Coehoorn, *Supermagnets, Hard Magnetic Materials*, 1990, p. 140, with kind permission from Kluwer Academic Publishers, Amsterdam.]

atomic spins S, and between atomic spins and conduction electron spins:

$$\mathcal{H} = -2 \sum_i^N \sum_j^z \mathcal{J}_{ij} \mathbf{S}_i \cdot \mathbf{S}_j - 2 \sum_{k<l} \mathcal{J}_{kl}^{\text{eff}} \mathbf{S}_k \cdot \mathbf{s}_l \tag{3.35}$$

where $\mathbf{s}$ are the conduction electron spins; this is a first approximation for the description of the magnetism of the rare earths.

The Zener model leads to a uniform polarization (or spin density) of the conduction electrons.

A more adequate description, however, allows the conduction (or itinerant) electrons to have a nonuniform spin density; this can be obtained with a susceptibility $\chi(\mathbf{r})$ that is nonlocal. This is equivalent to a susceptibility $\chi(\mathbf{q})$ dependent on the wavevector $\mathbf{q}$ ($|\mathbf{q}| = 2\pi/\lambda$). The polarization of the itinerant electrons, in this case, has the form (e.g., Martin 1967):

$$\rho(\mathbf{r})_\uparrow - \rho(\mathbf{r})_\downarrow = \sum_q [A(\mathbf{q}) \cos(\mathbf{q} \cdot \mathbf{r}) + B(\mathbf{q}) \sin(\mathbf{q} \cdot \mathbf{r})] \tag{3.36}$$

where $A(\mathbf{q})$ and $B(\mathbf{q})$ are the Fourier coefficients of the spin polarization. It should be noted that although the spin polarization varies spatially, the charge density is not affected.

The interaction leading to the preceding result is described by the hamiltonian

$$\mathcal{H} = -2 \sum_{i\langle j} \mathcal{J}_a(\mathbf{R}_i - \mathbf{R}_j) \mathbf{S}_i \cdot \mathbf{S}_j \tag{3.37}$$

with the indirect atomic exchange constant $\mathcal{J}_a$ given by the Fourier expansion

$$\mathcal{J}_a(\mathbf{R}_i - \mathbf{R}_j) = \sum_q \frac{\chi(\mathbf{q}) \mathcal{J}(\mathbf{q})^2}{4n^2 g^2 \mu_B^2} \cos[\mathbf{q} \cdot (\mathbf{R}_i - \mathbf{R}_j)] \tag{3.38}$$

This quantity exhibits an oscillatory behavior with the separation between the spins, and also an attenuation arising from the $\mathbf{q}$ dependence of the amplitude $\chi(\mathbf{q}) \mathcal{J}(\mathbf{q})^2$ (Fig. 3.3).

The susceptibility $\chi(\mathbf{q})$ is given as a function of the Pauli susceptibility function χ_P by (see Section 4.2)

$$\chi(\mathbf{q}) = \chi_P \left\{ \frac{1}{2} + \frac{4k_F^2 - q^2}{8k_F q} \ln \left| \frac{2k_F + q}{2k_F - q} \right| \right\} \tag{3.39}$$

where k_F is the value of the wavevector k at the Fermi level.

Making the approximation $\mathcal{J}(\mathbf{q}) \approx \mathcal{J}(0)$, we can obtain, from Eq (3.38), the

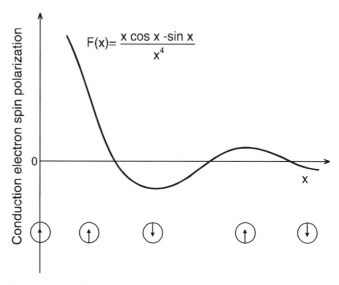

Figure 3.3 Dependence with the distance of the exchange integral $\mathcal{J}_a(\mathbf{R}_i - \mathbf{R}_j)$ in the indirect interaction in metals, according to the RKKY model.

result of Ruderman and Kittel (1954):

$$\mathcal{J}_a(\mathbf{R}_i - \mathbf{R}_j) = -\frac{\chi_P \mathcal{J}(0)^2}{4n^2 g^2 \mu_B^2} 12\pi n \, F(2k_F \, | \, \mathbf{R}_i - \mathbf{R}_j \, |) \quad (3.40)$$

with the function F given by

$$F(x) = \frac{1}{x^4}(x \cos x - \sin x) \quad (3.41)$$

Therefore, $\mathcal{J}_a(\mathbf{R}_i - \mathbf{R}_j)$ oscillates with the distance, with period $1/2k_F$, and its amplitude decreases as $|\mathbf{R}_i - \mathbf{R}_j|^{-3}$. Consequently, the conduction electron polarization presents the same oscillatory behavior; this is the most important result of the so-called RKKY (Ruderman–Kittel–Kasuya–Yosida) model.

The preceding results were obtained from a susceptibility $\chi(\mathbf{q})$ of electrons that do not interact among themselves. The existence of electron–electron Coulomb interactions increases the susceptibility. This fact may be taken into account in a simple way. The magnetization due to a field $\mathbf{H}(\mathbf{q})$ is

$$\mathbf{M}(\mathbf{q}) = \chi(\mathbf{q})\mathbf{H}(\mathbf{q}) \quad (3.42)$$

To include the electron–electron interaction, it suffices to assume that the electrons feel a molecular field due to their own magnetization:

$$\mathbf{M}(\mathbf{q}) = \chi(\mathbf{q})[\mathbf{H}(\mathbf{q}) + v\mathbf{M}(\mathbf{q})] \quad (3.43)$$

Table 3.I Enhancement factors $\{F = 1/[1 - v\chi(\mathbf{q})]\}$ for some metals

Element	Mo	Pd	Os
F	4.6	9.3	0.4

Source: Reprinted from Landolt-Börnstein, *Magnetic Properties of 3d Elements, New Series III/19a*, Springer-Verlag, New York, 1986, p. 39, with permission.

where $v = \lambda/\mu_0$ is a molecular field coefficient that measures the strength of the electron–electron interaction (μ_0 is the vacuum permeability).

Solving for magnetization, we obtain

$$\mathbf{M}(\mathbf{q}) = \frac{\chi(\mathbf{q})}{1 - v\chi(\mathbf{q})} \mathbf{H}(\mathbf{q}) \qquad (3.44)$$

from which one can derive a new susceptibility $\chi_e(\mathbf{q})$, called *enhanced susceptibility*, which includes the effect of electron–electron interactions:

$$\chi_e(\mathbf{q}) = \frac{\chi(\mathbf{q})}{1 - v\chi(\mathbf{q})} \qquad (3.45)$$

One therefore finds that the magnetic response of the electrons in the case where they interact with their own magnetization is amplified by an enhancement factor $F = 1/[1 - v\chi(\mathbf{q})]$. This factor attains a value of the order of 10 in the case of palladium; Table 3.I shows some values of F.

Experimentally, positive and negative effective exchange integrals are observed. Typical values of $\mathcal{J}$ for the 3d elements range from 10^{-21} to 10^{-20} J, corresponding to $\mathcal{J}/k$ varying from 10^2 to 10^3 K.

In metallic systems containing rare earths, one mechanism that may lead to negative values of $\mathcal{J}$ (Anderson and Clogston 1961) depends on the $s-f$ hybridization. In other words, it depends on the mixing of s and f character of the electrons, or on the virtual occupation of f states by the s electrons. An electron with wavevector $\mathbf{k}$ is absorbed in a nonoccupied $4f$ state, and reemitted with wavevector $\mathbf{k}'$. This process lowers the energy of the occupied $4f$ state and of the conduction electrons with spin parallel to the localized spin, therefore increasing the number of electrons of antiparallel spin to the $4f$ spin S_{4f}, and this is equivalent to a negative effective exchange parameter $\mathcal{J}$ (Fig. 3.4).

3.4 PAIR OF SPINS IN THE MOLECULAR FIELD (OGUCHI METHOD)

In the Weiss model each spin feels the result of the long-range magnetic order of the material through the molecular field; the role of the individual spins is to contribute to this grand average. The treatment is equivalent to taking into account only the time average of the projection of each spin. In real solids, however, the motion of a given spin shows strong correlation with the motion of its

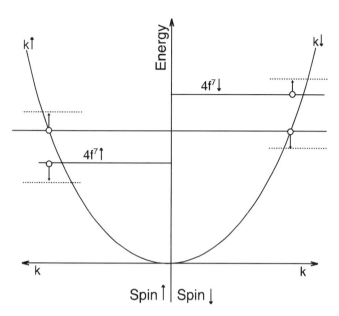

Figure 3.4 Nonmagnetized conduction bands of metallic Gd (spinup and spindown subbands represented by $k_\uparrow$ and $k_\downarrow$) showing the full $4f_\uparrow$ level (below E_F) and empty $4f_\downarrow$ (above E_F). The intraband mixing is responsible for the effective interaction, with $\mathcal{J} < 0$. [Reprinted from R. E. Watson, in *Hyperfine Interactions*, A. J. Freeman and R. B. Frankel, Eds., Academic Press, New York, 1967, p. 443.]

near-neighbor spins. In fact, even above the critical temperature T_C (the Curie temperature in a ferromagnet, i.e., the temperature above which there is no long- range magnetic order), some degree of local order is observed; in a small region around each spin, the moments remain correlated. A simple model to take into account this type of short-range correlation was proposed by Oguchi (1955). In this model one spin interacts with one of its neighbors, and this pair feels the effects of the other spins through a mean field.

The Weiss model considers a single magnetic moment that is coupled to the other moments through a mean field (or molecular field):

$$B_m = \lambda_m n \langle \mu_j^z \rangle_T \tag{3.46}$$

where $\langle \cdots \rangle_T$ denotes a thermal average.

This model does not take into account the correlation expected between the motions of neighbor magnetic moments. A measure of such correlation is given by the order parameter τ:

$$\tau = \frac{1}{J^2} \langle \mathbf{J}_i \cdot \mathbf{J}_j \rangle_T \tag{3.47}$$

where i and j label moments at neighbor sites.

In the Weiss approximation, the magnetization for external field $B_0 = 0$ and temperature T is given by

$$M_{0T} = n\langle \mu_J^z \rangle_T = g\mu_B n \langle J^z \rangle_T \qquad (3.48)$$

where n is the number of magnetic moments per unit volume and $\mathbf{M}_{0T}$ is the spontaneous magnetization, at temperature T:

$$\langle J^z \rangle_T = \frac{M_{0T}}{g\mu_B n} \qquad (3.49)$$

At $T = 0$

$$M_{00} = g\mu_B n J \qquad (3.50)$$

Noting that in the Weiss model, since the motion of each spin is independent of the other, $\langle \mathbf{J}_i \cdot \mathbf{J}_j \rangle = \langle \mathbf{J}_i \rangle \cdot \langle \mathbf{J}_j \rangle$, we can then obtain the values of the order parameter τ at different temperatures:

$$\begin{cases} 1 & \text{for } T = 0 \\ \left(\dfrac{M_{0T}}{g\mu_B n J}\right)^2 & \text{for } T < T_C \\ 0 & \text{for } T \geq T_C \text{ (since } M_{0T} = 0) \end{cases} \qquad (3.51)$$

Above T_C the order parameter is zero; in this region there is neither long-range order (or magnetization) nor short range order (or correlation between the spins in neighbor sites).

This is in disagreement with the usual experimental behavior of $1/\chi$ for $T > T_C$ (Fig. 3.5), and of the specific heat C_p; both quantities reflect the consequences of local order.

To describe this type of behavior, a model considering a coupled spin pair in a mean field was proposed (Oguchi 1955). The starting point is the hamiltonian (Smart 1966):

$$\mathcal{H} = -2J\mathbf{S}_i \cdot \mathbf{S}_j - g\mu_B(J_i^z + J_j^z)B \qquad (3.52)$$

where the first term describes the interaction between the two spins, and the second term the interaction of the pair with the total field $\mathbf{B} = B\mathbf{k}$. This is equivalent, in terms of the angular momentum operator $\mathbf{J}$, to

$$\mathcal{H} = -2\mathcal{J}(g-1)^2 \mathbf{J}_i \cdot \mathbf{J}_j - g\mu_B(J_i^z + J_j^z)B \qquad (3.53)$$

Defining the total angular momentum operator of the pair

$$\mathbf{J}' = \mathbf{J}_i + \mathbf{J}_j \qquad (3.54)$$

and following the same steps of the derivation of the Weiss model (Section 2.3),

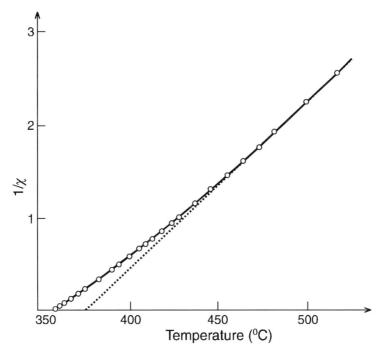

Figure 3.5 Curve of the inverse of the susceptibility of Ni as a function of temperature above T_C; the deviation from Curie–Weiss law results from the persistence of local order above this temperature.

we derive, for $J = \frac{1}{2}$, the value of the z component of $\mathbf{J}'$:

$$\langle J'^z \rangle_T = \frac{2\sinh(b)}{1 + \exp(-2j) + 2\cosh(b)} \qquad (3.55)$$

with

$$j = \frac{\mathcal{J}(g-1)^2}{kT}$$

$$b = \frac{g\mu_B B}{kT} \qquad (3.56)$$

The magnetic moment per ion (half the moment of the pair) is

$$\langle \mu^z \rangle_T = \frac{g\mu_B}{2} \langle J'^z \rangle_T = \frac{g\mu_B \sinh(g\mu_B B/kT)}{1 + \exp(-2\mathcal{J}(g-1)^2/kT) + 2\cosh(g\mu_B B/kT)} \qquad (3.57)$$

compared to the result of the Weiss model, which is, for spin $\frac{1}{2}$:

$$\langle \mu^z \rangle_T^W = \frac{1}{2} g\mu_B B_{1/2}(x) = \frac{1}{2} g\mu_B \tanh\left(\frac{g\mu_B B}{kT}\right) \tag{3.58}$$

To describe a ferromagnetic system, we will initially return to the Weiss model; this time we will distance ourselves from the approach of Section 2.6 and assume that the molecular field is due to the z near neighbors of the spin, as in Section 3.2. Then the molecular field constant becomes, from Eq. (3.34)

$$\lambda_m^W = \frac{2\mathcal{J}}{ng^2\mu_B^2}(g-1)^2 z \tag{3.59}$$

Analogously, the molecular field constant in the Oguchi model due to the $z-1$ neighbors (excluding the one forming the pair with the central ion) is

$$\lambda_m^O = \frac{2\mathcal{J}}{ng^2\mu_B^2}(g-1)^2(z-1) \tag{3.60}$$

In a similar way to the derivation of Section 2.6, we can obtain the transition temperature (T_C) in the Oguchi model, deriving the expression of the temperature at which the spontaneous magnetization vanishes, in zero external field. This is given by the equation

$$\exp(-2j_c) + 3 = 2(z-1)j_c \tag{3.61}$$

where z is the number of nearest neighbors and j_c is the quantity j defined above [Eq. (3.56)], for $T = T_C$. The value of T_C can be obtained, for different values of z, by solving Eq. (3.61). It turns out that the values of T_C given by the Oguchi model are lower than those given by the Weiss model, for the same $\mathcal{J}$ parameters. This is a general result; the incorporation of local order effects lowers the transition temperature of the magnetic system (Smart 1966).

To obtain the magnetic susceptibility we begin by computing the magnetization $\mathbf{M}_{BT}$ at temperature T and applied field B. The susceptibility per mole is given by

$$\chi_m = \frac{\partial \mathbf{M}_m}{\partial \mathbf{H}} = \mu_0 \frac{\partial \mathbf{M}_m}{\partial \mathbf{B}} \tag{3.62}$$

and it follows that

$$\chi_m = \frac{\mu_0 g^2 \mu_B^2 N}{kT(\exp[-2\mathcal{J}(g-1)^2/kT] + 3) - 2(z-1)\mathcal{J}(g-1)^2} \tag{3.63}$$

At high temperatures, one can approximate

$$\exp\left(-\frac{2\mathcal{J}(g-1)^2}{kT}\right) \cong 1 - \frac{2\mathcal{J}(g-1)^2}{kT} \tag{3.64}$$

and χ_m becomes

$$\chi_m = \frac{\mu_0 g^2 \mu_B^2 N/4k}{T - 2\mathcal{J}(g-1)^2 z/4k} \tag{3.65}$$

The numerator in the expression of χ_m is the Curie constant in the Weiss model (for $J = \frac{1}{2}$) [Eq. (2.64)], and we may rewrite Eq. (3.65) as

$$\chi_m = \frac{C}{T - \theta_p} \tag{3.66}$$

The paramagnetic Curie temperature θ_P in the Oguchi model is given by

$$\theta_P^O = \frac{\mathcal{J}z}{2k}(g-1)^2 \tag{3.67}$$

which is the same result of the Weiss model [from Eqs. (2.91) and (3.60)]. The conclusion is that at high temperatures the two models show quantitative agreement.

For intermediate temperatures, the approximation Eq. (3.64) is not valid, and we find that χ tends to infinity as T approaches T_C, but in this case $1/\chi$ is not proportional to $(T - \theta_P)$, that is, the dependence of the inverse susceptibility is not linear with T, as seen in Eq. (3.63).

Finally, the correlation function, or short-range order parameter, is given, for $J_1 = J_2 = \frac{1}{2}$ and $B_0 = 0$, by the statistical average:

$$\tau = 4\langle \mathbf{J}_i \cdot \mathbf{J}_j \rangle_T \tag{3.68}$$

$$\tau = \frac{4}{\mathcal{Z}} \sum_M \sum_J [J'(J'+1) - J_1(J_1+1) - J_2(J_2+1)] Tr[\exp(-\mathcal{H}_p/kT)] \tag{3.69}$$

with $\mathcal{Z}$ the partition function for the pair. The result is (Smart 1966):

$$\tau = \frac{(2\cosh(b)+1) - 3\exp(-2j)}{(2\cosh(b)+1) + \exp(-2j)} \tag{3.70}$$

with $b = g\mu_B B/kT$ and $j = \mathcal{J}(g-1)^2/kT$.

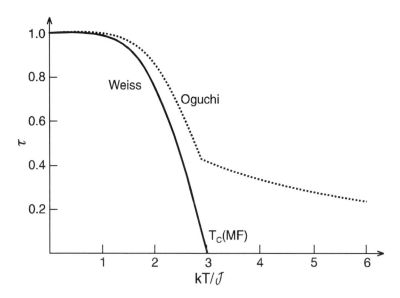

Figure 3.6 Dependence of the order parameter τ with the reduced temperature $kT/\mathcal{J}$ in the Weiss model, and in the Oguchi model, showing in the latter the persistence of local order above T_C ($z = 6$, $S = \frac{1}{2}$). [Reprinted from J. Smart, *Effective Field Theories of Magnetism*, Saunders, Philadelphia, PA, 1966, p. 42.]

The order parameter τ is given, noting that $B = 0$ for $T > T_C$:

$$\tau = \begin{cases} 1 & \text{for } T = 0 \quad \text{(as in the Weiss model)} \\ \dfrac{3[1 - \exp(-2j)]}{[3 + \exp(-2j)]} \neq 0 & \text{for } T > T_C \end{cases} \quad (3.71)$$

Thus, the model predicts that the local order subsists above T_C (Fig. 3.6), as observed experimentally by the dependence of $\chi(T)$ or by neutron scattering. The Oguchi method therefore overcomes this limitation in the Weiss model, of not accounting for local order effects.

There are other models that describe magnetic systems in terms of a pair of spins under the action of a molecular field; in the case of the constant coupling approximation, for example, this field is not proportional to the magnetization, and is obtained from statistical considerations. Instead of a pair of atoms, a larger cluster has also been considered; in the Bethe–Peierls–Weiss method the cluster has $z + 1$ spins under a molecular field.

3.5 SPIN WAVES: INTRODUCTION

In a ferromagnet at $T = 0$ K all the spins have the maximum projection S in the z direction; this is the ground state configuration. As the temperature is raised the

Figure 3.7 Schematic representation of (a) a spin wave propagating along a linear chain of spins in the x direction; (b) the same seen along the z direction. [Reprinted from C. Kittel, *Introduction to Solid State Physics*, 7th ed. Copyright © 1995, John Wiley & Sons, Inc., New York. Reprinted by permission of John Wiley & Sons, Inc.]

projections are reduced, and a classical image of this effect is shown in Fig. 3.7. A wavelike perturbation flows through the spin system: the spin wave. The spin wave theory leads to the description of the magnetism of ferromagnets at low temperatures, in the regime where $J^z \cong J$. We will introduce the spin waves first through a macroscopic description, and then show the relation to a simple microscopic model.

Let us consider the projection of the magnetization $M^z(x)$ varying continuously from point to point; if the magnetization deviates from its saturation value at a given point, a torque $\mathbf{M}(x) \times A\nabla^2\mathbf{M}(x)$ acts on the magnetization, and the equation of motion is (Martin 1967)[1]:

$$\frac{1}{\gamma}\dot{\mathbf{M}}(x) = \mathbf{M}(x) \times A\nabla^2\mathbf{M}(x) \qquad (3.72)$$

where $A = D/(\hbar\gamma M_0)$, with M_0 the saturation magnetization and D a parameter called "stiffness constant", which measures the strength of the tendency of aligning the local magnetization to recover its saturation value; γ is the gyromagnetic ratio of the atomic moments μ:

$$\boldsymbol{\mu} = \gamma\hbar\mathbf{S} = v\mathbf{M} \qquad (3.73)$$

In this expression, $\mathbf{M}$ is the magnetization and v is the volume occupied by one atom; dividing by v, we obtain the expression for the spin density $\mathcal{S}(x)$ in terms of the local magnetization $\mathbf{M}(x)$:

$$\mathcal{S}(x) = \frac{\mathbf{M}(x)}{\hbar\gamma} \qquad (3.74)$$

We look for the deviations $\mathbf{m}$ from the uniform magnetization $\mathbf{M}_0$:

$$\mathbf{m} = \mathbf{M} - \mathbf{M}_0 \qquad (3.75)$$

---

[1] The laplacian of a vector $\mathbf{M}$ is a vector of components $\nabla^2 M_x$, $\nabla^2 M_y$, and $\nabla^2 M_z$.

The solution of Eq. (3.72) gives

$$\mathbf{m} = m_0(\sin\omega t \mathbf{i} + \cos\omega t \mathbf{j})\sin(\mathbf{k}\cdot\mathbf{r}) \tag{3.76}$$

with

$$\omega = \frac{D}{\hbar}k^2 \tag{3.77}$$

where k is the modulus of the wavevector.

We will now discuss the spin waves within a microscopic model. The Heisenberg hamiltonian predicts that the lowest energy state (the ground state) of the spin system corresponds to a configuration with all spins aligned in parallel. It is easier to show this fact if we write the hamiltonian in terms of the operators

$$S^+ = S^x + iS^y \tag{3.78a}$$

$$S^- = S^x - iS^y \tag{3.78b}$$

where $i = \sqrt{-1}$. The hamiltonian becomes

$$\mathcal{H} = -2J\sum_{i<j}^{Nz}(\tfrac{1}{2}S_i^+ S_j^- + \tfrac{1}{2}S_i^- S_j^+ + S_i^z S_j^z) \tag{3.79}$$

Using the matrix form of the spin wavefunctions (the eigenvectors of S^z)

$$\chi_\alpha = \begin{pmatrix} 1 \\ 0 \end{pmatrix} \qquad \chi_\beta = \begin{pmatrix} 0 \\ 1 \end{pmatrix} \tag{3.80}$$

and of the spin operators (the Pauli matrices)

$$S^x = \frac{\hbar}{2}\begin{pmatrix} 0 & 1 \\ 1 & 0 \end{pmatrix} \qquad S^y = \frac{\hbar}{2}\begin{pmatrix} 0 & -i \\ i & 0 \end{pmatrix} \qquad S^z = \frac{\hbar}{2}\begin{pmatrix} 1 & 0 \\ 0 & -1 \end{pmatrix} \tag{3.81}$$

and recalling the rule of matrix multiplication

$$\begin{pmatrix} a & b \\ c & d \end{pmatrix}\begin{pmatrix} e \\ f \end{pmatrix} = \begin{pmatrix} ae + bf \\ ce + df \end{pmatrix} \tag{3.82}$$

we obtain

$$S^+\chi_\alpha = 0, \qquad S^-\chi_\beta = 0 \tag{3.83}$$

and the property that justifies the notation for the spin operators, with superscripts + and − is:

$$S^+ \chi_\beta = \hbar \chi_\alpha, \qquad S^- \chi_\alpha = \hbar \chi_\beta \qquad (3.84)$$

This means that the operator S^+ applied to a function corresponding to spin $-\frac{1}{2}$ transforms the function to that of spin $+\frac{1}{2}$; conversely, S^- inverts the spin from $+\frac{1}{2}$ to $-\frac{1}{2}$.

The total spin wavefunction for a system of N aligned atomic spins is the product of the individual functions:

$$\chi = \chi_\alpha(1)\chi_\alpha(2)\chi_\alpha(3)\cdots\chi_\alpha(N) \qquad (3.85)$$

Using the properties of the spin operators and χ functions described above, it is easy to demonstrate that the preceding wavefunction [Eq. (3.85)] satisfies Schrödinger's equation:

$$\mathcal{H}\chi = -\hbar^2 \frac{\mathcal{J}}{4} Nz\, \chi \qquad (3.86)$$

It can be demonstrated that this function corresponds to the minimum energy, that is, to the ground state. This is done by noting that the maximum value of $\langle \mathbf{S}_i \cdot \mathbf{S}_j \rangle$ is $\hbar^2/4$, and therefore the minimum of the energy

$$E = -2\mathcal{J} \sum_i^N \sum_j^z \langle \mathbf{S}_i \cdot \mathbf{S}_j \rangle \qquad (3.87)$$

is $-\hbar^2(\mathcal{J}/4)Nz$, in agreement with the preceding result [Eq. (3.86)]. We conclude that the perfectly aligned configuration is the ground state for $\mathcal{J} > 0$. A set of spins coupled ferromagnetically is aligned in parallel at 0 K.

We will now discuss the excited states of the spin system. A spin system in thermal contact with a thermal reservoir (e.g., the lattice in a solid) will not be in its ground state configuration, if $T \neq 0$. Assuming that the excited states are characterized only by changes in the orientational state of the spins, the excitation will imply a reduction in the spin projection along the quantization direction: as the temperature is raised, the z component of the magnetization $\mathbf{M}$ decreases.

We may describe this process with a classical image of spins precessing around the z direction with an angle θ that varies along the x direction. In energy terms, this form of excitation is less costly than the reduction of the magnetization through inversion of spins (Fig. 3.7).

We will consider a system of N spins, each one interacting with z neighbors, and in the presence of a magnetic field B. The hamiltonian is (assuming only spin angular momentum, i.e., $\mathbf{J} = \mathbf{S}$):

$$\mathcal{H} = -2J \sum_i^N \sum_j^z \mathbf{S}_i \cdot \mathbf{S}_j - g\mu_B \sum_i^N S_i^z B \qquad (3.88)$$

Let us take, for simplicity, a one-dimensional spin system; in this case $z = 2$. Neglecting the second term, that describes the interaction of the spins with the field $\mathbf{B}$ (Zeeman term), we obtain for the energy of the spins, in the classical limit (at $T = 0$):

$$E_1 = -2(N-1)JS^2 \qquad (3.89)$$

If instead of a spin system with all spins aligned, one had $N-1$ aligned spins and one antiparallel spin, the energy would be

$$E_2 = -2(N-3)JS^2 + 2 \times 2JS^2 \qquad (3.90)$$

This energy is larger than that in the ferromagnetic case (preceding case); the difference is

$$\Delta E = E_2 - E_1 = 8JS^2 \qquad (3.91)$$

We will show that the spin can take excited configurations with energy much lower than the preceding energy, if spin waves are created. The classical expression of the energy of the spin of number p, in a linear chain, interacting with two nearest neighbor atoms $[(p-1)$ and $(p+1)]$ is

$$E_p = -2J\mathbf{S}_{p-1} \cdot \mathbf{S}_p - 2J\mathbf{S}_p \cdot \mathbf{S}_{p+1} = -2J(\mathbf{S}_{p-1} + \mathbf{S}_{p+1}) \cdot \mathbf{S}_p \qquad (3.92)$$

which is equivalent to

$$E_p = -\frac{2J}{g\mu_B}(\mathbf{S}_{p-1} + \mathbf{S}_{p+1}) \cdot g\mu_B \mathbf{S}_p = -\mathbf{B}_p \cdot \boldsymbol{\mu}_p \qquad (3.93)$$

where $\mathbf{B}_p$ is the field due to the neighbors, acting on the moment p.

Equating the rate of change of the angular momentum $\hbar \mathbf{S}_p$ to the torque $\boldsymbol{\mu}_p \times \mathbf{B}_p$, one obtains

$$\hbar \dot{\mathbf{S}}_p = 2J\mathbf{S}_p \times (\mathbf{S}_{p-1} + \mathbf{S}_{p+1}) = 2J\mathbf{S}_p \times \sum_i^z \mathbf{S}_i \qquad (3.94)$$

where the sum is done over the z neighboring spins in the linear chain. Looking

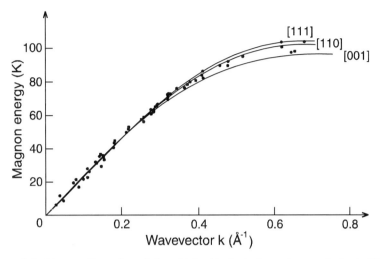

Figure 3.8 Magnon dispersion relation obtained by inelastic neutron scattering in RbMnFe$_3$ at 4.2 K; the curves are calculated for different directions. [Reprinted from C. Kittel, *Introduction to Solid State Physics*, 7th ed. Copyright © 1995, John Wiley & Sons, Inc., New York. Reprinted by permission of John Wiley & Sons, Inc.]

for solutions of the type

$$S_p^x = U \exp[i(pka - \omega t)] \tag{3.95a}$$

$$S_p^y = V \exp[i(pka - \omega t)] \tag{3.95b}$$

where a is the lattice spacing, U and V are constants, and p is an integer, we obtain the condition

$$\hbar\omega(k) = 4JS[1 - \cos(ka)] \tag{3.96}$$

The function $\omega(k)$ is called a *dispersion relation*; this is its expression for spin waves (Fig. 3.8). We made the approximation $S_p^x, S_p^y \ll S$ and $S_p^z \cong S$, valid for small deviations of the spins from the equilibrium position. We also obtain $V = -i\,U$, which shows that the motion of the spins is a precession around the z axis. The angular momenta precess around the direction z, and this excitation propagates along the chain in the plane (x, y).

In the limit of long wavelengths, since $k = 2\pi/\lambda$, $ka \ll 1$, $(1 - \cos ka) = 2\sin^2(ka/2) \approx \frac{1}{2}(ka)^2$ and the dispersion relation becomes

$$\hbar\omega = 2JSa^2k^2 \tag{3.97}$$

This is the same as (to be shown below):

$$\hbar\omega = Dk^2 \tag{3.98}$$

where D is the spin wave stiffness constant.

In the quantum description the total spin quantum number of the set of N spins may have values NS, $NS - 1$, $NS - 2$, and so forth.

Therefore, the z component of a spin is S^z given by [using S_p^x and S_p^y of Eqs. (3.95)]:

$$S^z = \sqrt{S^2 - S^{x2} - S^{y2}} = \sqrt{(S^2 - U^2)} \cong S - \frac{U^2}{2S} \tag{3.99}$$

for small values of U/S.

The number $N(S - S^z)$ that gives the reduction in the projection of the total spin in the z direction can attain only integer values. If this reduction is associated with the appearance of n_k spin waves of wavevector k, and each wave reduces the spin of one unit, we have

$$N(S - S^z) \cong N\frac{U_k^2}{2S} = n_k \tag{3.100}$$

or

$$U_k^2 = \frac{2Sn_k}{N} \tag{3.101}$$

The energy of interaction of N pairs of spins is

$$E = -2J\sum_i^N \mathbf{S}_p \cdot \mathbf{S}_i = -2JNS^2 \cos\phi \tag{3.102}$$

The angle ϕ is given (Fig. 3.9) by

$$\sin\frac{\phi}{2} = \frac{U\sin(ka/2)}{S} = \frac{U}{S}\sin\frac{ka}{2} \tag{3.103}$$

For $U/S \ll 1$, using $\cos x = 1 - 2\sin^2(x/2)$, we obtain

$$\cos\phi = 1 - 2\left(\frac{U}{S}\right)^2 \sin^2\frac{ka}{2} \tag{3.104}$$

and the energy is

$$E = -2JNS^2 + 4JNU^2 \sin^2\frac{ka}{2} = -2JNS^2 + 2JNU^2(1 - \cos ka) \tag{3.105}$$

Therefore the excitation energy of a spin wave is

$$\epsilon_k = 2JNU_k^2(1 - \cos ka) \tag{3.106}$$

86 INTERACTION BETWEEN TWO SPINS

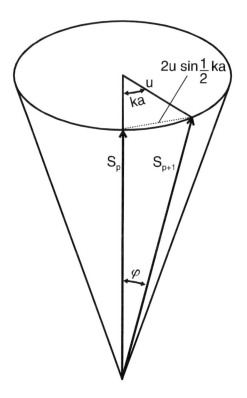

Figure 3.9 Neighbor spins in a ferromagnet along the direction of propagation of a spin wave. The projections of the spins form an angle of *ka* radians.

But, from (3.101):

$$\epsilon_k = 4JSn_k(1 - \cos ka) \tag{3.107}$$

From (3.96), it follows that the excitation energy of the spin waves is

$$\epsilon_k = n_k \hbar \omega_k \tag{3.108}$$

The spin waves are therefore quantized, and the quanta are called *magnons*: n_k is the number of magnons of wavevector k. The number of magnons n_k in thermal equilibrium at temperature T follows a Planck distribution:

$$\langle n_k \rangle = \frac{1}{\exp(\hbar \omega_k / kT) - 1} \tag{3.109}$$

Summing the reduction in magnetization due to all the magnons and dividing by the maximum value of the magnetization, we may obtain its relative variation

(Exercise 3.3):

$$\frac{\Delta M}{M_0} = \frac{\sum n_k}{NS} \propto \left(\frac{kT}{2JS}\right)^{3/2} \quad (3.110)$$

This characteristic dependence of the magnetization, proportional to $T^{3/2}$, is a result confirmed experimentally at low temperatures for many systems and is known as *Bloch's $T^{3/2}$ law* (Fig. 3.10).

One can establish a link between the preceding microscopic description of spin wave phenomena using the Heisenberg hamiltonian, and the phenomenological discussion in terms of the local magnetization introduced at the beginning of this section. To do this, we expand the spin density function in a Taylor series around the atom of order p:

$$\frac{1}{v}(\mathbf{S}_{p-1} + \mathbf{S}_{p+1}) = \mathbf{S}(x-a) + \mathbf{S}(x+a) \approx 2\mathbf{S}(x)_p + a^2\left(\frac{\partial^2 \mathbf{S}(x)}{\partial x^2}\right)_p + \cdots \quad (3.111)$$

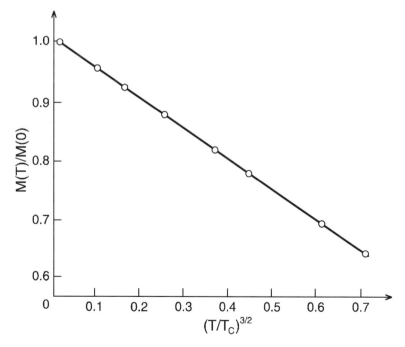

Figure 3.10 Variation of the spontaneous magnetization of Gd as a function of temperature, exhibiting a dependence of the form $\propto T^{3/2}$, characteristic of the contribution of spin waves. [Adapted from N. W. Ashcroft and N. D. Mermin, *Solid State Physics*. Copyright © 1976, Saunders College Publishing, Orlando, FL. Reproduced by permission of the publisher.]

or

$$\frac{1}{v}\sum_i^z \mathbf{S}_i \approx 2\mathbf{S}(x)_p + a^2 \nabla_p^2 \mathbf{S}(x) \qquad (3.112)$$

where $\mathbf{S}(x)_p$ is the spin density at the spin p, v is the volume that contains the atom with this spin, and a is the lattice parameter of the linear chain.

Substituting the expansion into the expression of the torque acting on the spin p [Eq. (3.94)], we obtain

$$\hbar \dot{\mathbf{S}}_p = 2J\mathbf{S}_p \times v[2\mathbf{S}(x)_p + a^2 \nabla_p^2 \mathbf{S}(x)] \qquad (3.113)$$

The first term of the vector product is zero since $\mathbf{S}(x)_p$ is parallel to $\mathbf{S}_p$; we then have:

$$\hbar \dot{\mathbf{S}}_p = 2J\mathbf{S}_p \times [a^2 v \nabla_p^2 \mathbf{S}(x)] \qquad (3.114)$$

or

$$\hbar \dot{\mathbf{S}}(x) = 2J\mathbf{S}(x) \times [a^2 v \nabla^2 \mathbf{S}(x)] \qquad (3.115)$$

Using (3.74)

$$\mathbf{S}(x) = \frac{\mathbf{M}(x)}{\hbar \gamma} \qquad (3.116)$$

we can express Eq. (3.114) in terms of $\mathbf{M}$; comparing with the equation of motion of $\mathbf{M}$ [Eq. (3.72)] the following relation is then apparent

$$D = 2Ja^2 S \qquad (3.117)$$

where S is the spin.

This equation embodies the connection between the two approaches to the spin wave problem: the microscopic description using the Heisenberg

Table 3.II Spin wave stiffness constants D for 3d metals at room temperature (in meV Å^2)[a]

Element	Fe	Co	Ni
D	280	510	455

[a] To convert to joules square meter (J m^2), multiply by 1.60219×10^{-42}.
Source: Reprinted from E. P. Wohlfarth, in *Ferromagnetic Materials*, Vol. 1, E. P. Wohlfarth, Ed., North-Holland, Amsterdam, 1980, with permission from Elsevier North-Holland.

hamiltonian and the phenomenological discussion through the local magnetization $\mathbf{M}(x)$; it shows that the stiffness constant D (see Table 3.II) is proportional to the exchange constant $\mathcal{J}$.

GENERAL READING

Craik, D., *Magnetism, Principles and Applications*, Wiley, Chichester, 1995.
Herring, C. in G. T. Rado and H. Suhl, Eds., *Magnetism*, Vol. IV, Academic Press, New York, 1964.
Martin, D. H., *Magnetism in Solids*, Iliffe, London, 1967.
Patterson, J. D., *Introduction to the Theory of Solid State Physics*, Addison-Wesley, Reading, MA, 1971.
Smart, J. S., *Effective Field Theories of Magnetism*, Saunders, Philadelphia, 1966.
Williams, D. E. G., *The Magnetic Properties of Matter*, Longmans, London, 1966.

REFERENCES

Anderson, P. W. and A. M. Clogston, *Bull. Am. Phys. Soc.* **2**, 124 (1961).
Ashcroft, N. W. and N. D. Mermin, *Solid State Physics*, Saunders College Publishing, Orlando, FL, 1976.
Coehoorn, R., *Supermagnets, Hard Magnetic Materials*, Kluwer Academic, Amsterdam, 1990.
Kittel, C., *Introduction to Solid State Physics*, 6th ed., Wiley, New York, 1986.
Landolt-Börnstein, *Magnetic Properties of 3rd Elements*, New Series III/19a, Springer-Verlag, New York, 1986.
Martin, D. H., *Magnetism in Solids*, Iliffe, London, 1967.
Oguchi, T., *Progr. Theor. Phys.* (Kyoto) **13**, 148 (1955).
Patterson, J. D., *Introduction to the Theory of Solid State Physics*, Addison-Wesley, Reading, MA, 1971.
Ruderman, M. A. and C. Kittel, *Phys. Rev.* **96**, 99 (1954).
Smart, J. S., *Effective Field Theories of Magnetism*, Saunders, Philadelphia, 1966.
Watson, R. E., "Conduction Electron Charge and Spin Density Effects due to Impurities and Local Moments in Metals," in A. J. Freeman and R. B. Frankel, Eds., *Hyperfine Interactions*, Academic Press, New York, 1967, p. 413
Wohlfarth, E. P., in E. P. Wohlfarth, Ed., *Ferromagnetic Materials*, Vol. 1, North-Holland, Amsterdam, 1980, p. 1.
Zener, C., *Phys. Rev.* **81**, 440 (1951).

EXERCISES

3.1 *Magnon Dispersion Relation.* Show that for a simple cubic lattice with $z = 6$ the magnon dispersion relation [Eq. (3.96)] becomes

$$\hbar\omega = 2\mathcal{J}S\left[z - \sum_{\boldsymbol{\delta}} \cos(\mathbf{k}\cdot\boldsymbol{\delta})\right]$$

where δ is the vector that connects the central atom to each nearest neighbor. Show that for $ka \ll 1$,

$$\hbar\omega \approx 2JSa^2k^2$$

where a is the lattice parameter.

3.2 *Magnon Specific Heat.* The total magnon energy is given by the sum of the energies of each magnon multiplied by the number of magnons in the state k:

$$U = \sum_k n_k \hbar\omega = \frac{V}{(2\pi)^3} \int d^3k \frac{\hbar\omega}{e^{\hbar\omega/kT} - 1}$$

Using the approximate relation $\omega = Ak^2$, evaluate U for low temperatures and show that

$$C_v = \frac{\partial U}{\partial T} \propto T^{3/2}$$

3.3 *Bloch $T^{3/2}$ Law.* The thermal excitation of spin waves reduces the saturation value of the magnetization according to

$$M(T) = M(0)\left(1 - \frac{1}{NS}\sum_k n_k\right)$$

Show that at low temperatures, when $\omega(k) \approx Ak^2$

$$\frac{M(T) - M(0)}{M(0)} \propto T^{3/2}$$

4

MAGNETISM ASSOCIATED WITH THE ITINERANT ELECTRONS

4.1 INTRODUCTION

The hypothesis of itinerancy of the electrons may be used to describe the magnetic properties of the metals. Itinerant electrons are electrons that do not remain bound to a given atom, but instead move across the whole matrix. This description applies to the behavior of electrons in metals. Within this hypothesis one may explain, for example, the temperature-independent paramagnetism (Pauli paramagnetism) of the alkali metals (Li, Na, K, Rb, and Cs), and the ferromagnetism of the metals of the $3d$ transition series (Fe, Ni, and Co) and their alloys.

The itinerant electrons occupy states with a (quasi-) continuous distribution of energy; these states appear as we form a metal by putting together the isolated atoms, as illustrated in Fig. 4.1. Initially ($r = \infty$) there are only atomic states; as the atoms approach each other, the originally sharp atomic energy states broaden. For the equilibrium atomic separation ($r = r_0$) there is a superposition of the energy range of the $4s$ and $3d$ electrons (in the example in Fig. 4.1, which shows schematically the situation of metallic Fe), forming bands. These band electrons are delocalized, in the sense that they are shared by all atoms of the crystal.

In the elements of the iron group, the $3d$ electrons are responsible for the magnetism; the $4s$ electrons give a smaller contribution to the magnetic properties; this is evident, for example, from the values of the corresponding magnetic moments per atom (Table 4.I).

The itinerant character of the $3d$ electrons, responsible for the magnetism of the elements of the iron group, contrasts with the localized behavior of the $4f$ electrons, which play the same role in the rare earths. In the actinides, whose

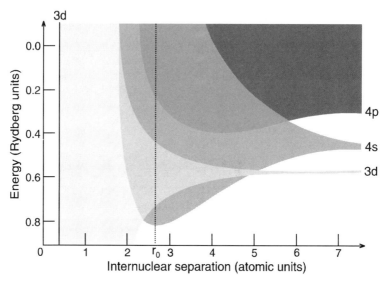

Figure 4.1 Schematic representation of the broadening of the energy states of 4s and 3d electrons in metallic Fe, as a function of the separation r between the atoms. As the atoms approach each other to form the crystal (with the equilibrium distance $r = r_0$), the atomic states give way to states that overlap in energy, forming bands.

magnetism arises from the incomplete 5f shell, the situation is more complex—the degree of localization varies along the series. For a comparison of the spatial behavior of the 3d, 4f and 5f shells, see Fig. 4.2.

The band structure differs for normal metals, noble metals, and transition metals (Section 2.2.1). This can be seen in the schematic representation of the curve of energy–wavevector k (the dispersion curve) and density of states–energy curve. This is shown in Fig. 4.3; the closed shells are atomic-like states that appear in the low energy part of the plots (the straight lines in the graphs). The conduction electrons appear in the upper part, forming parabolic bands in the density of state graphs. The d electrons, represented by a sharper peak in the $n(E)$ curves, are located at low energy in the normal metals, at intermediate energies in the noble metals, and at higher energy in the transition metals. The d band is split for the noble and transition metals, and the upper subbands overlap

Table 4.1 Magnetic moments of iron, cobalt and nickel, and *d* and *s* contributions measured from the diffraction of polarized neutrons

	Fe	Co	Ni
Total magnetic moment (μ_B)	2.216	1.715	0.616
Moment of 3d electrons (μ_B)	2.39	1.99	0.620
Moment of 4s electrons (μ_B)	−0.21	−0.28	−0.105

Source: Reprinted from E. P. Wohlfarth, in *Ferromagnetic Materials*, Vol. 1, E. P. Wohlfarth, Ed., North-Holland, Amsterdam, 1980, p. 34, with permission from Elsevier North-Holland.

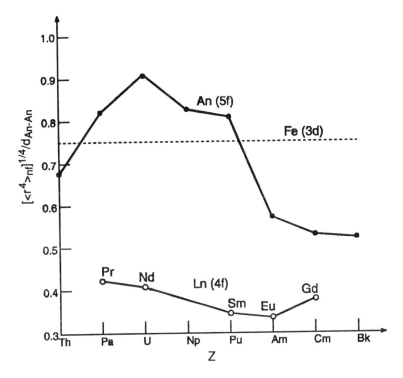

Figure 4.2 Ratio of the average radius of the incomplete shells to interatomic separation for 3d electrons of Fe, for 4f electrons of the lanthanides (Ln), and 5f electrons of the actinides (An), versus Z. Note the large difference in the ratio for Fe and for the rare earths, and the marked dependence of the ratio with Z, for the actinides. [Reprinted from Landolt-Börnstein, *Magnetic Properties of Metals*, New Series III/19fl, Springer-Verlag, New York, 1991, p. 2, with permission.]

with the conduction electron energy in both cases. The upper subband is at the Fermi level in the case of the transition metals.

The dispersion relation for noble metals shows deviations from the unperturbed shapes where the two bands cross, a phenomenon known as *hybridization*.

4.2 PARAMAGNETIC SUSCEPTIBILITY OF FREE ELECTRONS

The simplest itinerant electron model is that of a gas of free electrons, that is, a gas of electrons that interact neither with the atomic cores nor among themselves. The expression for the total density of energy states of this gas (Fig. 4.4), having only the constraint of being contained in the volume V, is obtained from the Schrödinger equation, and is given by (Exercise 4.5)

$$N(E) = 4\pi V \left(\frac{2m_e}{h^2}\right)^{3/2} E^{1/2} \qquad (4.1)$$

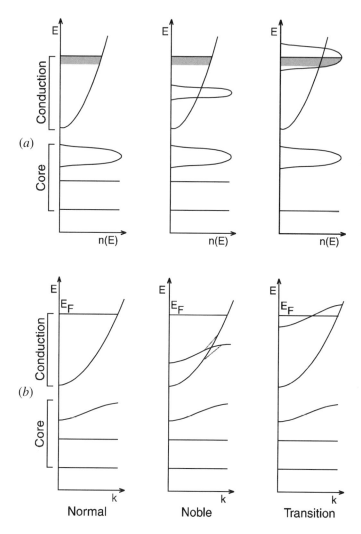

Figure 4.3 Schematic representation of (a) the density of electronic states $n(E)$ and (b) curve of $E(k)$ (dispersion curve) of a normal metal, a noble metal, and a transition metal. [Reprinted from M. Gerl, in *Métaux et Alliages*, C. Janot and M. Gerl, Eds., Masson et Cie, Paris, 1973, p. 91.]

where m_e is the electron mass and E is the energy. Each state may be occupied by at most two electrons, one with spinup ($m_s = +\frac{1}{2}$) and another with spindown ($m_s = -\frac{1}{2}$). At $T = 0$ K, all the states up to E_F, the maximum energy (called the *Fermi energy*), are occupied by two electrons each, and the total number of electrons in the volume V (free-electron gas) is

$$N = \int_0^{E_F} N(E)dE = 4\pi V \left(\frac{2m_e}{h^2}\right)^{3/2} \int_0^{E_F} E^{1/2} dE = \frac{8\pi V}{3}\left(\frac{2m_e}{h^2}\right)^{3/2} E_F^{3/2} \quad (4.2)$$

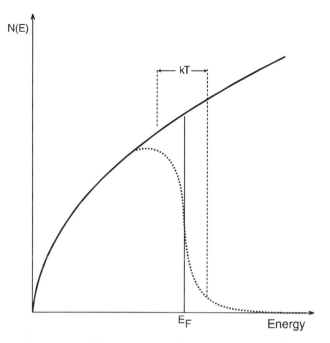

Figure 4.4 Density of states $N(E)$ as a function of the energy, for a gas of free electrons at 0 K [Eq. (4.1)], and at a temperature T. The states are occupied up to the Fermi energy E_F.

Substituting the expression of N [Eq. (4.2)] into (4.1), we may write the density of states as a function of the total number of electrons N:

$$N(E) = 4\pi V \left(\frac{2m_e}{h^2}\right)^{3/2} E^{1/2} = \frac{3}{4}\left(\frac{N}{E_F^{3/2}}\right) E^{1/2} \qquad (4.3)$$

The density of states curves for real metals can be much more complicated than indicated in Eq. (4.3); a curve computed for bcc iron is given as an example in Fig. 4.5.

The probability that a state of energy E is occupied by an electron at a temperature T is $f(E)$, the Fermi–Dirac function

$$f(E) = \frac{1}{\exp[(E-\mu)/kT] + 1} \qquad (4.4)$$

and $\mu = \mu(T)$ is the chemical potential, which at $T = 0$ K is identical to the maximum energy E_F (Fermi energy) (Fig. 4.4, Fig. 4.6). E_F is of the order of a few electronvolts, and kT at the usual temperatures, is of the order of 10^{-2} eV; at room temperature ($T = 300$ K), we have $kT = \frac{1}{40}$ eV.

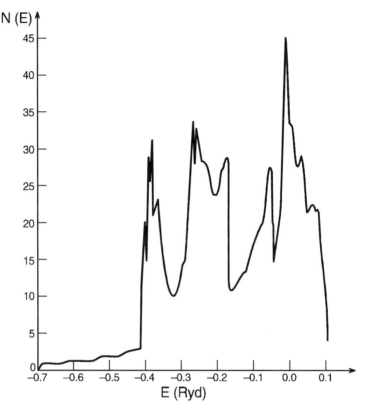

Figure 4.5 Computed density of states $N(E)$ per atom, for bcc iron, as a function of the energy, in Rydberg units (1 Ryd = 2.18×10^{-18} J). The energy origin is at the Fermi level. [Reprinted from E. P. Wohlfarth, in *Ferromagnetic Materials*, Vol. 1, E. P. Wohlfarth, Ed., North-Holland, Amsterdam, 1980, p. 7, with permission from Elsevier North-Holland.]

In the expression of the total density of states $N(E)$, the electrons have only kinetic energy, since we are dealing with a gas of free electrons—the potential energy is zero. If we apply an external magnetic field of induction B_0, a term of magnetic energy appears, corresponding to $-\mu_B B_0$ for the electrons with magnetic moment up, and $+\mu_B B_0$ for the magnetic moment down electrons (the unbound electrons have only spin moments). The total number of electrons per unit volume $n = n_\uparrow + n_\downarrow$, of course, does not vary, only $n_\uparrow$ is now different from $n_\downarrow$. Noting the definition of E_F of the figure (Fig. 4.7), we may observe, using $n(E) = N(E)/V$, that

$$n_\uparrow = \tfrac{1}{2}\int_{-\mu_B B_0}^{E_F} n(E + \mu_B B_0)dE = \tfrac{1}{2}\int_{0}^{E_F + \mu_B B_0} n(E)dE \quad (4.5a)$$

and

$$n_\downarrow = \tfrac{1}{2}\int_{\mu_B B_0}^{E_F} n(E - \mu_B B_0)dE = \tfrac{1}{2}\int_{0}^{E_F - \mu_B B_0} n(E)dE \quad (4.5b)$$

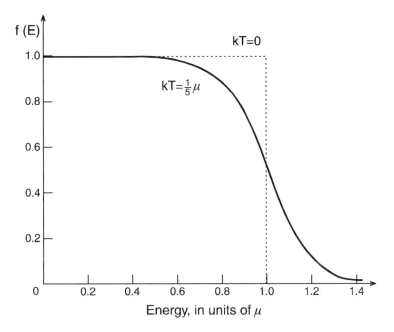

Figure 4.6 Fermi–Dirac distribution [Eq. (4.4)], which gives the occupation of the electronic states at a temperature T, drawn for $T = 0$ K and $T \neq 0$ K.

The resulting magnetization is given by

$$M = \mu_B(n_\uparrow - n_\downarrow) \tag{4.6}$$

which is equal to

$$\tfrac{1}{2}\mu_B \left\{ \int_0^{E_F+\mu_B B_0} n(E)dE + \int_{E_F-\mu_B B_0}^0 n(E)dE \right\} = \tfrac{1}{2}\mu_B \int_{E_F-\mu_B B_0}^{E_F+\mu_B B_0} n(E)dE \tag{4.7}$$

By the fundamental theorem of integral calculus, the preceding integral, between $E_F - \varepsilon$ and $E_F + \varepsilon$ (where $\varepsilon = \mu_B B_0$), in the limit $\varepsilon \to 0$, is equal to the integrand (at the point $E = E_F$) times $2\varepsilon = 2\mu_B B_0$. Thus

$$M = \mu_B^2 B_0 n(E_F) \tag{4.8}$$

and the susceptibility at 0 K, given by $\partial M/\partial H = \mu_0 \partial M/\partial B_0$, is then

$$\chi_0 = \mu_0 \mu_B^2 n(E_F) \tag{4.9}$$

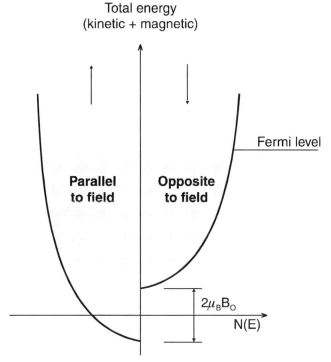

Figure 4.7 Density of states of a gas of free electrons with magnetic moments parallel (↑) and magnetic moments antiparallel (↓), to a magnetic field B_0. Note that $n_\uparrow \neq n_\downarrow$, from which follows that the system has a net magnetization $M = \mu_B(n_\uparrow - n_\downarrow)$.

This is the Pauli spin susceptibility of an electron gas at absolute zero. This susceptibility is proportional to the electron density of states at the Fermi level.

At temperatures above 0 K, the proportion of occupied electron states as a function of temperature; in other words, the statistics, has to be taken into account and we have to make the convolution of the density of states function with the Fermi–Dirac distribution $f(E)$, which gives the probability of occupation of the states at temperature T. The number of electrons with magnetic moments parallel to the magnetic field (up) and antiparallel (down), per unit volume, is now [making the change in variables as in Eqs. (4.5)]

$$n_\uparrow = \tfrac{1}{2}\int_0^\infty n(E) f(E - \mu_B B_0) dE \qquad (4.10a)$$

$$n_\downarrow = \tfrac{1}{2}\int_0^\infty n(E) f(E + \mu_B B_0) dE \qquad (4.10b)$$

In the particular case of a gas of free electrons, the density of states is given by the expression (4.1). Substituting $n(E) = N(E)/V$ in Eqs. (4.10), and using

Eqs. (4.3) and (4.4):

$$n_\uparrow = \frac{3}{4}\left(\frac{n}{E_F^{3/2}}\right)\int_0^\infty \frac{E^{1/2}dE}{\exp[(E - \mu_B B_0 - \mu)/kT] + 1} \quad (4.11a)$$

or

$$n_\uparrow = \frac{3}{4}\left(\frac{n}{E_F^{3/2}}\right)(kT)^{3/2}\int_0^\infty \frac{x^{1/2}dx}{\exp(x - \varepsilon) + 1} \quad (4.11b)$$

with $x = E/kT$ and $\varepsilon = (\mu_B B_0 + \mu)/kT$. Writing

$$F_{1/2}(\varepsilon) = \int_0^\infty \frac{x^{1/2}dx}{\exp(x - \varepsilon) + 1} \quad (4.12)$$

the total number of electrons per unit volume becomes

$$n = n_\uparrow + n_\downarrow = \frac{3n}{4}\left(\frac{kT}{E_F}\right)^{3/2}\left[F\left(\frac{\mu_B B_0 + \mu}{kT}\right) + F\left(\frac{-\mu_B B_0 + \mu}{kT}\right)\right] \quad (4.13)$$

and the magnetization becomes

$$M = \mu_B(n_\uparrow - n_\downarrow) = \frac{3n\mu_B}{4}\left(\frac{kT}{E_F}\right)^{3/2}\left[F\left(\frac{\mu_B B_0 + \mu}{kT}\right) - F\left(\frac{-\mu_B B_0 + \mu}{kT}\right)\right] \quad (4.14)$$

The integrals $F(n)$ may be calculated numerically, and were tabulated by McDougall and Stoner (1938). From the expression of M it can be shown, expanding in series the function F, that the susceptibility becomes (Exercise 4.1)

$$\chi = \mu_0\mu_B^2 n(E_F)\left[1 - \left(\frac{\pi^2}{12}\right)\left(\frac{kT}{E_F}\right)^2 + \cdots\right] \quad (4.15)$$

where $\mu_0\mu_B^2 n(E_F) = \chi_0$ [Eq. (4.9)].

Since $kT \ll E_F$, one can see from Eq. (4.15) that the susceptibility χ for a gas of free electrons (called *Pauli susceptibility*) is practically independent of temperature, and is given by

$$\chi = \mu_0\mu_B^2 n(E_F) \quad (4.16)$$

The Pauli susceptibility is small, of the order of the diamagnetic susceptibility.

4.3 FERROMAGNETISM OF ITINERANT ELECTRONS

A simple model for the description of transition metal ferromagnetism is the Stoner (1938) model, which treats the electron–electron interactions within the mean field approximation.

Analogously to the treatment of the magnetism of localized electrons, one can obtain the magnetization of the itinerant electrons, as in the previous section, and then add another magnetic field (the molecular field) to $\mathbf{B}_0$. Before we do that, however, we will discuss the condition the conduction band parameters have to satisfy to order magnetically; this is the Stoner criterion.

4.3.1 Magnetization at $T = 0$ K: The Stoner Criterion

A split band with $n_\uparrow$ electrons up and $n_\downarrow$ down has a magnetization given by Eq. (4.6). Its interaction with a molecular field $\mathbf{B}_m$ is described by

$$\mathcal{H}_m = -\tfrac{1}{2}\boldsymbol{\mu} \cdot \mathbf{B}_m = -\tfrac{1}{2}\mu_B(n_\uparrow - n_\downarrow)\lambda_m \mu_B(n_\uparrow - n_\downarrow) \tag{4.17}$$

or

$$\mathcal{H}_m = -\tfrac{1}{2}\lambda_m \mu_B^2 (n_\uparrow - n_\downarrow)^2 = -\tfrac{1}{2}\lambda_m \mu_B^2 (n^2 - 4n_\uparrow n_\downarrow) \tag{4.18}$$

where we have used $n = n_\uparrow + n_\downarrow$. The factor $\tfrac{1}{2}$ in (4.17) arises from the fact that $\mathcal{H}_m$ describes the interaction of the magnetization with a molecular field produced by the same magnetization.

We can see that $\mathcal{H}_m$ has two terms: one constant (n^2) and another in $n_\uparrow n_\downarrow$. We will retain only the last term; then

$$\mathcal{H}_m = 2U n_\uparrow n_\downarrow \tag{4.19}$$

where $U = \lambda \mu_B^2$ is the Stoner, or Stoner-Hubbard parameter.

Since $n_\uparrow = n_\downarrow = n/2$ for the nonmagnetized band, the variation in magnetic energy E_m as the band is magnetized is

$$\Delta E_m = 2U n_\uparrow n_\downarrow - 2U \tfrac{1}{4} n^2 = -U \tfrac{1}{2} n^2 \left(\frac{n_\uparrow - n_\downarrow}{n} \right)^2 \tag{4.20}$$

The magnetic moment per electron, or relative magnetization (in μ_B) is ζ:

$$\zeta = \frac{n_\uparrow - n_\downarrow}{n} \tag{4.21}$$

and we have

$$\Delta E_m = -U\tfrac{1}{2}n^2\zeta^2 \qquad (4.22)$$

and

$$n_\uparrow = \frac{n}{2}(1+\zeta) \qquad (4.23a)$$

$$n_\downarrow = \frac{n}{2}(1-\zeta) \qquad (4.23b)$$

When the electron gas is magnetized, the split subbands ↑ and ↓ are shifted $2\delta E$ in relation to one another. The change in kinetic energy corresponds, as shown in Fig. 4.8, to lifting the shaded region of the moment down subband to occupy the position of the shaded region in the moment up subband. The area of each region is $\tfrac{1}{2}(n_\uparrow - n_\downarrow)$, and the vertical displacement is δE.

The total variation in kinetic energy ΔE_k is then

$$\Delta E_k = \frac{1}{2}(n_\uparrow - n_\downarrow)\delta E = \frac{n\zeta\delta E}{2} \qquad (4.24)$$

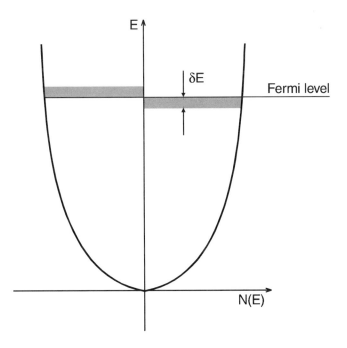

Figure 4.8 Density of states of a gas of free electrons with parallel moments (↑) and antiparallel moments (↓). Under an applied magnetic field, the area of the shaded region in the down subband (right-hand side) is transferred to the top of the up subband; its final position in the latter subband is also shown.

Therefore, the total variation in energy as the band is magnetized will be

$$\Delta E_T = \Delta E_m + \Delta E_k = -U\frac{1}{2}n^2\zeta^2 + \frac{n\zeta\delta E}{2} \qquad (4.25)$$

Since

$$n(E_F)\delta E = (n_\uparrow - n_\downarrow) = n\zeta \qquad (4.26)$$

we substitute δE into Eq. (4.25) and obtain

$$\Delta E_T = -U\frac{1}{2}n^2\zeta^2 + n\zeta\frac{n\zeta}{2n(E_F)}$$

or
$$\qquad (4.27)$$

$$\Delta E_T = \frac{n^2}{2n(E_F)}\zeta^2[1 - Un(E_F)]$$

From this equation, one can derive the following condition:

$$\begin{cases} \text{If } [1 - Un(E_F)] > 0, \text{ then } E_T \text{ is minimum for zero magnetization } (\zeta = 0) \\ \text{If } [1 - Un(E_F)] < 0, \text{ then } E_T \text{ is minimum for nonzero magnetization } (\zeta \neq 0) \end{cases}$$

$$(4.28)$$

This means that the condition for spontaneous magnetic order (i.e., for $\zeta \neq 0$) is

$$[1 - Un(E_F)] < 0 \qquad (4.29)$$

a condition known as the *Stoner criterion for ferromagnetism*. From this condition, one sees that ferromagnetism is favored for strong electron–electron interaction (i.e., large U) and high density of states $n(E_F)$ at the Fermi level. Computed values of $[1 - Un(E_F)]$ give -0.5 to -0.7 for Fe, -1.1 for Ni, and $+0.2$ for Pd (Wohlfarth 1980).

The ferromagnetic transition metals may have different degrees of occupation of the spinup and spindown subbands: the strong itinerant ferromagnets have only one incomplete subband, and the other totally filled (e.g., nickel); weak ferromagnets (e.g., iron) have both subbands incomplete (Fig. 4.9).

4.3.2 Magnetization at $T \neq 0$ K

In the Stoner model the electrons with moments up and moments down are under the effect of magnetic fields $B_\uparrow$ and $B_\downarrow$ that include the external field B_0 and

FERROMAGNETISM OF ITINERANT ELECTRONS

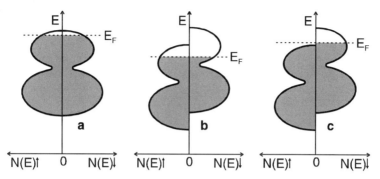

Figure 4.9 Density of states curves for 3d electrons with spinup (↑) and spindown (↓), in the following situations: (a) paramagnet, (b) weak ferromagnet, and (c) strong ferromagnet.

a molecular field $\lambda_m M$:

$$B_\uparrow = B_\downarrow = B_0 + \lambda M \qquad (4.30)$$

In the absence of an external field, using the magnetization per electron $\zeta = (n_\uparrow - n_\downarrow)/n = M/n\mu_B$

$$B_\uparrow = B_\downarrow = \lambda_m M = \lambda_m n \mu_B \zeta \qquad (4.31)$$

Using θ', a molecular field parameter (proportional to λ_m) with dimension of temperature

$$\theta' = \frac{\lambda_m n \mu_B^2}{k} \qquad (4.32)$$

with k the Boltzmann constant, the magnetic field becomes

$$B_\uparrow = B_\downarrow = \frac{k\theta'\zeta}{\mu_B} \qquad (4.33)$$

The energy of one electron in each subband in the molecular field is

$$E_\uparrow = E_k - k\theta'\zeta \qquad (4.34a)$$

$$E_\downarrow = E_k + k\theta'\zeta \qquad (4.34b)$$

where E_k is the kinetic energy. Equations (4.10), in the Stoner model for ferromagnetism, become

$$n_\uparrow = \tfrac{1}{2}\int_0^\infty n(E)f(E - \mu_B B_0 - k\theta'\zeta)dE \qquad (4.35a)$$

$$n_\downarrow = \tfrac{1}{2}\int_0^\infty n(E)f(E + \mu_B B_0 + k\theta'\zeta)dE \qquad (4.35b)$$

In the case of $B_0 = 0$, and for free electrons, we have, using the preceding expressions for n and M [Eqs (4.13 and 4.14)] with $\varepsilon = (k\theta'\zeta + \mu)/kT$

$$n = \frac{3}{4}n\left(\frac{kT}{E_F}\right)^{3/2}\left[F\left(\frac{k\theta'\zeta + \mu}{kT}\right) + F\left(\frac{-k\theta'\zeta + \mu}{kT}\right)\right] \qquad (4.36)$$

$$M = \frac{3}{4}n\mu_B\left(\frac{kT}{E_F}\right)^{3/2}\left[F\left(\frac{k\theta'\zeta + \mu}{kT}\right) - F\left(\frac{-k\theta'\zeta + \mu}{kT}\right)\right] \qquad (4.37)$$

Solving numerically the integrals for $B_0 \neq 0$, we obtain the (volume) magnetization M, or the magnetization per unit mass $\sigma(B_0, T) = M/\rho$, where ρ is the density, or the reduced magnetizations $\zeta = \sigma(B_0, T)/\sigma(0,0)$ and the susceptibilities.

The results obtained are shown in Figs. 4.10 and 4.11. The magnetization curves obtained reproduce reasonably well the experimental results (e.g. of the Cu–Ni alloys) (Fig. 4.11), where such curves for several concentrations may be fitted with ζ computed for different values of $k\theta'$.

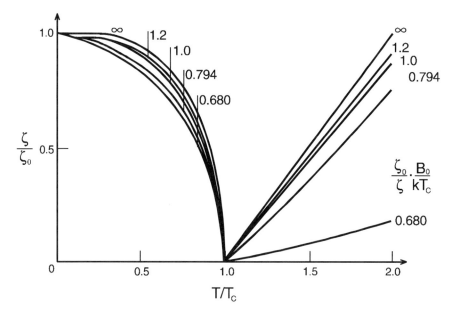

Figure 4.10 Spontaneous reduced magnetization (ζ/ζ_0), and inverse of the reduced susceptibility versus reduced temperature T/T_C, for different values of $k\theta'/E_F$ in the Stoner model.

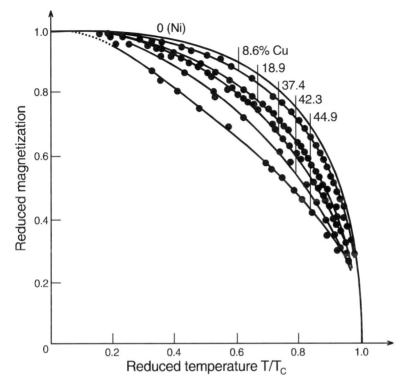

Figure 4.11 Magnetization (ζ) in the Stoner model versus reduced temperature T/T_C for different values of the parameter $k\theta'/E_F$, and experimental values of the spontaneous magnetization of the Cu_xNi_{1-x} alloys.

Contrary to what we had found in the case of localized magnetism (Weiss model), in the Stoner model there is a critical value of the parameter θ', that is, there is a value of $k\theta'/E_F$ below which there is no magnetic order (Fig. 4.12). In the localized case, T_C is proportional to λ_m or $\mathcal{J}$ [Eq. (2.83)], so that for any value of $\mathcal{J}$ there exists ferromagnetic order. Another difference in the itinerant electron case is that even when magnetic order exists, ζ_0 (i.e., ζ at 0 K) may or may not reach its maximum value (= 1).

The critical values of $k\theta'/E_F$ may be computed from the integral equations that define n and M [Eqs. (4.13) and (4.14)]. When $T \to 0$ K, the Fermi–Dirac function $f(E)$ [Eq. (4.4)] in Eq. (4.11a) tends to 1 for $E < E_F$, and the functions $F[(k\theta'\zeta + \mu)/kT]$ and $F[(-k\theta'\zeta + \mu)/kT]$ become $\frac{2}{3}[(k\theta'\zeta + \mu)/kT]^{3/2}$ and $\frac{2}{3}[(-k\theta'\zeta + \mu)/kT]^{3/2}$, respectively. The expressions for n and M are obtained from the Eqs. (4.36) and (4.37) and are written

$$n = \left(\frac{n}{E_F^{3/2}}\right)\left(\frac{kT}{E_F}\right)^{3/2}\frac{2}{3}\left\{\left[\left(\frac{k\theta'\zeta + \mu}{kT}\right)^{3/2}\right] + \left[\left(\frac{-k\theta'\zeta + \mu}{kT}\right)^{3/2}\right]\right\} \quad (4.38)$$

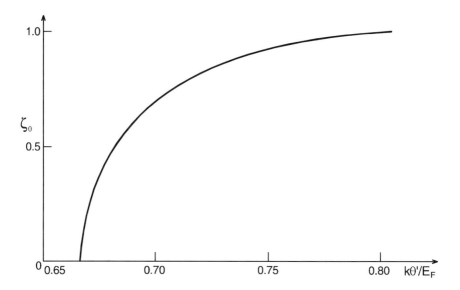

Figure 4.12 Spontaneous magnetization (ζ_0) at $T = 0$ K versus the parameter $k\theta'/E_F$ in the Stoner model.

$$M = \left(\frac{n}{E_F^{3/2}}\right)\mu_B \left(\frac{kT}{E_F}\right)^{3/2} \frac{2}{3}\left\{\left[\left(\frac{k\theta'\zeta + \mu}{kT}\right)^{3/2}\right] - \left[\left(\frac{-k\theta'\zeta + \mu}{kT}\right)^{3/2}\right]\right\} \quad (4.39)$$

Dividing the first equation by n and the second equation by $n\mu_B$, we obtain, by summing and subtracting

$$(1 + \zeta) = \frac{1}{E_F^{3/2}}(E_F + k\theta')^{3/2} \quad (4.40a)$$

$$(1 - \zeta) = \frac{1}{E_F^{3/2}}(E_F + k\theta')^{3/2} \quad (4.40b)$$

Taking to the power $\frac{2}{3}$ and subtracting the second equation from the first, we have

$$(1 + \zeta)^{2/3} - (1 - \zeta)^{2/3} = \frac{2k\theta'}{E_F}\zeta \quad (4.41)$$

The value of $k\theta'/E_F$ below which no magnetic order can occur is given, rewriting Eq. (4.41) for small values of ζ. In this case

$$(1 + \zeta)^{2/3} \cong 1 + \frac{2}{3}\zeta \quad (4.42)$$

and Eq. (4.41) becomes

$$\frac{2}{3} \cong \frac{k\theta'}{E_F} \qquad (4.43)$$

Consequently, below $k\theta'/E_F = 2/3$ there is no ferromagnetic order; this condition is equivalent to the usual Stoner criterion [Eq. (4.29)] (see Exercise 4.3).

The value of $k\theta'/E_F$ for which ζ_0 reaches its maximum value ($= 1$) may be calculated, making ζ of the order of 1; in this case

$$(1+\zeta)^{2/3} \cong 2^{2/3} \qquad (4.44)$$

and

$$(1-\zeta)^{2/3} \cong 0 \qquad (4.45)$$

Substituting into Eq. (4.41), we find that below a value of $k\theta'/E_F$ given by

$$\frac{k\theta'}{E_F} = 2^{-1/3} \qquad (4.46)$$

there is no saturation, in other words, ζ has, at $T = 0$ K, a value $\zeta_0 < 1$.

Figure 4.12 shows the variation of ζ_0 versus $k\theta'/E_F$; one can see the threshold value of $k\theta'/E_F$ for magnetic order (0.667) and the value below which the magnetization at $T = 0$ K is not saturated ($k\theta'/E_F = 2^{-1/3} = 0.794$).

4.4 COUPLED LOCALIZED ITINERANT SYSTEMS

The collective model of Stoner provides a simple description and reproduces several aspects of the behavior of metals, as, for example, the variation of the magnetization as a function of temperature in the metals of the d group, such as nickel. A model containing both localized magnetic moments and itinerant moments has sometimes been used to describe the same ferromagnetic metals. The question of the coexistence of localized and itinerant moments in these metals is a controversial point.

On the other hand, such a mixed model seems particularly appropriate for certain metallic systems; these are the intermetallic compounds containing rare earths and d transition metals. These compounds present properties that are characteristic of localized systems (e.g., Curie–Weiss-type dependence of the susceptibility), side by side with others associated with itinerant magnetism (e.g., Slater–Pauling-type dependence for the d magnetic moments). There are compounds that order magnetically, even when the rare earth present is nonmagnetic (e.g., RFe_2, R_2Fe_{17}); the magnetism here results from the d–d

interaction, and is a band phenomenon. Other compounds order only if the transition metal is combined with a magnetic rare earth (e.g., RNi$_2$); in these, the order arises from the interaction between localized moments through the conduction electrons. A third group shows mixed features, with both the rare earth and the d electrons contributing to T_C.

Characterizing the d–d interactions with the parameter $k\theta'$ and the interaction between conduction electrons and local moments (angular momentum J) with the parameter $\mathcal{J}$, we may group these systems into four classes, exemplified by rare-earth transition metal intermetallic compounds of the AB$_2$ series:

1. $k\theta'$ small, $J\mathcal{J} = 0$; that is, LuNi$_2$, $T_C \approx 0$ K
2. $k\theta'$ large, $J\mathcal{J} = 0$; that is, LuFe$_2$, $T_C \approx 600$ K
3. $k\theta'$ small, $J\mathcal{J} \neq 0$; that is, GdNi$_2$, $T_C \approx 80$ K
4. $k\theta'$ large, $J\mathcal{J} \neq 0$; that is, GdFe$_2$, $T_C \approx 800$ K

The study of rare-earth 3d compounds has generated much interest due to their huge importance as materials for permanent magnets. In this application, the high magnetic ordering temperatures associated with the 3d elements are combined with the strong anisotropies characteristic of the rare earths (see Chapters 1 and 5).

To study these systems, we will consider a model in which there are two coupled sublattices, sublattice i (ion) and the sublattice e (electron). Since the superposition of the 4f orbitals of the ions is negligible, these orbitals interact only with the conduction electrons: the conduction electrons, however, interact with the ions, and among themselves. The molecular fields that act on the ions and on the electrons are (e.g., Iannarella et al. 1982):

$$B_i = B_0 + \frac{1}{\mu_B} J(g-1)\mathcal{J}\zeta_e \quad (4.47a)$$

$$B_e = B_0 + \frac{1}{\mu_B}[J(g-1)\mathcal{J}\zeta_i + k\theta'\zeta_e] \quad (4.47b)$$

with $\zeta_e = M_e/(n_e\mu_B)$ and $\zeta_i = M_i/(g\mu_B Jn_i)$

The Stoner equations [Eqs. (4.13) and (4.14)] are rewritten with B_e in the place of B_0

$$n = n_\uparrow + n_\downarrow = \frac{3n}{4}\left(\frac{kT}{E_F}\right)^{3/2}\left[F\left(\frac{E+\mu_B B_e}{kT}\right) + F\left(\frac{E-\mu_B B_e}{kT}\right)\right] \quad (4.48)$$

$$M = \mu_B(n_\uparrow - n_\downarrow) = \frac{3n}{4}\mu_B\left(\frac{kT}{E_F}\right)^{3/2}\left[F\left(\frac{E+\mu_B B_e}{kT}\right) - F\left(\frac{E-\mu_B B_e}{kT}\right)\right] \quad (4.49)$$

In the limit $T \to 0$, analogously to Eq. (4.41), we obtain

$$(1+\zeta_e)^{2/3} - (1-\zeta_e)^{2/3} = 2\frac{k\theta'}{E_F}\zeta_e + 2J(g-1)\left(\frac{J}{E_F}\right) \quad (4.50)$$

To obtain the magnetizations $\zeta_e(T)$ and $\zeta_i(T)$, we solve [Eqs. (4.48) and (4.49)] numerically. The resulting curves, of magnetization, and of the inverse of the susceptibility, are presented in Fig. 4.13 (for the narrowband limit).

The curves of electronic magnetization and susceptibility are similar to those obtained with the simple Stoner model. It can be seen (Fig. 4.13) that $\zeta_e(T)$ is not zero, even in the case when $k\theta'/E_F = 0$. The ionic magnetization $\zeta_i(T)$ always reaches its maximum value ($= 1$) at $T = 0$. For $T \gg T_C$, the inverse of the susceptibility, for ions and electrons, follows a linear dependence on temperature.

In contradistinction with the behavior of the Stoner model, there is spontaneous magnetic order for any value of $k\theta'$, provided J is not zero.

The electronic magnetization at $T = 0$ depends on the value of $k\theta'/E_m$, for different values of J. The magnetic behavior as a function of the parameters $k\theta'/E_F$ and J may be well illustrated by equi-T_C and equi-ζ_e curves (Iannarella et al. 1982). If, in the Stoner model, we make the width of the band (E_F) tend to zero, we will find an almost perfect equivalence with the localized system. The model of two coupled systems becomes, in this case, practically equivalent to two interacting localized moments.

4.5 MAGNETIC PHASE TRANSITIONS: ARROTT PLOTS

The free energy of a sample with small magnetization, described within the molecular field approximation, can be expressed as a Landau expansion in powers of magnetization $M(H,T)$ (Landau and Lifshitz 1968). The magnetic contribution to the free energy f_m in a field H is written

$$f_m = \frac{A}{2}M^2(H,T) + \frac{B}{4}M^4(H,T) + \cdots - \mu_0 M(H,T)H \quad (4.51)$$

To obtain the equilibrium magnetization in the presence of H, we find the minimum of the free energy f_m as a function of M. This gives, ignoring higher-order terms:

$$M^2(H,T) = -\frac{A}{B} + \left(\frac{\mu_0}{B}\right)\frac{H}{M(H,T)} \quad (4.52)$$

This result shows that under these conditions the square of the magnetization depends linearly with the variable H/M. A graph of isothermal values of M^2 versus H/M is known as an *Arrott plot* (Fig. 4.14).

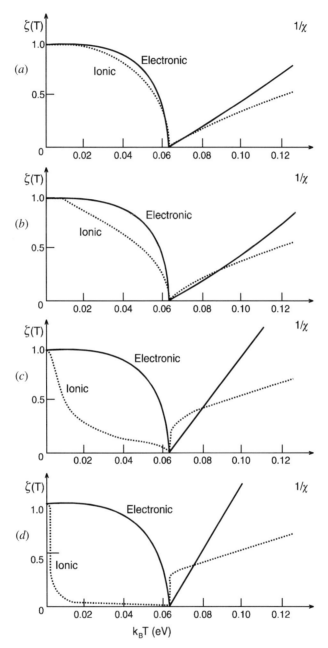

Figure 4.13 Ionic magnetization $\zeta_i(T)$, electronic magnetization $\zeta_e(T)$, and inverse of the susceptibilities in the ion-electron coupled system (in the limit of narrow band), for different values of the pair $[k\theta', J(g-1)\mathcal{J}]$. (a) for $k\theta' = 0$ and $J(g-1)\mathcal{J} = 0.134$ eV; and (b) $k\theta' = 0.08$ eV and $J(g-1)\mathcal{J} = 0.08$ eV; (c) $k\theta' = 0.123$ eV and $J(g-1)\mathcal{J} = 0.0125$ eV; and (d) $k\theta' = 0.124$ eV and $J(g-1)\mathcal{J} = 0.001614$ eV, for $kT_C = 0.062$ eV and $J = \frac{7}{2}$. [Reprinted from L. Iannarella, A. P. Guimarães, and X. A. Silva, *Phys. Stat. Sol. (b)* **114**, 259 (1982). Reprinted by permission of Wiley-VCH Verlag, Weinheim.]

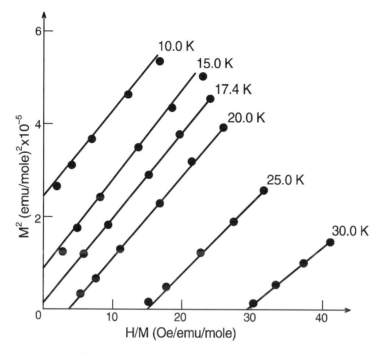

Figure 4.14 Plot of M^2 versus H/M (Arrott plot) for the itinerant ferromagnet ZrZn$_2$. H is the internal magnetizing field. [Reprinted from E. P. Wohlfarth, in *Magnetism—Selected Topics*, S. Foner, Ed., Gordon and Breach, New York, 1976, p. 74.]

In the case of very weak itinerant ferromagnets, the Stoner model [from Eqs. (4.23) and (4.39)] leads to (e.g., Wohlfarth 1976)

$$M^2(H,T) = M^2(0,0)\left[1 - \left(\frac{T}{T_C}\right)^2\right] + 2n\chi_0 M^2(0,0)\frac{H}{M(H,T)} \quad (4.53)$$

which has the same form as Eq. (4.52). One can thus identify the terms A and B that appear in (4.51):

$$A = \frac{\mu_0}{2n\chi_0}\left[\left(\frac{T}{T_C}\right)^2 - 1\right] \quad (4.54)$$

$$B = \frac{\mu_0}{2n\chi_0 M^2(0,0)} \quad (4.55)$$

From Eq. (4.54) one can see that the line in an Arrott plot representing the measurements made at $T = T_C$ passes through the origin. This fact is used to determine T_C experimentally from such plots.

A modified version of the Arrott plot has been used to study phase transitions in systems that cannot be described by the molecular field approximation (Seeger and Kronmüller 1989). It consists of a graph of $M^{1/\beta}$ versus $(H/M)^{1/\gamma}$, where β and γ are critical exponents.

GENERAL READING

Crangle, J., *Solid State Magnetism*, Van Nostrand Rheinhold, London, 1991.
Gignoux, D., "Magnetic Properties of Metallic Systems," in R. W. Cahn, P. Haasen, and E. J. Kramer, Eds., *Materials Science and Technology*, Vol. 3A, Part I, VCH, Weinheim, 1992.
Kittel, C., *Introduction to Solid State Physics*, 6th ed., Wiley, New York, 1986.
Shimizu, M., *Rep. Prog. Phys.* **44**, 329 (1981).
Williams, D. G. E., *The Magnetic Properties of Matter*, Longmans, London, 1966.
Wohlfarth, E. P., "Band Magnetism and Applications," in S. Foner, Ed., *Magnetism— Selected Topics*, Gordon and Breach, New York, 1976, p. 59.

REFERENCES

Gerl, M., in C. Janot and M. Gerl, Eds., *Métaux et Alliages*, Masson et Cie., Paris, 1973, p. 93.
Iannarella, L., A. P. Guimarães, and X. A. da Silva, *Phys. Stat. Sol. (b)* **114**, 255 (1982).
Landau, L. D. and E. M. Lifshitz, *Statistical Physics*, 2nd ed., Pergamon Press, New York, 1968.
Landolt-Börnstein, *Magnetic Properties of Metals*, Landolt-Börnstein Tables, New Series III/19f1, Springer-Verlag, New York, 1991.
McDougall, J. and E. C. Stoner, *Phil. Trans. Roy. Soc.* **A237**, 67 (1938).
Seeger, M. and H. Kronmüller, *J. Mag. Mag. Mat.* **78**, 393 (1989).
Stoner, E. C., *Proc. Roy. Soc.* (London) **A165**, 372 (1938).
Wohlfarth, E. P., "Band Magnetism and Applications," in S. Foner, Ed., *Magnetism— Selected Topics*, Gordon and Breach, New York, 1976, p. 59.
Wohlfarth, E. P., in E. P. Wohlfarth, Ed., *Ferromagnetic Materials*, Vol. 1, North-Holland, Amsterdam, 1980, p. 1.

EXERCISES

4.1 *Pauli Susceptibility (1).* Show that if T is small compared to the Fermi temperature, the Pauli susceptibility is given by

$$\chi(T) \approx \chi_0 \left\{ 1 - \frac{\pi^2}{6}(kT)^2 \left[\left(\frac{n'(E_F)}{n(E_F)}\right)^2 - \frac{n''(E_F)}{n(E_F)} \right] \right\}$$

where n, n', and n'' are the density of states and their derivatives at the Fermi surface. Show that in this case, for free electrons this expression is reduced to

$$\chi(T) \approx \chi_0 \left[1 - \frac{\pi^2}{12}\left(\frac{kT}{E_F}\right)^2\right]$$

4.2 *Pauli Susceptibility (2).* The spin susceptibility of an electron gas at $T = 0$ may be discussed in the following way: let

$$N_\uparrow = \tfrac{1}{2}N(1+\zeta); \ N_\downarrow = \tfrac{1}{2}N(1-\zeta)$$

be the moment up and moment down electron concentrations, respectively.

(a) Show that in a magnetic field B the total energy of the 'moment up' band in the model of the free-electron gas is

$$E_\uparrow = E_0(1+\zeta)^{5/2} - \tfrac{1}{2}n\mu B(1+\zeta)$$

where $E_0 = (\tfrac{3}{10})nE_F$. Find an analogous expression for $E_\downarrow$.

(b) Minimize $E_{\text{total}} = E_\uparrow + E_\downarrow$ in relation to ζ and solve for ζ in the approximation $\zeta \ll 1$. Show that the magnetization is given by $M = 3n\mu_B^2 B/2E_F$.

4.3 *Stoner Criterion for Ferromagnetism.* Show that the Stoner criterion given by Eq. (4.29) is equivalent to that of Eq. (4.43).

4.4 *Ferromagnetism of Conduction Electrons.* The effect of the exchange interaction among conduction electrons may be approximated, assuming that the electrons with parallel spins interact among themselves with energy $-V$, with $V > 0$, and electrons with antiparallel spin do not interact. Use the results of the exercise 4.2 and show that the energy of the moment up subband is given by

$$E_\uparrow = E_0(1+\zeta)^{5/3} - \tfrac{1}{8}Vn^2(1+\zeta)^2 - \tfrac{1}{2}n\mu B(1+\zeta)$$

(a) Find a similar expression for $E_\downarrow$.

(b) Minimize the total energy and show that the magnetization is

$$M = \frac{3n\mu^2}{2E_F - \tfrac{3}{2}Vn}B$$

that is, the interaction increases the susceptibility.

(c) Show that with $B = 0$ the total energy is unstable to $\zeta = 0$ when $V > 4E_F/3n$. If this condition is satisfied, a ferromagnetic state ($\zeta \neq 0$) will have a lower energy than the paramagnetic state. Since $\zeta \ll 1$, this is a sufficient condition for ferromagnetism, but it may not be necessary.

4.5 *Density of States of a Free Electron Gas.* Show that a gas of free electrons contained in a volume V has a density of states given by Eq. (4.1).

5

THE MAGNETIZATION CURVE

The magnetic characterization of materials is done primarily from the graph of their magnetization **M** as a function of the intensity of the external magnetic field **H**. This is their magnetization curve, or **M–H** curve. From the **M–H** (or **B–H**) plot, many important parameters of the magnetic material can be measured; some of them are defined in Section 5.5, and include the saturation magnetization, coercivity, and retentivity.

The magnetic materials present a large diversity of shapes of magnetization curves; these reflect complex phenomena that take place in the materials, such as the motion of domain walls, the rotation of domains, and changes in the direction of magnetization. Before discussing some of these processes, we will examine the shapes of the magnetization curves of some idealized materials.

5.1 IDEAL TYPES OF MAGNETIC MATERIALS

We will consider four types of ideal materials that, under the influence of magnetic fields, approximate the behavior of a large range of real materials. It is instructive to discuss the shapes of the $M-H$ curves for these materials. It is certainly more rewarding to start with these, rather than with the more complex real magnetic materials.

We may describe all known materials, in an approximate way, from four classes of ideal materials (Herrmann 1991): (1) the ideal nonmagnetic materials, (2) the ideal magnetically hard materials, (3) the ideal magnetically soft magnetic materials, and (4) the ideal diamagnet.

116   THE MAGNETIZATION CURVE

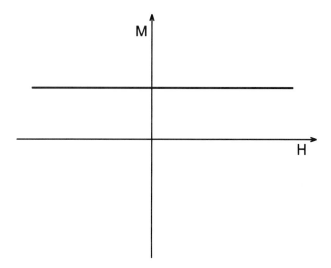

Figure 5.1 Magnetization curve of a magnetized sample of an ideal hard magnetic material. [Reprinted from F. Herrmann, *Am. J. Phys.* **59**, 448 (1991).]

In the ideal nonmagnetic materials, the application of an external field does not result in any magnetization. The magnetization is zero for any value of **H**, and the curve of magnetization versus field coincides with the axis of **H**. Paramagnetic and diamagnetic materials may be identified in some circumstances with these ideal nonmagnetic materials. For example, when considering the magnetization of a material containing a mixture of phases, we may take the magnetization of paramagnetic or diamagnetic impurities as zero; they are then identified with this ideal nonmagnetic material.

In the ideal hard magnetic material, the magnetization is not affected by the external field **H**; it remains constant for any value of **H**. This is the property that makes the hard magnetic materials useful for the manufacture of permanent magnets. The magnetization curve of a magnetized sample of such material is a horizontal line, parallel to the *H* axis (Fig. 5.1); this ideal behavior is inspired in the relatively flat magnetization curve of hard magnetic materials (see also Section 5.5). In the ideal hard magnetic material, in opposition to the soft magnetic material, the external field penetrates completely the sample: $\mathbf{H}_{int} \approx \mathbf{H}$ (see Chapter 1).

In the soft magnetic material, the magnetization increases rapidly as the external magnetic field is increased. In the limit of an ideal soft magnetic material, the magnetization curve is a vertical straight line that coincides with the **M** axis (Fig. 5.2); the ideal soft magnetic material is a medium that can be magnetized with an arbitrarily small magnetic field intensity. If the geometry is such that the demagnetizing factor $N_d \neq 0$, the external magnetic field is completely shielded, so that the internal field **H** is zero. Therefore in these ideal materials the magnetic fields do not penetrate the samples; this effect is the

IDEAL TYPES OF MAGNETIC MATERIALS   117

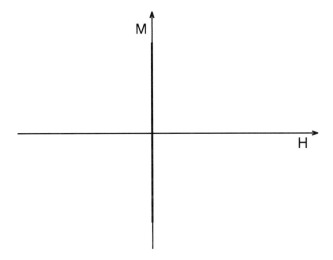

Figure 5.2 Magnetization curve of an ideal soft magnetic material.

result of the arrangement of magnetic dipoles at the surface of the sample, in such a way that its interior is completely shielded from the influence of the external fields. For their ability to impede the penetration of external magnetic fields, the magnetically soft materials are very useful, among other things, as magnetic shields.

These ideal materials are the magnetic analogs of perfect electric conductors, which do not allow the penetration of electric lines of force in their interior (Fig. 5.3). The lines of force outside the ideal soft material are equivalent to those that would be observed in the presence of an opposite magnetic pole, located inside the material.

The ideal diamagnetic material has zero induction **B** for any value of applied external magnetic field **H**. Since **B** remains zero as **H** increases, $|\mathbf{M}|$ has to

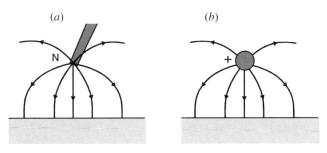

Figure 5.3 (a) Magnetic north pole near the surface of an ideal soft magnetic material, and (b) positive charge near the surface of an electric conductor.

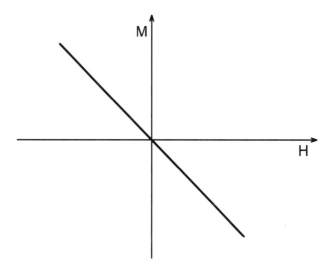

Figure 5.4 $M-H$ curve for an ideal diamagnet; the magnetization varies linearly with H, corresponding to a susceptibility $\chi = -1$.

increase at the same rate to compensate [from Eq. (1.5)], and the susceptibility $\chi = M/H$ is equal to -1. Since $\mathbf{B} = \mu\mathbf{H}$, in this ideal material the magnetic permeability μ has a value of zero. The magnetization curve for an ideal diamagnet is shown in Fig. 5.4. Superconductors behave like ideal diamagnets, for applied fields of magnitude smaller than the critical field H_c (or smaller than H_{c1} in the case of type II superconductors). The lines of force of the field $\mathbf{H}$ (or $\mathbf{B}$)

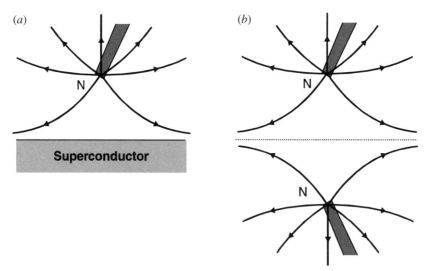

Figure 5.5 (a) North magnetic pole near a superconducting surface (ideal diamagnetic material); (b) same, showing the equivalent magnetic pole inside the material. [Reprinted from F. Herrmann, Am. J. Phys. **59**, 448 (1991).]

cannot penetrate this material. However, differently from what occurs in the magnetically soft material, the shielding mechanism involves the presence of electric currents at the surface, instead of magnetic poles. The lines of force of **H** are parallel to the surface (Fig. 5.5); the effect of the ideal diamagnetic material on the lines of force is equivalent to that of a magnetic pole of the same polarity, located inside the material.

5.2 CONTRIBUTIONS TO THE ENERGY IN MAGNETIC MATERIALS

Inside a magnetic material, the magnetic moments are subject to several interactions (Kittel 1949), such as (1) the magnetostatic energy, that is, the magnetic energy in the demagnetizing field; (2) the magnetic anisotropy; (3) the exchange interaction, responsible for the magnetic order; and (4) the magneto-elastic interaction, relevant in the phenomenon of magnetostriction. We will now focus on these different terms for the energy inside the magnetic domains.

5.2.1 Magnetostatic Energy

We can consider a magnetic dipole as formed by two fictitious entities called *magnetic poles*, of magnetic strengths $+p$ and $-p$, separated by a distance d; the pole strength p is measured in amperes meter (SI). Two such poles, $+p(\mathbf{r}_1)$ and $-p(\mathbf{r}_2)$, separated by a distance $r = |\mathbf{r}_1 - \mathbf{r}_2|$, apply on one another a force given by Coulomb's law

$$F = \frac{\mu_0}{4\pi} \frac{p^2}{r^2} \tag{5.1}$$

where μ_0 is the vacuum permeability, whose value is $\mu_0 = 4\pi \cdot 10^{-7} \mathrm{H/m}$. The force on each pole is $-dU/dr$, where U is the potential due to the other pole:

$$U = -\frac{\mu_0}{4\pi} \frac{p}{r} \tag{5.2}$$

The force that acts on a pole in an applied field H is $F = \mu_0 p H$. The magnetic dipole moment of the pair of poles p, separated by a distance r is $m = pr$.

We can calculate the work necessary to form a magnetic dipole by separating these two poles for a distance d; this work is identical to the energy required to magnetize a bar of unit section:

$$\int dW = \int_0^d F(r)dr = \mu_0 \int_0^d pH \, dr = \mu_0 \int_0^M H \, dM \tag{5.3}$$

One can thus compute with this integral the work necessary to magnetize an originally unmagnetized sample, which is measured by the area between the

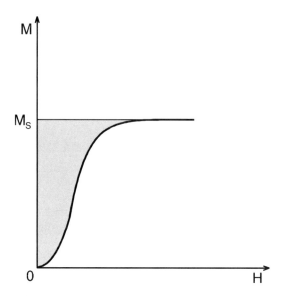

Figure 5.6 Area that measures the work required to magnetize a sample.

curve $M(H)-H$ and the M axis, in Fig. 5.6; this work is converted into potential energy and heat.

If we vary the applied field in such a way as to produce a full magnetization loop (see Section 5.5), we return to the same point in the $M-H$ diagram, and the variation in potential energy is zero. The area of the loop then represents the energy dissipated as heat—known as the *hysteresis loss*.

The *magnetostatic energy* is the energy of a magnetized material, in the absence of an applied magnetic field; in this case, the only magnetic field that acts is the demagnetizing field. The energy of the material in its demagnetizing field is also called *self-energy*. A sample of magnetic material that is taken to saturation by the application of a field $\mathbf{H}_{ext}$ of increasing amplitude will, in general, keep a certain magnetization as it is removed from the field.

The graph of magnetization versus the internal field $\mathbf{H}$ ($= \mathbf{H}_{ext} + \mathbf{H}_d$) will be as shown in Fig. 5.7. At the maximum external field, the curve will reach the point A, and the magnetization will reach the saturation value M_s. As the field is lowered to zero, the magnetization will evolve to an equilibrium value M_e (point C); at this point, the only magnetic field on the sample will be the demagnetizing field H_d, an internal field along the negative H axis. This curve is called the *demagnetization curve*, and point C is the intersection of the straight line $M = -H/N_d$ with the curve for M. For each value of M, the corresponding value of H given by this line is the demagnetizing field; if one adds this value of $|H_d|$ at each abscissa H of the curve, one recovers the curve of $M - H_{ext}$.

The magnetostatic energy E_{ms} per unit volume, at point C, may be computed from Eq. (5.3), as the work done against the demagnetizing field. Since in this

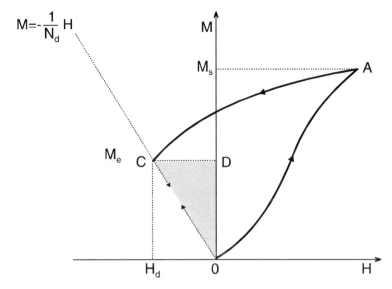

Figure 5.7 Magnetization of a magnetized material versus internal field H; this is the demagnetization curve. Point C corresponds to the equilibrium magnetization of the sample under the action of the demagnetizing field H_d. The marked area measures the magnetostatic energy (divided by μ_0).

case the energy is the self-energy, a factor of $\frac{1}{2}$ appears:

$$E_{ms} = -\tfrac{1}{2}\mu_0 \int_0^{M_s} H_d \, dM - \tfrac{1}{2}\mu_0 \int_{M_s}^{M_e} H_d \, dM$$

$$= -\tfrac{1}{2}\mu_0 \int_0^{M_e} H_d \, dM = \frac{\mu_0}{2} N_d M_e^2 \tag{5.4}$$

where M_s is the saturation magnetization and M_e is the equilibrium magnetization, that is, the magnetization for zero external field. E_{ms} is given by the shaded area in Fig. 5.7.

For a given geometry, and therefore a given N_d, a permanent magnet will always operate on the same line OC, called the *load line*. If, for example, the demagnetizing curve is obtained with the sample at a higher temperature, the magnetization will be smaller, but the working point (C') will still fall on the same line OC. The modulus of the slope of the load line in the $B \times H$ curve ($\equiv -B_m/H_m$) is called the *permeance coefficient*.

5.2.2 Magnetic Anisotropy

The shape of the curve of magnetization versus applied field **H** in ferromagnetic single crystals depends on the direction of application of **H**. This can be seen in Fig. 5.8 for crystals of iron and nickel. The origin of this effect lies in the fact that

122 THE MAGNETIZATION CURVE

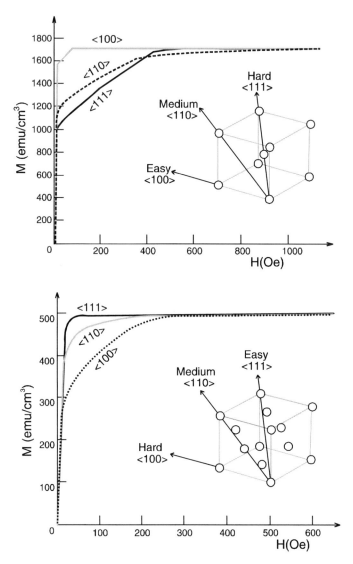

Figure 5.8 Dependence of the magnetization with the magnetic field **H** applied along different crystal directions, for single crystals of Fe and Ni. The easy directions of magnetization are ⟨100⟩ for Fe and ⟨111⟩ for Ni.

the magnetic moments inside the magnetic material do not point indifferently to any direction in relation to the crystalline axes. There exists, for each crystal, a preferred direction, known as the "direction of easy magnetization," or the "easy direction."[1]. For example, in metallic iron, the easy direction is [100] (and the

---
[1] In many cases, as will be exemplified later on, the directions on a plane are equivalent, and instead of an easy direction, one has an easy plane.

equivalent directions are [010] and [001]). Applying a magnetic field along these directions, one reaches the maximum magnetization (or saturation magnetization) with lower values of **H**. The direction along which a multidomain sample is easier to magnetize is the same direction of magnetization of the individual undisturbed domains.

There are several contributions to the magnetic anisotropy; for instance, the magnetocrystalline anisotropy (or crystal anisotropy) is the main source of intrinsic anisotropy. The extrinsic contributions are related to the shape of the samples, their state of mechanical stress, and so on (see Table 5.III).

The magnetocrystalline anisotropy energy (or crystal anisotropy energy) arises mainly from the interaction of the electronic orbital angular momenta with the crystalline field, that is, with the electric field at the site of the magnetic ions.

The exchange interaction is isotropic, and therefore, cannot be responsible for this effect; the microscopic origin of crystal anisotropy lies in the interaction of the atomic orbital momentum with the charges of the lattice. The spin momentum of the atoms, in its turn, is involved in this interaction through spin–orbit coupling.

The magnetic anisotropy energy E_K per unit volume may be derived in the case of a single-domain perfect crystal. This energy is written as a function of the direction cosines α_1, α_2, and α_3, defined in relation to the axes of the crystal. Since the energy is only a function of the angle with the easy axis (and indifferent to the direction along this axis), it must remain the same when we change the sign of these cosines, and therefore, odd powers of the cosines cannot appear in its expression. Also, the permutations among the cosines must leave the energy E_K invariant.

The most general form that the energy may have in terms of the powers of the direction cosines α_i for a cubic crystal is

$$E_K = K_0 + K_1(\alpha_1^2\alpha_2^2 + \alpha_2^2\alpha_3^2 + \alpha_3^2\alpha_1^2) + K_2(\alpha_1\alpha_2\alpha_3)^2 + \cdots \quad (5.5)$$

Substituting into E_K the direction cosines of the directions [100], [110], and [111], symmetry directions in the cubic system, we obtain the expression of the energy for these three directions:

$$E_{100} = K_0 \quad (5.6a)$$

$$E_{110} = K_0 + K_1/4 \quad (5.6b)$$

$$E_{111} = K_0 + K_1/3 + K_2/27 \quad (5.6c)$$

The anisotropy constants K_0, K_1, and K_2 may then be derived from the areas of the magnetization curves obtained for each direction, since the anisotropy energy for each direction is given by the area between the curve and the M axis, as in Fig. 5.6. The K values vary with temperature, tending to zero at the transition temperature T_C. The anisotropy constants are measured in units of joules per cubic meter (SI), or ergs per cubic centimeter (CGS); the SI value is obtained by multiplying the CGS value by 10^{-1}. When K_2 can be neglected, K_1 defines the direction of easy magnetization—for $K_1 > 0$, the easy direction is [100] ($E_{100} < E_{110} < E_{111}$); for negative K_1, the easy direction is [111] ($E_{111} < E_{110} < E_{100}$).

For uniaxial crystals the description is simplified; for hexagonal crystals, for example, the anisotropy energy is usually written in terms of the sine of the angle θ between the c axis and the direction of magnetization as

$$E_K = K_0 + K_1 \sin^2 \theta + K_2 \sin^4 \theta + \cdots \tag{5.7}$$

In many cases, $K_1 > 0$ and $K_2 > -K_1$, and the anisotropy energy is a minimum for $\theta = 0$; the magnetization points along the c axis of the crystal. This occurs, for example, in cobalt metal, and in barium ferrite ($BaFe_{12}O_{19}$). In these cases the anisotropy is uniaxial, since E_K does not depend on the angle with the directions of the basal plane, in the hexagonal crystal. In the simplest situation, $|K_1| \gg |K_2|$, and the anisotropy energy may be written (ignoring the constant term K_0):

$$E_K = K_1 \sin^2 \theta \tag{5.8}$$

Some values of the anisotropy constant K_1 are given in Table 5.I.

We may, for certain purposes, assume that the uniaxial magnetic anisotropy is due to the action of an equivalent field, H_a (anisotropy field), with direction equal to that of the easy magnetization axis. Its expression can be obtained computing the value of the magnetic field that produces on the magnetization the same torque of the anisotropy interaction, for small angles ($\sin \theta \approx \theta$). The torque τ is $\mathbf{M} \times \mathbf{B}$, and the energy is $-\mathbf{M} \cdot \mathbf{B}$; therefore, for small angles, $dE = \tau \, d\theta$. For uniaxial anisotropy, for example, the condition of equal torque is written, from Eq. (5.8), as a function of the saturation magnetization M_s:

$$\mu_0 H_a M_s \theta = 2 K_1 \theta \tag{5.9}$$

from what results, for the expression of the anisotropy field,

$$H_a = \frac{1}{\mu_0} \frac{2 K_1}{M_s} \tag{5.10}$$

Table 5.I Anisotropy constants K_1 of some cubic metals and intermetallic compounds at room temperature

Crystal	Fe	Ni	ErFe$_2$	DyFe$_2$	TbFe$_2$	HoFe$_2$
$K_1(10^3$ J m$^{-3})$	45	−5	−330	2100	−7600	580

One may show that this gives also the magnitude of the external magnetic field that produces saturation when applied in a direction perpendicular to the axis of easy magnetization.

The shape of a sample affects its magnetic anisotropy energy. As we have seen in Section 1.2, the demagnetizing field depends on the shape of the sample, and on the direction of the applied field. The demagnetizing field is smaller along the longer dimension of the sample, and larger in the opposite case. For this reason, if one wants to induce the appearance of an internal magnetic field inside a given sample, a less intense field is required if applied along a larger dimension. In other words, the direction of larger dimension is an easy axis of magnetization, in the cases where this easy axis is determined by shape anisotropy.

To obtain the expression for the shape anisotropy energy, we use the magnetostatic energy, $-(\mu_0/2)\mathbf{M} \cdot \mathbf{H}_d$ [from Eq. (5.4)]. The shape anisotropy energy for an ellipsoid of major axis c and minor axes $a = b$ can be computed by projecting the components of the magnetization $\mathbf{M}$ along the three axes. The sum of these contributions is

$$E_K = \frac{\mu_0}{2} N_c M^2 + \frac{\mu_0}{2}(N_a - N_c) M^2 \sin^2 \theta \tag{5.11}$$

where θ is the angle between the c axis and the direction of magnetization, and the N terms are the corresponding demagnetizing factors (Fig. 5.9).

In the case of a spherical sample, $a = b = c$, $N_a = N_c$, and the shape anisotropy energy is

$$E_K = \frac{\mu_0}{2} N M^2 = \frac{\mu_0}{6} M^2 \tag{5.12}$$

where we have used $N = \frac{1}{3}$ for the sphere. Thus, in the case of the sphere, the shape anisotropy energy is not zero, but it is isotropic; that is, it does not depend on θ.

The expression for a bidimensional magnetic sample, applicable to a magnetic thin film, can be obtained from Eq. (5.11), in the limit of a flat oblate ellipsoid. If

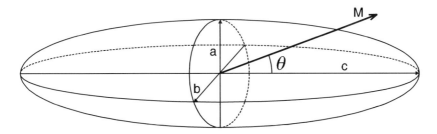

Figure 5.9 Sample of ellipsoidal shape with magnetization along a direction forming an angle θ with the major axis.

the length of the c axis tends to zero, then $N_c \to 1$, $N_a \to 0$, and we obtain (ignoring a term that does not contain θ)

$$E_K = -\frac{\mu_0}{2} M^2 \sin^2 \theta \tag{5.13}$$

where θ is now the angle formed between **M** and the normal to the plane of the sample (Fig. 5.10).

Equation (5.13) shows that, in terms of shape anisotropy, any direction on the plane is an easy direction—we may then speak of an easy plane; this equation can be seen as describing a form of uniaxial anisotropy [Eq. (5.8)], with uniaxial anisotropy constant $K_u = -(\mu_0/2)M^2$.

In thin films, the breaking of local symmetry associated with the presence of the interface gives origin to another contribution, the magnetic surface anisotropy (Néel 1954), which amounts to a term $\sigma = K_s \cos^2 \theta$ added to the surface energy. This corresponds to an anisotropy energy per volume

$$E_s = \frac{1}{d} K_s \cos^2 \theta \tag{5.14}$$

where K_s is called out of plane surface anisotropy constant, and d is the thickness of the film (Gradman 1993); K_s is in the range $0.1-1.0 \times 10^{-3}$ J m^{-2}.

5.2.3 Exchange Interaction

The interaction between the atomic spins responsible for the establishment of

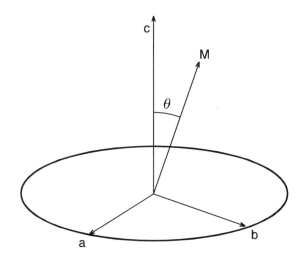

Figure 5.10 Planar sample with magnetization along a direction forming an angle θ with the normal.

magnetic order is the exchange interaction, an interaction of electrical origin (see Chapter 3). It may be written, for a pair of neighbor atoms, as a function of their spin operators $\mathbf{S}_i$ and $\mathbf{S}_j$ [Eq. (3.25)]

$$\mathcal{H}_{\text{exch}} = -2\mathcal{J}\mathbf{S}_i \cdot \mathbf{S}_j \tag{5.15}$$

where $\mathcal{J}$ is the exchange parameter. We can therefore write in approximate form, for the exchange energy of a pair of atoms of spin S in a ferromagnetic material, as a function of the angle ϕ between the spins:

$$E_{\text{exch}}^{\text{pair}} = -2\mathcal{J}S^2 \cos\phi \tag{5.16}$$

Expanding $\cos\phi$:

$$\cos\phi = 1 - \frac{\phi^2}{2} + \frac{\phi^4}{24} - \cdots \tag{5.17}$$

Taking into account the terms up to second order and substituting into the expression of E_{exch}, we obtain, neglecting the term that does not depend on ϕ:

$$E_{\text{exch}}^{\text{pair}} = \mathcal{J}S^2\phi^2 \tag{5.18}$$

For a one-dimensional chain of $N+1$ neighbor atoms, the exchange energy is

$$E_{\text{exch}} = \mathcal{J}S^2 \sum_{i}^{N} \phi_i^2 \tag{5.19}$$

where the sum is made on the N pairs of nearest neighbors.

5.2.4 Magnetoelastic Energy and Magnetostriction

A solid under tension has an elastic energy that is expressed as a function of the strains ϵ_{ij}; this notation represents a strain arising from a force applied in the direction along the axis i, to a surface of normal in the direction parallel to the j axis. In the case of a magnetic material, the elastic energy has an additional term that results from the interaction between the magnetization and the strains; this is the magnetoelastic energy. Its expression may be derived through the computation of the anisotropy energy in the presence of the stresses (Kittel 1949). The magnetoelastic energy is the contribution to the anisotropy energy that arises in a solid under stress.

The stress σ at a point P is defined as the force divided by the area $\Delta F/\Delta A$ in the limit as $\Delta A \to 0$. A solid body under the effect of a stress σ undergoes a strain ϵ. The strain ϵ is dimensionless, and is given by the relative change in length $(\delta l/l)$. Within certain limits, this deformation is linear, and the strain is proportional to the stress σ (an empirical result known as *Hooke's law*):

$$\epsilon = \frac{\sigma}{E} \qquad (5.20)$$

where E is Young's modulus.

A solid sample under tension along a given direction also reduces its transverse dimension, or width w_0, and the Poisson ratio ν (dimensionless) is defined from the variation δw of this dimension:

$$\frac{\delta w}{w_0} = -\nu \frac{\delta l}{l} \qquad (5.21)$$

For most materials the Poisson ratio has a value between 0.1 and 0.3.

Hooke's law can be rewritten in terms of the components ϵ_{ij} of the strain and σ_{ij} of the stress (Landau and Lifshitz 1959):

$$\epsilon_{xx} = \frac{\sigma_{xx} - \nu(\sigma_{yy} + \sigma_{zz})}{E} \qquad (5.22)$$

$$\epsilon_{ij} = \frac{(1+\nu)\sigma_{ij}}{E} \quad (i,j) = x,y,z \quad (i \neq j) \qquad (5.23)$$

The elastic energy per unit volume is given in the cubic system by (Landau and Lifshitz 1959)

$$E_{el} = \tfrac{1}{2}c_{11}(\epsilon_{xx}^2 + \epsilon_{yy}^2 + \epsilon_{zz}^2) + \tfrac{1}{2}c_{44}(\epsilon_{xy}^2 + \epsilon_{yz}^2 + \epsilon_{zx}^2)$$
$$+ c_{12}(\epsilon_{yy}\epsilon_{zz} + \epsilon_{xx}\epsilon_{zz} + \epsilon_{xx}\epsilon_{yy}) \qquad (5.24)$$

where the c_{ij} are the elastic moduli.

The magnetoelastic energy may be obtained expanding in Taylor series the expression of the anisotropy energy [Eq. (5.5)], as a function of the strains ϵ_{ij}:

$$E_K = E_K(0) + \sum_{i,j} \left(\frac{\partial E_K}{\partial \epsilon_{ij}}\right)_0 \epsilon_{ij} + \cdots \qquad (5.25)$$

The anisotropy energy is then a sum of a term for zero strain—the anisotropy energy proper—and additional terms that involve the strains ϵ_{ij}. These terms of the expansion constitute the magnetoelastic energy E_{ME}:

$$E_K = E_K(0) + E_{ME} \qquad (5.26)$$

Expanding the series, and using the expression of E_K for an unstrained crystal [Eq. (5.5)], we obtain the following equation for the magnetoelastic energy in a cubic crystal (Kittel 1949):

$$E_{ME} = B_1(\alpha_1^2\epsilon_{xx} + \alpha_2^2\epsilon_{yy} + \alpha_3^2\epsilon_{zz}) + B_2(\alpha_1\alpha_2\epsilon_{xy} + \alpha_2\alpha_3\epsilon_{yz} + \alpha_3\alpha_1\epsilon_{zx}) \quad (5.27)$$

The B factors are known as *magnetoelastic coupling constants*; the α terms are the direction cosines.

The equilibrium configurations of a magnetized cubic crystalline solid are given by the tensor of strains ϵ_{ij} that minimizes the total energy, which is expressed by [using the second term of Eqs. (5.5), and Eqs. (5.24) and (5.27)]:

$$\begin{aligned} E = E_K + E_{el} + E_{ME} = \\ K(\alpha_1^2\alpha_2^2 + \alpha_2^2\alpha_3^2 + \alpha_3^2\alpha_1^2) \\ + \tfrac{1}{2}c_{11}(\epsilon_{xx}^2 + \epsilon_{yy}^2 + \epsilon_{zz}^2) + \tfrac{1}{2}c_{44}(\epsilon_{xy}^2 + \epsilon_{yz}^2 + \epsilon_{zx}^2) \\ + c_{12}(\epsilon_{yy}\epsilon_{zz} + \epsilon_{xx}\epsilon_{zz} + \epsilon_{xx}\epsilon_{yy}) \\ + B_1(\alpha_1^2\epsilon_{xx} + \alpha_2^2\epsilon_{yy} + \alpha_3^2\epsilon_{zz}) \\ + B_2(\alpha_1\alpha_2\epsilon_{xy} + \alpha_2\alpha_3\epsilon_{yz} + \alpha_3\alpha_1\epsilon_{zx}) \end{aligned} \quad (5.28)$$

The solutions are of the form $\epsilon_{ij} = \epsilon_{ij}(K, B_1, B_2, c_{mn})$.

A sample of magnetic material changes its dimensions as it is magnetized; this phenomenon is called *magnetostriction*. In more general terms, magnetostriction is the occurrence of variations of the mechanical deformation of a magnetic sample due to changes in the degree of magnetization, or in the direction of magnetization. Materials have positive magnetostriction when they exhibit a linear expansion as they are magnetized (e.g., the alloy Permalloy); and negative magnetostriction, in the opposite situation (e.g., nickel metal). Its microscopic origin involves the interaction of the orbital atomic moment with the electric charges in the crystalline lattice (the crystal field).

Magnetostriction is defined quantitatively as the relative linear deformation

$$\lambda = \frac{\delta l}{l_0} \quad (5.29)$$

where $\delta l = l - l_0$ is the variation in the linear dimension l of the sample.

This effect is very small; λ is normally of the order of 10^{-5} to 10^{-6}. The magnetostriction λ has the same dimension as the strains ϵ caused by a mechanical tension. Thus, a crystal of a ferromagnetic material that is perfectly cubic above the ordering temperature T_C will present a small distortion when cooled below this temperature.

There exists also the inverse magnetostrictive effect, that is, the effect of the change in the magnetization through the action of an applied stress. Magnetostriction is also observed when a magnetic field is applied to a magnetized sample (called in this case *forced magnetostriction*).

Table 5.II Saturation magnetostriction of some polycrystalline materials, at room temperature

Material	λ_s ($\times 10^6$)
Ni	−33
Co	−62
Fe	−9
Ni$_{60}$Fe$_{40}$ (Permalloy)	+25
Fe$_3$O$_4$	+40
TbFe$_2$	+1753

The magnetostriction constants usually quoted are the saturation magnetostriction constants λ_s, specifically, the values of $\delta l/l_0$ for samples taken from the unmagnetized state to magnetic saturation. Some values of λ_s for polycrystalline materials at room temperature are shown in Table 5.II. The magnetostriction constants fall with increasing temperature, tending to zero at the Curie temperature.

Let us assume a sample of cubic crystal structure that changes from a demagnetized state to a state of magnetic saturation. Its saturation magnetostriction λ_s, along a direction defined by the direction cosines β_1, β_2 and β_3 relative to the crystal axes, is (e.g., Kittel 1949):

$$\lambda_s(\alpha,\beta) = \tfrac{3}{2}\lambda_{100}(\alpha_1^2\beta_1^2 + \alpha_2^2\beta_2^2 + \alpha_3^2\beta_3^2 - \tfrac{1}{3})$$
$$+3\lambda_{111}(\alpha_1\alpha_2\beta_1\beta_2 + \alpha_2\alpha_3\beta_2\beta_3 + \alpha_3\alpha_1\beta_3\beta_1) \quad (5.30)$$

where α_1, α_2, and α_3 are the direction cosines of the direction of magnetization; λ_{100} and λ_{111} are the saturation magnetostrictions along the directions [100] and [111], and are related to the magnetoelastic coupling constants (B_1 and B_2), more fundamental quantities that appear in Eq. (5.27), and to the elastic moduli c_{ij}.

Calling θ the angle between the direction $[\beta_1\beta_2\beta_3]$ along which the magnetostriction is being measured, and the magnetic field (parallel to $[\alpha_1\alpha_2\alpha_3]$), we have

$$\cos\theta = \alpha_1\beta_1 + \alpha_2\beta_2 + \alpha_3\beta_3 \quad (5.31)$$

If we make the approximation of considering the magnetostriction isotropic, we will then have $\lambda_{100} = \lambda_{111} = \lambda_s$. Substituting into Eq. (5.30), the magnetostriction becomes

$$\lambda(\theta) = \tfrac{3}{2}\lambda_s(\cos^2\theta - \tfrac{1}{3}) \quad (5.32)$$

which can be written, substituting $\cos^2\theta = 1 - \sin^2\theta$ and neglecting the constant term:

$$\lambda(\theta) = -\tfrac{3}{2}\lambda_s \sin^2\theta \qquad (5.33)$$

This expression for the magnetostriction does not depend on the crystal directions, only on the angle between the magnetization and the direction along which the magnetostriction is measured.

The relation between magnetostriction and magnetoelastic energy can be seen for a solid of length l_0 submitted to a tensile stress σ, in a direction forming an angle θ with the magnetization; its dimension is altered by $dl(\theta)$. The increase in energy (per volume) is the work $-\sigma\, dl(\theta)/l_0$ done by the stress σ (noting that σ is equivalent to a negative pressure):

$$dE = -\sigma \frac{dl(\theta)}{l_0} = -\sigma\, d\lambda(\theta) \qquad (5.34)$$

For a solid magnetized to saturation, the total energy is the work done as the solid is deformed by magnetostriction; this is the magnetoelastic energy. Therefore, in the case of isotropic magnetostriction in a cubic crystal submitted to a stress σ, the magnetoelastic energy is written [using Eqs. (5.33) and (5.34)]:

$$E_{ME} = \tfrac{3}{2}\lambda_s \sigma \sin^2\theta \qquad (5.35)$$

Comparing thus with Eq. (5.8), we see that this expression has the form of an anisotropy energy. We can conclude that the magnetostriction, through the magnetoelastic energy, is equivalent to an uniaxial anisotropy, with anisotropy constant

$$K_u = \tfrac{3}{2}\lambda_s \sigma \qquad (5.36)$$

The different kinds of contributions to magnetic anisotropy are shown in Table 5.III.

For the magnetostriction measured in the same direction as the applied field, $\alpha_i = \beta_i$, and substituting into Eq. (5.30), we obtain

$$\lambda_s(\alpha) = \tfrac{3}{2}\lambda_{100}(\alpha_1^4 + \alpha_2^4 + \alpha_3^4 - \tfrac{1}{3}) + 3\lambda_{111}(\alpha_1^2\alpha_2^2 + \alpha_2^2\alpha_3^2 + \alpha_3^2\alpha_1^2) \qquad (5.37)$$

where the α_i are the direction cosines of this direction.

Using the expansion

$$(\alpha_1^2 + \alpha_2^2 + \alpha_3^2)^2 = (\alpha_1^4 + \alpha_2^4 + \alpha_3^4) + 2(\alpha_1^2\alpha_2^2 + \alpha_2^2\alpha_3^2 + \alpha_3^2\alpha_1^2) \equiv 1 \qquad (5.38)$$

we simplify and obtain

$$\lambda_s(\alpha) = \lambda_{100} + 3(\lambda_{111} - \lambda_{100})(\alpha_1^2\alpha_2^2 + \alpha_2^2\alpha_3^2 + \alpha_3^2\alpha_1^2) \qquad (5.39)$$

Table 5-III Axial anisotropy constants and anisotropy mechanisms

Anisotropy	Mechanism	Uniaxial Constant
Crystalline	Crystal field	$K_u = K_1$
Shape	Magnetostatic	$K_u = K_s = \mu_0/2(N_a - N_c)M^2$
Stress	Magnetoelastic	$K_u = K_\sigma = \frac{3}{2}\lambda_s \sigma$
Néel	Surface	$K_u = K_s$

Source: Based on B. D. Cullity, *Introduction to Magnetic Materials*, Addison-Wesley, Reading, MA, 1972, p. 272.

which gives the saturation magnetostriction of a cubic crystal measured along the same direction as the applied magnetic field, of direction cosines α_i.

In systems with weak crystalline anisotropy, the anisotropy term derived from the magnetoelastic energy may dominate. The definition of the easy direction of magnetization will then be determined by the stress σ. The local effect of the application of a stress to a demagnetized sample will be the growth of domains in the preferred directions, both parallel and antiparallel (for $\lambda_s > 0$). The magnetization will remain zero, but there will be motion of domain walls. If the crystalline anisotropy is very low, the walls remaining after the application of the tension may be removed with the application of a negligible magnetic field.

Depending on the shape, a magnetized sample may undergo a deformation due to the tendency to minimize the magnetoelastic energy (shape effect); this effect is different from the action of magnetostriction (see Cullity 1972).

Measurements of anisotropy energies or magnetostriction in polycrystalline samples yield average values of these quantities. For example, the saturation magnetostriction of a cubic random polycrystal can be obtained by averaging Eq. (5.30) and is given by

$$\overline{\lambda_s} = \tfrac{2}{5}\lambda_{100} + \tfrac{3}{5}\lambda_{111} \tag{5.40}$$

Polycrystals in which the individual crystallites show preferred orientation are said to present texture, and their magnetic properties cannot be described by simple expressions such as Eq. (5.40).

5.3 MAGNETIC DOMAINS

Samples of ferromagnetic materials seldom have a nonzero total magnetic moment; that is, they do not behave as magnetized objects. This is the case, for example, for an ordinary iron object at room temperature. Why are all samples of ferromagnetic materials not magnets? The explanation is that the ferromagnetic samples are divided into small regions, called *domains*, each one with its magnetization pointing along a different direction, in such a way that the resulting magnetic moment (and the average magnetization) remains nearly zero. Inside each domain the magnetization has its saturation value.

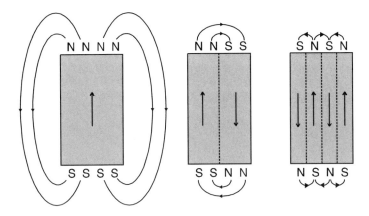

Figure 5.11 Division of a single magnetic domain, minimizing the magnetostatic energy.

Domains are created because their existence reduces the magnetostatic energy. This can be illustrated in the case of a sample of rectangular cross section (Fig. 5.11); as the original single domain splits, the magnetic energy of the system goes down. This energy is reduced even more with the formation of closure domains (not shown), with magnetization perpendicular to that of the other domains.

Between two adjacent magnetic domains with magnetization directions differing by an angle θ, there is an intermediate region of finite width, known as a *domain wall*. If a 180° domain wall has a thickness of N atoms, each one with spin S, the average angle between neighbor spins is π/N, and the energy per pair of neighbors is $E_{\text{exch}}^{\text{pair}} = \mathcal{J} S^2 (\pi/N)^2$ [from Eq. (5.18)]. A line of atoms with $N+1$ neighbors perpendicular to the domain wall has an energy

$$E_{\text{exch}} = N E_{\text{exch}}^{\text{pair}} = \frac{\mathcal{J} S^2 \pi^2}{N} \tag{5.41}$$

The condition for the energy E_{exch} to be minimum is that N grows indefinitely; however, if N increases, the anisotropy energy increases, since the number of spins not aligned to the direction of easy magnetization also increases. If the separation between the atoms is a, a unit length of the domain wall crosses $1/a$ lines of atoms; a unit area of wall is crossed by $1/a^2$ lines. The exchange energy per unit area is then

$$e_{\text{exch}} = \pi^2 \frac{\mathcal{J} S^2}{N a^2} \tag{5.42}$$

The anisotropy energy per unit volume of a uniaxial crystal is $E_K = K \sin^2 \theta$ [Eq. (5.8)]. Since a wall of unit area has a volume Na, the anisotropy energy per unit area is

$$e_K = \overline{K\sin^2\theta} Na \approx KNa \qquad (5.43)$$

The condition that minimizes the total energy per unit area $e = e_{\text{exch}} + e_K$ (exchange plus anisotropy) is given by

$$\frac{de}{dN} = \frac{-\pi^2 \mathcal{J} S^2}{N^2 a^2} + Ka = 0 \qquad (5.44)$$

and the wall that satisfies this condition has a number of atoms given by

$$N = \frac{\pi S}{a^{3/2}} \sqrt{\frac{\mathcal{J}}{K}} \qquad (5.45)$$

Therefore the thickness of this wall is

$$\delta = Na = \frac{\pi S}{a^{1/2}} \sqrt{\frac{\mathcal{J}}{K}} \qquad (5.46)$$

The domain wall thickness is therefore directly proportional to $\sqrt{\mathcal{J}}$ and inversely proportional to $\sqrt{K}$.

The subdivision into domains does not proceed indefinitely, again for energy

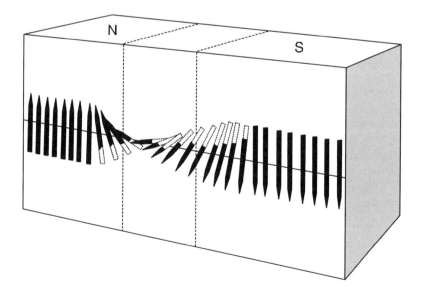

Figure 5.12 Magnetic moments inside a 180° domain wall (Bloch wall).

considerations. The formation of interfaces (walls) between the domains leads to an increase in energy due to magnetic anisotropy, and to the exchange interaction. This occurs because the anisotropy energy is minimum for a direction parallel to the original direction of magnetization of the domains, while the exchange energy is minimum for a parallel alignment of the moments. The width of the domain wall is defined by the competition between the anisotropy energy and the exchange energy; the former is reduced for narrow walls and the latter, for thick walls. As example of domain wall widths, we have 50 nm for the 90° domain walls in iron, and 15 nm for the 180° walls in cobalt.

The domain walls are also called Bloch walls, although this denomination is used more specifically for a type of wall in which the magnetization turns outside the plane of the magnetizations of the neighbor domains (Fig. 5.12). The wall whose moments turn in the same plane of the domain moments is called a *Néel wall*.

Our discussion of the formation of magnetic domains is applicable to single crystalline magnetic samples. For polycrystals, the individual crystals (or grains) normally present a multidomain structure, if their sizes are larger than a critical dimension, given roughly by Eq. (5.46) (see Coey 1996).

5.4 REVERSIBLE AND IRREVERSIBLE EFFECTS IN THE MAGNETIZATION

A small external magnetic field applied to a single domain along an arbitrary direction produces a torque that tends to turn the magnetic moments, causing them to deviate from the direction of easy magnetization. This effect produces a reversible increase of the component of the magnetization in the direction of the applied field. The angle of rotation, and consequently the increase in the magnetization (or the susceptibility), depends on the competition between the value of the anisotropy field and the intensity of the external field. For a field applied according to an angle θ_0 with the direction of uniaxial anisotropy, and forming an angle θ with the magnetization, the energy will be

$$E = -K_u \cos^2(\theta - \theta_0) - \mu_0 M_s H \cos\theta \tag{5.47}$$

where K_u is the parameter of uniaxial anisotropy.

For larger magnetic fields applied to a single-domain particle, irreversible processes occur, arising from the irreversible rotation of the magnetization. For example, the magnetization of the single domain in Fig 5.13 rotates from its original direction through the action of the field H. As the intensity of H increases, **M** eventually flips to a direction opposite to the positive c axis, shown in the figure. If, after that, H is reduced, the magnetization does not return to its original direction, but instead aligns with $-c$; this change in magnetization is therefore irreversible.

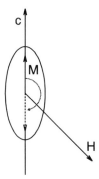

Figure 5.13 Single domain rotating its direction of magnetization under an applied magnetic field, in an irreversible process.

In real multidomain crystals, the energy of a domain wall is dependent on its position, due to the interaction with impurities and defects. This dependence may be, for example, as shown in Fig. 5.14. Small wall displacements around the position $x = s_0$, shown in the figure, are reversible, and this makes the corresponding variations of the magnetization also reversible.

The interaction of a domain wall with defects or impurities hinders its motion; a domain wall that is immobilized by this interaction is said to be pinned. If the edge of a domain wall is pinned, but its surface is allowed to move, another form of reversible magnetization results from this motion under the external field. With an increase in the H field, this wall deforms as a membrane under pressure. With this deformation, or bowing, its area, and consequently, its energy, also increase.

In the displacement Δs of a 180° domain wall, the magnetization of the

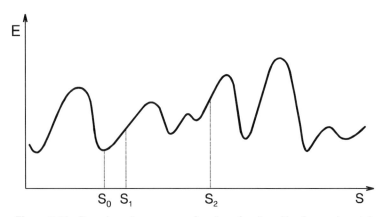

Figure 5.14 Domain wall energy as a function of wall position in a real crystal.

sample is increased by $\Delta M_s = 2M_s \, \Delta s$. The increase in magnetic energy, Eq. (5.47), is:

$$\Delta E = -2\mu_0 M_s H \Delta s \cos\theta \qquad (5.48)$$

and the force per unit area, or pressure, is given, in the limit as $\Delta s \to 0$ by:

$$\frac{\partial E}{\partial s} = -2\mu_0 M_s H \cos\theta \qquad (5.49)$$

The most important irreversible magnetization mechanism in magnetically ordered solids is related to the irreversible displacement of domain walls. This may be illustrated by Fig. 5.14, which shows the action of a magnetic field pushing the wall, for example, from the position s_0 to s_1. If at s_1 the derivative of the energy reaches a local maximum, an increase in field **H** will make the wall jump to s_2. The wall stops at s_2 because at this point the equivalent force (or pressure) exerted by **H** is again balanced by the restoring force, which is proportional to the derivative of the energy at the point s_2 [Eq. (5.49)]. If, from this moment onward, the field **H** is canceled, the wall will move to the nearest minimum, and consequently the magnetization will not return to the original value corresponding to the point s_0.

The jumps of the domain walls (e.g., from s_1 to s_2 in Fig. 5.14) can be detected through the discrete changes in magnetic flux through a coil wound around the sample. The discontinuous change in magnetization with constantly increasing **H** is known as the *Barkhausen effect*, and the steps in the induced electromotive force (e.m.f.) are called *Barkhausen noise*.

5.5 THE MAGNETIZATION PROCESS

The magnetic characterization of a sample can be made by plotting its magnetization in a graph, against the applied field **H**, generally in the form of a (1) virgin curve and (2) magnetization curve or hysteresis cycle. The *virgin curve* is the curve of magnetization versus H for an originally unmagnetized sample. The *hysteresis cycle* or *hysteresis loop* is the full magnetization curve, traced from $H = H_{\max}$ to $H = -H_{\max}$ and back (Fig. 5.15).

The variation of the magnetization of a material as a function of the intensity of the applied field H is a complex phenomenon that reflects the action of several microscopic mechanisms. A sample of magnetic material is formed, in general, by an ensemble of magnetic domains that may, under the influence of the applied field, change volume, or turn their magnetization directions away from the easy directions. The shape of the magnetization curve is affected by the presence of local impurities, defects, and grain boundaries; these are relevant for the appearance of domains with opposite magnetization (nucleation), for the pinning of domain walls, and so on.

138 THE MAGNETIZATION CURVE

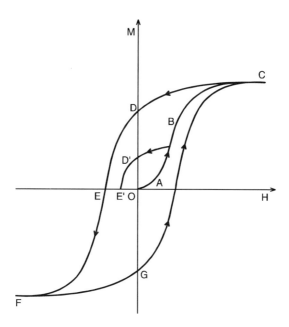

Figure 5.15 Initial (or virgin) magnetization curve (*OABC*) and magnetization curve, or hysteresis loop (*CDEFGC*). The curve *BD'E'* is followed if *H* is decreased from a point below saturation (corresponding to point *C*).

Starting from an unmagnetized sample, the magnetization curve, or M–H loop, has the general shape shown in Fig. 5.15, where we may distinguish three different regions. In the OA region, the magnetization increases slowly with the application of the external magnetic field; in the region AB this occurs more rapidly, and in the region BC the magnetization tends to a value of saturation. If the applied field does not grow until the magnetization reaches its maximum value, but instead, starts to decrease after reaching an intermediate value, the magnetization traces a curve that is, in general, different from the curve OC. Only for small fields, and consequently small magnetizations, this effect is not observed; for example, the curve OA may be traced in two senses: with increasing or decreasing field. If the magnetic field increases until the magnetization reaches the point B (Fig. 5.16), and is later reduced, the magnetization falls, for example, until B'; if H starts to increase again from this point, the magnetization follows the closed curve limited by B and B'. Curves of this type are called *minor loops* (Fig. 5.16).

From the virgin $B \times H$ curve (Fig. 5.17), we can evaluate the initial magnetic permeability μ_i (the derivative of the curve at the origin) and the maximum permeability μ_m (tangent of the largest angle formed by the straight line that is tangent to the curve and passes through the origin).

The complete magnetization curve is traced when the field H increases up to $H_{\max}$, decreases to $-H_{\max}$, and returns to the maximum value. Figure 5.15 shows

THE MAGNETIZATION PROCESS **139**

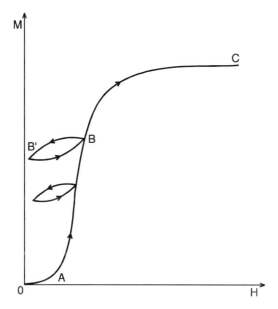

Figure 5.16 Initial magnetization curve (*OABC*), showing minor loops (e.g., *BB'B*).

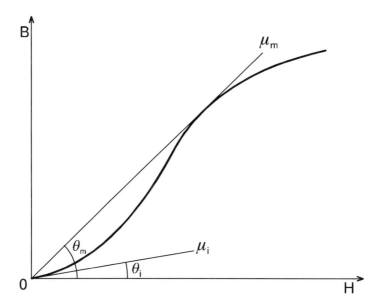

Figure 5.17 Initial curve of magnetic induction **B**, showing the angles that define the maximum permeability μ_m ($\equiv \tan\theta_m$) and the initial permeability μ_i ($\equiv \tan\theta_i$).

a typical magnetization curve with a value of H_{max} sufficient to reach the saturation of the magnetization. Several aspects may be stressed in relation to this curve: (1) as mentioned above, the decreasing field traces a curve that is different from the initial (or virgin) curve OC; (2) when the field reaches the value $H = 0$, the magnetization is not zero, but it has a finite value (OD, called the *retentivity*), and (3) the field for which the magnetization reaches zero is a negative field, whose modulus ($\overline{OE}$) is called the *coercive field* (or *coercivity*). When the $M-H$ loop is traced without reaching the saturation magnetization, the magnetization for zero field is called *remanent magnetization* (or *remanence*) ($\overline{OD'}$ in Fig. 5.15), and the field for $M = 0$ is called the *coercive force* ($\overline{OE'}$ in Fig. 5.15).

We can also describe the magnetic behavior of a sample through the graph of its magnetic induction B versus field H ($B-H$ curve). This curve is equivalent, but not identical, to the $M-H$ curve, because B and M are connected through Eq. (1.5):

$$\mathbf{B} = \mu \mathbf{H} = \mu_0(\mathbf{H} + \mathbf{M}) = \mathbf{B}_{ext} + \mu_0 \mathbf{M} \tag{5.50}$$

(we have assumed the internal field $\mathbf{H} = \mathbf{H}_{ext}$); μ_0 is the vacuum permeability and μ is the magnetic permeability of the medium. The polarization $\mathbf{J}$ is defined as $\mu_0 \mathbf{M}$. The $B-H$ curve differs from the $M-H$ curve, since the former does not present saturation; as H increases to the point of saturating M, the induction B will still continue to increase linearly with H [from Eq. (5.50)]. It is usual to distinguish the coercivity obtained from the curve of induction ($_B H_c$) from that obtained from the magnetization curve, or from the polarization curve ($_J H_c$).

Note that the internal field $\mathbf{H}$ is in general a sum of the applied field $\mathbf{H}_{ext}$ and the demagnetizing field $\mathbf{H}_d$, the latter depending on the shape of the sample, and direction of the applied field. To obtain a $M-H$ loop that is independent of these factors, we should subtract from H_{ext} the quantity $|H_d|$, thus obtaining a graph of M versus H (the internal field).

The physical quantities whose values are obtained from the virgin magnetization curve, and from the hysteresis curve, are listed in Table 5.IV.

Table 5.IV Magnetic parameters derived from the hysteresis curve (*M–H*) and from the virgin magnetization curve (see Figs 5.15 and 5.17)

Quantity	Symbol	Representation	Unit (SI)	Unit (CGS)
Saturation magnetization	M_s	OC	$A\,m^{-1}$	G
Coercivity, or coercive field	H_c	OE	$A\,m^{-1}$	Oe
Coercive force (without saturation)	H_c	OE'	$A\,m^{-1}$	Oe
Retentivity	M_r	OD	$A\,m^{-1}$	G
Remanence (without saturation)	M_r	OD'	$A\,m^{-1}$	G
Maximum permeability (virgin curve)	μ_m	$\tan(\theta_m)$	–	–
Initial permeability (virgin curve)	μ_i	$\tan(\theta_i)$	–	–
Energy product	$(BH)_{max}$		$J\,m^{-3}$	GOe

As mentioned in Section 1.5, a magnetic material adequate for the construction of permanent magnets must have elevated values of the coercive field and of the retentivity (or of the remanence); this reflects the fact that (1) a large (negative) field is required to take the magnetically saturated assembly of domains to the condition of zero net magnetization, and that (2) a high degree of alignment of the domains remains when the external field is removed. These favorable properties may be measured by a quantity known as the *energy product*, $(BH)_{max}$, [see Eq. (1.17)], which is equal to the area of the largest rectangle than can be inscribed in the second quadrant of the $B-H$ curve. Therefore, to optimize $(BH)_{max}$, a magnetic material must have, in principle, the maximum retentivity and the maximum coercivity.

The magnetic properties of some materials at room temperature, for different degrees of magnetic hardness, are presented in Tables 5.V (soft magnetic materials), 5.VI (intermediate magnetic materials), and 5.VII (hard magnetic materials); see Fig. 5.18 (see also Table 1.II and Fig. 1.15).

The fact that the magnetization follows two distinct curves, one for increasing fields and another for decreasing fields, is called *hysteresis*; for this reason the magnetization curve is also called *hysteresis curve*, or *hysteresis loop*. We can, under special conditions, obtain a magnetization curve without hysteresis (anhysteretic). For this, it is necessary to apply, for each value of H, a superposed oscillating magnetic field of decreasing intensity, with an initial amplitude sufficiently large to saturate the sample. The anhysteretic magnetization curve is traced by recording the magnetization when this amplitude reaches zero, versus H.

The work necessary to change the magnetization of an element of volume of a magnetic material, from M_1 to M_2, under an applied field H, is given by

$$\delta W = \mu_0 \int_{M_1}^{M_2} H \, \delta M \qquad (5.51)$$

The integral between $M_1 = 0$ and $M_2 = M_s$ is a measure of the area between the magnetization axis and the curve in Fig. 5.6. As we go through a full hysteresis cycle, beginning with $H = H_{max}$ and going back to this value, the variation in potential energy must be zero, and therefore, the energy corresponding to the area of the hysteresis curve is dissipated as heat. This energy converted into heat is the hysteresis loss.

The variation of magnetization as a function of magnetic field H is the result of several different processes operating in the sample. For small values of the field (curve OA, Fig. 5.15), the magnetization increases mostly through reversible motion of the walls, in such a way that the domains whose magnetization have projections along the same direction of H increase their size at the expense of the others (changing from Fig. 5.19a to Fig. 5.19b). In this region the magnetization also increases due to moment rotation inside the domains, against the anisotropy field.

For intermediate values of the field H, the magnetization increases via the irreversible displacement of the domain walls (Fig. 5.19c). In this process, the saturation magnetization is reached; its value corresponds to the value of

Table 5.V Properties of some soft magnetic materials

Main Elements besides Fe	Saturation Polarization $J = \mu_0 M$ (T)	Saturation Magnetostriction (in 10^{-6})	Coercivity (DC) (A m^{-1})	Relative Permeability ($\times 10^{-3}$)	Electrical Resistivity ($10^{-4}\,\Omega\cdot$cm)
				($H = 4m\,\text{A cm}^{-1}$)[a]	
		Commercial Alloys			
72-83 Ni + Mo, Cu, Cr	0.75–0.95	≈1	0.3–4	30–250	0.55–0.6
35-40 Ni	1.30–1.40	22–25	20–40	3–9	0.55–0.6
		Other Special Alloys			
6.5 Si	1.8	≈1	8–20	≈10	0.8
16 Al	0.8–0.9	15	2–5	4–8	1.45
		Powder Core Materials		($B = 40\,\text{mT}$)	
80 Ni, Mo	0.5–0.85	Depends	10–100	30–250	> 10^{10}
50 Ni	1.2	on alloy	200	30–150	> 10^{10}
		Amorphous Alloys		($H = 4m\,\text{A m}^{-1}$)[b]	
Fe$_{78}$Si$_9$B$_{13}$	1.55	27	3	8	1.37
Co$_{74}$Fe$_2$Mn$_4$Si$_{11}$B$_9$	1.0	<0.2	1.0	2	1.15
		Nanocrystalline Alloys			
Fe$_{73.5}$Cu$_1$Nb$_3$Si$_{13.5}$B$_9$	1.25	+2	1	100	1.35

Source: Reprinted from R. Boll, in *Materials Science and Technology*, K. H. J. Bushow, Ed., Vol. 3B, Part II, p. 439. Copyright © 1994, Wiley-VCH Verlag, Weinheim. Reprinted by permission of John Wiley & Sons, Inc.
[a] For alloys with round loops.
[b] Alloys with round or fat loops, $f = 50$ Hz.

Table 5.VI Magnetic properties of some intermediate magnetic materials (or semihard magnetic materials)[a]

	γ-Fe$_2$O$_3$	Fe$_2$O$_3$–Fe$_3$O$_4$	Co-γ-Fe$_2$O$_3$	CrO$_2$	Ba Ferrite
B_r (T)	0.11	0.15	0.15	0.15	0.12
H_c (kA m^{-1})	26	37	52	45	64

Source: Reprinted from J. Evetts, Ed., *Concise Encyclopedia of Magnetic and Superconducting Materials*, Pergamon, London, 1992, p. 223, with permission from Elsevier Science.
[a] All examples are from magnetic recording materials.

Table 5.VII Magnetic properties of some commercially available permanent magnet materials

Material	T_C (°C)	$(BH)_{max}$ (kJ m^{-3})	B_r T	$_JH_c$ (kA m^{-1})	$_BH_c$ (kA m^{-1})
Ferroxdure (SrFe$_{12}$O$_{19}$)	450	28	0.39	275	265
Alnico 4	850	72	1.04	–	124
SmCo$_5$	720	130–180	0.8–0.91	1100–1500	600–670
Sm(CoFeCuZr)$_7$	800	200–240	0.95–1.15	600–1300	600–900
NdFeB (sintered magnet)	310	200–350	1.0–1.3	750–1500	600–850

Source: Reprinted from K. H. J. Bushow, in *Materials Science and Technology*, K. H. J. Bushow, Ed., Vol. 3B, Part II, p. 475. Copyright ©1994, Wiley-VCH Verlag, Weinheim. Reprinted by permission of John Wiley & Sons, Inc.

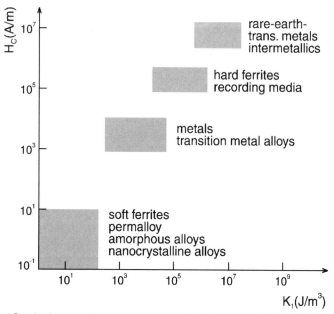

Figure 5.18 Graph of ranges of coercivity H_c and of anisotropy constant K_1 for different types of magnetic materials. [Reprinted from H. Kronmüller, *J. Mag. Mag. Mat.* **140–144**, 26 (1995), with permission from Elsevier North-Holland, NY.]

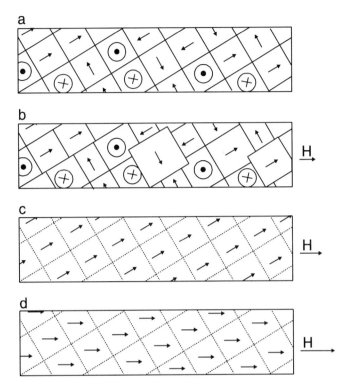

Figure 5.19 Schematic representation of the effect of an applied magnetic field on the domain structure of a ferromagnetic material; (a) before application of the field; (b) with the applied field, the domains with magnetization parallel to H increase at the expense of the other domains; (c) the technical saturation of the sample, which becomes practically one single domain (d) for higher fields, the magnetization rotates inside the domain. In c and d the dotted squares represent microscopic regions of the sample (not the domains).

the magnetization inside the domains at the temperature of the experiment; this is called *technical saturation*. Finally, for high values of H, the increase in M arises from (reversible) rotations of the magnetization of the domains, which tend to align with H (Fig. 5.19d). The magnetization grows still further, through the increase in the degree of alignment of the magnetic moments inside the domains; this is called *forced magnetization*.

The shapes of the $M-H$ curves are generally dependent on the direction of the applied field **H** relative to the crystal axes, due to the effect of crystalline anisotropy. This may be illustrated in an idealized sample with two magnetic domains, uniaxial anisotropy, no irreversible effects in the magnetization and high wall mobility. We also neglect shape anisotropy. If one applies a magnetic field parallel to the anisotropy axis (Fig. 5.20a) the domain wall will move and the magnetization will reach saturation for a negligible field. The $M-H$ curve will be as a shown in Fig. 5.20b.

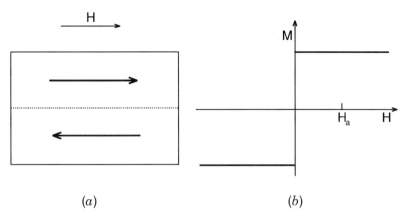

Figure 5.20 (a) Sample of magnetic material with two domains, under the action a field **H** parallel to the direction of easy magnetization; (b) shape of M–H curve; note the saturation for low fields.

If, on the other hand, the applied magnetic field **H** points along a direction perpendicular to the anisotropy axis, the wall will not move, and the magnetization will gradually turn inside the domains, as this field overcomes the effect of the anisotropy field $\mathbf{H}_a$. The M–H curve will be a sloping straight line, reaching saturation for the field $\mathbf{H} = \mathbf{H}_a$ (Figs. 5.21a and 5.21b).

It is instructive to follow the direction of magnetization of the domains inside the material, at different points in the hysteresis curve. In the first place, it should be noted that different domain configurations may correspond to the same value of magnetization. The configurations are shown in Fig. 5.22. In particular, for a null magnetization there is more than one possible configuration of the domains; the ideal demagnetized state is usually taken as that in which the

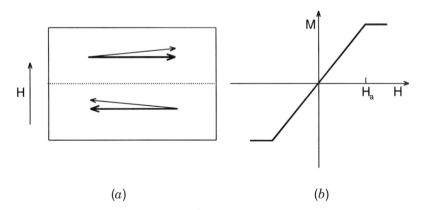

Figure 5.21 (a) Sample of magnetic material with two domains, under the action of a field **H** perpendicular to the direction of easy magnetization; (b) Shape of M–H curve; note the gradual increase of the magnetization, reaching saturation for $\mathbf{H} = \mathbf{H}_a$.

146 THE MAGNETIZATION CURVE

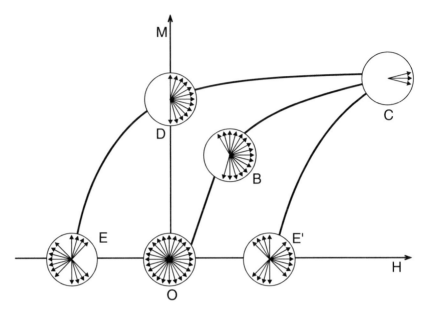

Figure 5.22 Distribution of the directions of magnetization of the domains at different points of the magnetization curve.

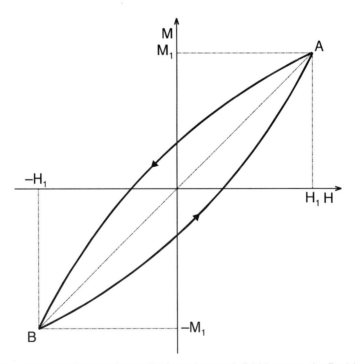

Figure 5.23 Magnetization curve for small values of magnetic field, known as the *Rayleigh curve*.

volume of the sample is equally divided among the possible types of magnetic domains.

The magnetization curves obtained with very low values of H (of the order of 100 A m^{-1} or 1 Oe) have a special shape, called the *Rayleigh curve* (Fig. 5.23). The magnetic permeability for small field intensity is a linear function of H and can be written in the form

$$\mu = \mu_i + \nu H \tag{5.52}$$

where μ_i is the initial permeability, which is also given by the tangent of the angle of the straight line tangent at the origin of the curve of B versus H, and ν is the Rayleigh coefficient; these two quantities are characteristic of each material. The magnetization curve as a function of H for low values of H [from (5.50)] is a parabola, of the form

$$M = aH + bH^2 \tag{5.53}$$

with $a = (\mu_i - \mu_0)/\mu_0$ and $b = \nu/\mu_0$. The area limited by the hysteresis curve in this case is proportional to H^3.

As explained in the previous section, permanent magnet materials are hard magnetic materials, with high retentivity M_r (or high remanence) and high

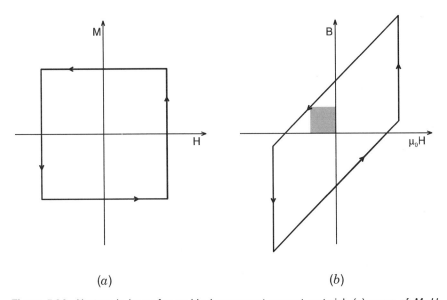

(a) (b)

Figure 5.24 Hysteresis loops for an ideal permanent magnet material: (a) curve of M–H, showing a square loop; (b) curve $B - \mu_0 H$, showing the square whose area is a measure of the ideal energy product $(BH)_{\text{max}}^{\text{limit}}$ (see text).

coercivity H_c. The single characteristic that best describes the suitability of a given material for the use in permanent magnets is the maximum energy product $(BH)_{max}$.

An ideal permanent magnet material has a perfectly square $M-H$ loop, which means that its magnetization remains at the saturation level for any value of H, from the maximum applied field to the coercive field (H_c) (Fig. 5.24a). Therefore, the magnetization at the remanence point (M_r) has the same value of the saturation magnetization M_s. Drawing the hysteresis loop as B versus $\mu_0 H$ (Fig. 5.24b), one can use the same units in both axes (in tesla). For this ideal magnet material, the graph in the second quadrant is a straight line connecting the point $(0, B_r)$ to the point $(\mu_{0B} H_c, 0)$, since the only change in B in this range of magnetic field arises from the variation in H itself.

The maximum energy product $(BH)_{max}$ is the area of the square in the hysteresis loop, plotted as $B \times \mu_0 H$ (Fig 5.24b). From this curve, noting that $\mu_{0B} H_c = B_r$ and $B_r = \mu_0 M_s$, it is easy to estimate the value of the energy product in this ideal case:

$$(BH)_{max}^{limit} = \mu_0 \frac{M_s^2}{4} = \frac{B_r^2}{4\mu_0} \qquad (5.54)$$

This is therefore the upper theoretical limit for this quantity; for example, the measured energy product for a sample of NdFeB with induction at remanence $B_r = 1.35\,T$ is $320\,kJ\,m^{-3}$, corresponding to approximately 90% of the value of $363\,kJ\,m^{-3}$ predicted from the above expression.

Permanent magnet materials are frequently multiphase, or heterogeneous, consisting of different components that have different magnetic properties, such as magnetic hardness. Also, the domain structure is complicated by the presence of both multidomain and single-domain grains. Therefore, the analyses of the processes responsible for the shape of the hysteresis loops are correspondingly more complex (see Givord 1996).

5.6 DYNAMIC EFFECTS IN THE MAGNETIZATION PROCESS

There is a class of magnetic phenomena associated with the time dependence of the response to external applied magnetic fields. In the discussion of the magnetization process, we have not yet considered the form of dependence of $H(t)$, assuming implicitly that at each moment the system is in equilibrium. In this section we will briefly treat these phenomena, limiting our scope to time effects observed in ferromagnetic materials. This restriction excludes some important time effects observed, for example, in spin glasses.

These dynamic effects can be divided into aftereffects and resonances. The application of a magnetic field **H** of sufficient intensity to take a sample to magnetic saturation does not induce the instantaneous appearance of a magnetization M_s, for two main reasons: (1) for very short times, eddy currents appear in the sample that oppose the growth of the induction **B** (or of **M**); (2) because the several microscopic processes underlying the magnetization process take finite characteristic times to be completed. The delay in the growth of the magnetization due to this last cause is called *magnetic aftereffect*, and may vary from a fraction of a second to many hours. In these processes we will ignore time dependences associated with nonreversible causes, due to the action of the magnetic fields, such as structural changes, or aging of the material. In the case of Fe–C alloys, for example, the aftereffects were attributed to the diffusion of carbon atoms that occupy interstitial sites, producing deformations that change the energy of the domain walls and can lead to their displacement. A thermal fluctuation aftereffect arises from the thermal fluctuation of the magnetization direction in small single-domain particles (or in pinned domain walls); this is a strongly temperature-dependent process, the rate of change falling with the temperature. This effect usually leads to a linear variation of the sample magnetization with the logarithm of the time, and this property is known as magnetic viscosity; it is expressed quantitatively as (see Givord 1996):

$$S = -\frac{dM}{d(\ln t)} \tag{5.55}$$

where M is the magnetization and t is the time.

One of these time effects is the disaccommodation, which consists in the

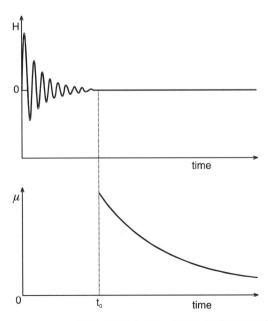

Figure 5.25 The phenomenon of disaccommodation—the magnetic permeability decreases with time, after the application of an oscillating magnetic field of decreasing amplitude.

variation of the magnetic susceptibility (or of the permeability) of a material, after the application of a magnetic field; in this case it is the magnetic response, not the magnetization, that changes. It can be observed after the sample is demagnetized through the action of an oscillating field of decreasing intensity (Fig. 5.25). In ferrites this effect was attributed to changes in the magnetic properties brought about by the migration of vacancies. In FeSi alloys this effect arises from a linear ordering of atoms of the constituent elements.

Eddy currents appear each time a conducting sample is subjected to a time varying magnetic field $\mathbf{H}(t)$. The variation of the magnetic flux $\phi = BA$ (A is the area) creates an electromotive force ϵ that produces currents in the material being magnetized (Faraday law):

$$\epsilon = -\frac{d\phi}{dt} = -\frac{d(BA)}{dt} \tag{5.56}$$

The currents generate a magnetic field that opposes the rate of variation $d(BA)/dt$ of the flux (Lenz's law); they have the effect, therefore, of hindering the increase of the magnetization at the same rate of change as the external field $\mathbf{H}(t)$. The eddy currents are proportional to the square of the frequency, and, of course, inversely proportional to the electrical resistivity of the material. These currents are more amplified yet in the domain walls. They are an important source of energy loss inside magnetic materials, an effect that is particularly relevant in power transformers.

The most important losses in ferromagnetic materials, however, are hysteresis losses (Section 5.5); losses by microscopic eddy currents and by other mechanisms associated with irreversible processes contribute to these losses. At high frequencies, on the other hand, domain wall motion is reduced, and the losses by microscopic eddy currents dominate. The presence of hysteresis, of eddy currents, and other mechanisms leads to the appearance, in an oscillating magnetic field, of an imaginary term in the magnetic permeability and in the magnetic susceptibility.

Under an oscillating magnetic field of angular frequency ω, given by

$$H = H_0 e^{i\omega t} \tag{5.57}$$

a retarded flux density B arises, with phase difference δ:

$$B = B_0 e^{i(\omega t - \delta)} \tag{5.58}$$

and the magnetic permeability is given by

$$\mu = \frac{B}{H} = \frac{B_0 e^{i(\omega t - \delta)}}{H_0 e^{i\omega t}} = \frac{B_0}{H_0} e^{-i\delta} \tag{5.59}$$

DYNAMIC EFFECTS IN THE MAGNETIZATION PROCESS

The permeability μ can be written in complex form

$$\mu = \mu' - i\mu'' \qquad (5.60)$$

where the normal permeability (in phase with H) is given by the real part of μ:

$$\mu' = \frac{B_0}{H}\cos\delta \qquad (5.61)$$

and the out-of-phase part (the imaginary part), which is related to the dissipative processes, is given by

$$\mu'' = \frac{B_0}{H}\sin\delta \qquad (5.62)$$

The loss factor is given by the ratio

$$\frac{\mu''}{\mu'} = \tan\delta \qquad (5.63)$$

The magnetization process due to the increase in volume of the domains cannot be instantaneous, since the walls move in a magnetic medium with a finite velocity. Observations made in different materials record velocities between 1 and 10^4 cm s^{-1}. Although there is no displacement of mass in the motion of a domain wall, there is inertia against this motion, which results from the torques applied by the angular momenta associated with the atomic magnetic moments.

The variation in the energy of a domain wall of area A, under the influence of a field **H** that produces a displacement x, is

$$E = -2\mu_0 A M_s H x \qquad (5.64)$$

and the force per unit area is

$$F = -\left(\frac{1}{A}\right)\frac{dE}{dx} = 2\mu_0 M_s H \qquad (5.65)$$

The equation of motion of the wall can now be written

$$\frac{d^2 x}{dt^2} + \beta\frac{dx}{dt} + \alpha x = 2\mu_0 M_s H \qquad (5.66)$$

The effective mass of the wall is $m = \mu_0 \sigma / 2\gamma^2 A'$, where σ is the energy per unit area, γ is the gyromagnetic ratio of the spins, and A' is the exchange stiffness, a coefficient proportional to the exchange energy ($A' = JS^2/a$, where a is the interatomic spacing).

Let us assume magnetizations pointing along the axes x and $-x$ in two adjacent domains separated by a 180° wall, with the normal to the domain wall in the z direction (Fig. 5.26). A field $\mathbf{H} = H\mathbf{i}$ will produce a torque on the moments localized in the wall, that will push them out of the planes of $\mathbf{H}$ and $\mathbf{M}$, that is, upwards, as shown in Fig. 5.26. The z component of the magnetization in the wall will create a demagnetizing field:

$$H_d = -N_d M_z \tag{5.67}$$

and its action on the moments will be perpendicular to z, and will cause the moments to turn in the plane xy. It is this effect that leads to the displacement of the wall; the final result is the volume increase of the domains with magnetization parallel to H, and a reduction of the volume of the antiparallel domains.

Several resonance phenomena are observed in solids submitted to oscillating electromagnetic fields; in general, their observation requires the simultaneous

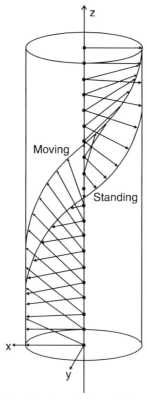

Figure 5.26 Spin structure inside a domain wall, shown in two configurations: in the stationary state, and moving under an applied magnetic field. The plane of the domain wall is perpendicular to the z axis.

application of an external static magnetic field. The atomic magnetic moments, the magnetic moments of the conduction electrons, and the magnetic moments of the nuclei can absorb energy from an oscillating field. These phenomena are usually described in terms of classical equations of motion of the magnetization, as the Bloch equations, the Landau–Lifshitz equations, or the Gilbert equation [see, e.g., Morrish (1965)]. These equations of motion are discussed in Sections 7.2, 7.3, and 8.4.

Among the resonances that may be observed, one includes (1) cyclotron resonance, (2) electron paramagnetic or spin resonance (EPR or ESR), (3) ferromagnetic resonance (FMR) and antiferromagnetic resonance (AFMR), (4) spin wave resonance, and (5) nuclear magnetic resonance (NMR). This last phenomenon will be discussed in more detail in Chapters 7 and 8; a brief introduction to FMR will be presented in Section 8.4.

Cyclotron resonance (or Azbel'–Kaner resonance) is observed in metals and semiconductors, under the action of microwaves; in nonmagnetic materials it requires the application of an external magnetic field. It results from the interaction of the electric field of the microwaves with the charge of the electrons, keeping them in circular orbits; therefore, it is not in fact a magnetic resonance. The resonance condition is the same as the operating regime of the cyclotrons, given by

$$\omega = \frac{eB}{m} \quad (5.68)$$

where B is the magnetic induction and m is the effective mass of the electrons; this frequency is twice the expression of the Larmor frequency [Eq. (2.10)].

In electron paramagnetic resonance, transitions are observed between the energy states of the unpaired electrons in the atoms or molecules, in the presence of an external magnetic field. For a magnetic induction ranging in tenths of tesla (or in the range of kilogauss), the resonance frequencies are in the region of gigahertz. A special type of EPR is observed in diamagnetic metals, due to conduction electrons: the conduction electron spin resonance (CESR). It is characterized by very broad lines, due to the fast relaxation rates of the conduction electrons (see Chapter 8).

In ordered magnetic systems, the atomic spins precess in phase, under the influence of an applied magnetic field. With the incidence of microwaves, we may obtain, according to the case, ferromagnetic resonance (FMR), or antiferromagnetic resonance (AFMR) (see Section 8.4). These resonances are usually observed in the same way as is EPR, although in the case of FMR the atomic magnetic moments are also under the influence of the demagnetizing fields. The resonance condition for ferromagnetic resonance where the ions feel an anisotropy field $\mathbf{B}_a$, is

$$\omega = \gamma(B_0 + B_a) \quad (5.69)$$

where γ is the gyromagnetic ratio of the atomic moments and $\mathbf{B}_0$ is the applied magnetic field (Section 8.4). As in other types of resonance where the magnetic moments are electronic moments (as in EPR), for applied fields in the range of tenths of tesla (or kilogauss), the resonance frequencies are usually in the gigahertz range.

Finally, the spin wave resonance is a special type of FMR, which may be observed in thin samples of magnetically ordered materials. These samples, under a magnetic field applied along the normal to the plane of the film, present an amplitude of precession of the atomic moments that varies along the same normal. For some spin wave wavelengths, the turning magnetization is maximum, and there is resonance. The resonance condition is defined by

$$\hbar\omega = g\mu_B(B_0 - \mu_0 M) + DK^2 \qquad (5.70)$$

where

$K = n\pi/L$ (L is the smallest dimension of the sample)
n = an integer
D = the spin wave stiffness constant (see Section 3.5)
g = the g factor
M = magnetization.

The observation of the different resonances succinctly described above presupposes the possibility of penetration of the electromagnetic waves into the solids. In the case of metallic solids, this penetration is limited because of the skin effect; the intensity of the electromagnetic field inside the conducting samples falls exponentially, decreasing to $1/e$ of the value on the surface for a penetration depth s, given by

$$s = \sqrt{\frac{2\rho}{\omega\mu}} \qquad (5.71)$$

where

ρ = the electric conductivity
ω = the angular frequency of the electromagnetic wave
μ = the magnetic permeability at the applied frequency ω.

A case of large practical importance is the penetration of the fields at the ac frequency of the electricity network (60 or 50 Hz) into electric conductors and transformer cores. For a copper conductor, at this frequency, for example, s is approximately 1 cm; for an iron core, $s = 1$ mm. The skin effect is also relevant in the phenomenon of giant magnetoimpedance, which consists in the large variation of impedance of a sample as a function of applied magnetic field.

At higher frequencies, the penetration is a function of the permeability $\mu(\omega)$.

For a frequency around 100 MHz, the penetration depth in a metal is smaller than 10 μm.

GENERAL READING

Barbara, B., D. Gignoux, and C. Vettier, *Lectures on Modern Magnetism*, Science Press and Springer-Verlag, Beijing and Berlin, 1988.
Boll, R., "Soft Magnetic Metals and Alloys," in K. H. J. Buschow, Ed., *Materials Science and Technology*, Vol. 3B, Part II, VCH, Weinheim, 1994, p. 399.
Buschow, K. H. J., "Permanent Magnetic Materials," in K. H. J. Buschow, Ed., *Materials Science and Technology*, Vol. 3B, Part II, VCH, Weinheim, 1994, p. 451.
Carr, W. J., Jr., "Principles of Ferromagnetic Behavior," in A. E. Berkowitz and A. E. Kneller, Eds., *Magnetism and Metallurgy*, Vol. 1, Academic Press, New York, 1969, p. 45.
Chikazumi, S., *Physics of Magnetism*, 2nd ed., Clarendon Press, Oxford, 1997.
Coey, J. M. D., Ed., *Rare-Earth Iron Permanent Magnets*, Clarendon Press, Oxford, 1996.
Crangle, J., *Solid State Magnetism*, Van Nostrand Rheinhold, London, 1991.
Culken, J. R., A. E. Clark, and K. B. Hathaway, "Magnetostrictive Materials," in K. H. J. Buschow, Ed., *Materials Science and Technology*, Vol. 3B, Part II, VCH, Weinheim, 1994, p. 529.
Cullity, B. D., *International J. Mag.* **1**, 323 (1971).
Cullity, B. D., *Introduction to Magnetic Materials*, Addison-Wesley, Reading, 1972.
Givord, D. and M. F. Rossignol, "Coercivity," in J. M. D. Coey, Ed., *Rare-Earth Iron Permanent Magnets*, Clarendon Press, Oxford, 1996.
Grandjean, F. and G. J. Long, in *Supermagnets, Hard Magnetic Materials*, Kluwer Academic Press, Amsterdam, 1990, p. 27.
Ingram, D. J. E., "Magnetic Resonance," in G. M. Kalvius and R. S. Tebble, Eds., *Experimental Magnetism*, Wiley, Chichester, 1979.
Jiles, D., *Introduction to Magnetism and Magnetic Materials*, Chapman & Hall, London, 1991.
Kittel, C., *Rev. Mod. Phys.* **21**, 541 (1949).
Landau, L. D. and E. M. Lifshitz, *Theory of Elasticity*, Pergamon Press, London, 1959.
Lee, E. W., "Magnetostriction," in G. M. Kalvius and R. S. Tebble, Eds., *Experimental Magnetism*, Vol. 1, Wiley, Chichester, 1979.
Martin, D. H., *Magnetism in Solids*, Iliffe, London, 1967.
McCurrie, R. A., *Ferromagnetic Materials*, Academic Press, London, 1994.
Morrish, A. H., *Physical Principles of Magnetism*, Wiley, New York, 1965.
Parker, R J., *Advances in Permanent Magnets*, Wiley, New York, 1990.
Rathenau, G. W. and G. De Vries, "Diffusion," in A. E. Berkowitz and A. E. Kneller, Eds., *Magnetism and Metallurgy*, Vol. 2, Academic Press, New York, 1969.
Schultz, S., "Microwave Resonance in Metals," in E. Passaglia, Ed., *Techniques of Metal Research*, Vol. VI, Part 1, Interscience, New York, 1973, p. 338.
Williams, D. E. G., *The Magnetic Properties of Matter*, Longmans, London, 1966.

REFERENCES

Boll, R., "Soft Magnetic Metals and Alloys," in K. H. J. Buschow, Ed., *Materials Science and Technology*, Vol. 3B, Part II, Wiley-VCH, Weinheim, 1994, p. 399.

Buschow, K. H. J., "Permanent Magnet Materials," in K. H. J. Buschow, Ed., *Materials Science and Technology*, Vol. 3B, Part II, Wiley-VCH, Weinheim, 1994, p. 451.

Coey, J. M. D., Ed., *Rare-Earth Iron Permanent Magnets*, Clarendon Press, Oxford, 1996.

Cullity, B. D., *Introduction to Magnetic Materials*, Addison-Wesley, Reading, MA, 1972.

Evetts, J., Ed., "Magnetic Materials: An Overview," in *Concise Encyclopedia of Magnetic and Superconducting Materials*, Pergamon Press, London, 1992.

Givord, D. and M. F. Rossignol, "Coercivity," in J. M. D. Coey, Ed., *Rare-Earth Iron Permanent Magnets*, Clarendon Press, Oxford, 1996.

Gradman, U., in K. H. J. Buschow, Ed., *Handbook of Magnetic Materials*, Vol. 7, North-Holland, Amsterdam, 1993, p. 1.

Herrmann, F., *Am. J. Phys.* **59**, 447 (1991).

Kittel, C., *Rev. Mod. Phys.* **21**, 541 (1949).

Kittel, C., *Introduction to Solid State Physics*, 6th ed., Wiley, New York, 1986.

Kronmüller, H., *J. Mag. Mag. Mat.* **140–144**, 25 (1995).

Landau, L. D. and E. M. Lifshitz, *Theory of Elasticity*, Pergamon Press, London, 1959.

Morrish, A. H., *Physical Principles of Magnetism*, Wiley, New York, 1965.

Néel, L., *J. Phys. Radium* **15**, 225 (1954).

EXERCISES

5.1. *Anisotropy Energy.* For a crystal with cubic symmetry, show that the fourth order term $\alpha_1^4 + \alpha_2^4 + \alpha_3^4$ does not appear in the expression of the anisotropy energy, although it satisfies the symmetry requirements.

5.2 *Magnetoelastic Coupling.* In a cubic crystal, the density of elastic energy in terms of the components of the tensor e_{ij} is:

$$U_{el} = \tfrac{1}{2}C_{11}(\epsilon_{xx}^2 + \epsilon_{yy}^2 + \epsilon_{zz}^2) + \tfrac{1}{2}C_{44}(\epsilon_{xy}^2 + \epsilon_{yz}^2 + \epsilon_{zx}^2)$$
$$+ C_{12}(\epsilon_{yy}\epsilon_{zz} + \epsilon_{xx}\epsilon_{zz} + \epsilon_{xx}\epsilon_{yy})$$

and the dominant term in the anisotropy energy is

$$U_K = K_1(\alpha_1^2\alpha_2^2 + \alpha_2^2\alpha_3^2 + \alpha_3^2\alpha_1^2)$$

The magnetoelastic coupling may be formally taken into account with the introduction of the term

$$U_a = B_1(\alpha_1^2\epsilon_{xx} + \alpha_2^2\epsilon_{yy} + \alpha_3^2\epsilon_{zz}) + B_2(\alpha_1\alpha_2\epsilon_{xy} + \alpha_2\alpha_3\epsilon_{yz} + \alpha_3\alpha_1\epsilon_{zx})$$

where B_1 and B_2 are the magnetoelastic coupling constants. Show that the total energy is minimum when

$$\epsilon_{xx} = \frac{-B_1[C_{11}^2\alpha_1^2 + C_{12}^2\alpha_2^2 - C_{11}C_{12}\alpha_3^2]}{C_{11}^3 + C_{12}^3}$$

$$\epsilon_{xy} = -\frac{B_2\alpha_1\alpha_2}{C_{44}}$$

with similar expressions for the other components of ϵ_{ij}.

5.3 *Equilibrium Configuration of a Bloch Wall.* Let $U_{\text{exch}}(\phi, \phi')$ and $U_K(\phi)$ be the exchange and anisotropy energies, respectively, at a point x along a Bloch wall. $\phi = \phi(x)$ is the angle that the magnetization at the point x makes with the anisotropy field (same direction of the magnetization in the domains), and ϕ' is its first derivative in relation to x. If the extreme values of ϕ are 0 and ϕ_0, the total energy of the wall may be written as

$$J = \int_0^{\phi_0} [U_K + U_{\text{exch}}(\phi, \phi')]dx$$

The equilibrium condition of the wall can be obtained from the variational principle:

$$\delta J = 0$$

(a) Compute δJ for the case in which $U_{\text{exch}} = h(\phi)\phi'^2$, where $h(\phi)$ is an arbitrary function of ϕ. Substitute

$$\phi'\,\delta\phi' = \frac{d}{dx}(\phi'\,\delta\phi) - \phi''\,\delta\phi$$

and show that

$$\delta J = \int_0^{\phi_0} \left[\left(\frac{dU_K}{d\phi} + \frac{dh}{d\phi}\phi'^2 - 2h\phi''\right)\delta\phi + 2h\frac{d}{dx}(\phi'\,\delta\phi)\right]dx$$

(b) Recalling that $\delta\phi$ must be zero in the extremes of the variation, integrate the last term of the preceding expression and show that

$$\int_0^{\phi_0} 2h(\phi)\frac{d}{dx}(\phi'\,\delta\phi)dx = -2\int_0^{\phi_0} \phi'^2\frac{dh}{d\phi}\delta\phi\,dx$$

(c) Substitute this result into δJ and, from the condition $\delta J = 0$, obtain a differential equation connecting these quantities.

(d) Use $\phi' = d\phi/dx$ as integrating factor and show that

$$U_K(\phi) = h(\phi)\phi'^2$$

Thus, the equilibrium configuration is that in which the energies of exchange and anisotropy are equal *at every point of the wall*.

6

HYPERFINE INTERACTIONS

6.1 INTRODUCTION

The electric charges present in the nucleus interact with the electrons that surround it; in an analogous way, the electric currents (or the magnetic moments) associated with the electrons and the nuclei also interact. The magnetic and electrostatic interaction between nuclei and electrons may be written in a general way as a sum of products

$$\mathcal{H} = \sum_l C^N(l) \times C^e(l) \tag{6.1}$$

where $C^N(l)$ and $C^e(l)$ are nuclear and electronic operators corresponding to the multipolar electric terms [of parity $p = (-1)^l$] or magnetic terms [of parity $p = (-1)^{l+1}$], where l is an integer.

The main contributions to the interaction associated with the following nuclear moments are: (1) *electric part*—nuclear electric monopole moment (which is simply the nuclear charge) and nuclear electric quadrupole moment and (2) *magnetic part*—nuclear magnetic dipole moment. In some cases the magnetic octupole interaction can also be detected, but usually it can be neglected.

The interaction of the electric monopole moment of the nucleus with the electric field due to the electrons is the Coulomb interaction, and does not concern us here.

The other interactions between nuclei and electrons are called *hyperfine interactions.* In magnetic materials, the main hyperfine interaction is the interaction of magnetic origin; the electrostatic interaction is usually smaller.

Experimentally it is observed that the hyperfine interactions are much weaker than the exchange interactions or the interactions of the ionic moment with the crystal field; the latter, in turn, are much weaker than the spin–orbit interactions ($\mathcal{H}_{LS}$) (in the rare earths). That is, illustrating the case of the rare earths,

$$\mathcal{H}_{LS} \gg \mathcal{H}_{\text{exch}} + \mathcal{H}_{\text{cf}} \gg \mathcal{H}_{\text{hf}} \tag{6.2}$$

Typical values of these interaction energies for the rare earth ions are $E_{LS}/k \sim 10^4$ K, $E_{\text{exch}}/k \sim 10^3$ K, $E_{\text{cf}}/k \sim 10^2$ K, and $E_{\text{hf}}/k \sim 10^{-4}$ K.

The atomic nuclei are characterized by the atomic number Z and by the mass number A: Z is the number of protons, and A the number of nucleons (protons + neutrons) present. The angular momenta of the nucleons couple in such way as to produce zero total angular momentum I in the cases when both Z and $(A - Z)$ are even. In every other case, $I \neq 0$, and it is either integer (a multiple of $\hbar$) or half–integer (multiple of $\hbar/2$). The nuclei having nonzero angular momentum have an associated magnetic dipole moment given by

$$\boldsymbol{\mu} = g_I \mu_N \mathbf{I} \tag{6.3}$$

where g_I is the nuclear g-factor and μ_N is the nuclear magneton, given by (m_p = proton mass):

$$\mu_N = \frac{e\hbar}{2m_p} = \frac{\mu_B}{1836} \tag{6.4}$$

where μ_B is the Bohr magneton. The nuclear magnetic moment is also written $\mu = \gamma \hbar \mathbf{I}$, as a function of the gyromagnetic ratio γ.

Since $\mu_N \ll \mu_B$ and the g factors (g_I) of the nucleus are of the order of 1, therefore comparable to the electronic g factors, it follows that the nuclear magnetic moments are much smaller than the ionic moments. For this reason, the nuclear magnetism of matter produces more subtle effects than the electronic (or ionic) magnetism. In general, the magnetic interaction energy of the nuclei ($\mu_I B$) is much smaller than kT, for usual values of B. To find effects comparable to those of the electronic magnetization, we need to reach temperatures three orders of magnitude lower.

Every nucleus with $I \neq 0$ has a magnetic dipolar moment. The nuclei that have $I > \frac{1}{2}$ also possess an electric quadrupole moment Q, since their charge distribution lacks spherical symmetry.

6.2 ELECTROSTATIC INTERACTIONS

The nuclei located in a solid interact with the electric charges of the electrons bound to the same atom, of electrons of neighbor atoms, and of conduction electrons (the latter, in the case of metals and semiconductors).

The interaction energy of a distribution of charges $\rho(\mathbf{r})$ limited in space and submitted to a potential $V(\mathbf{r})$ is given by

$$W = \int \rho(\mathbf{r})V(\mathbf{r})dv \tag{6.5}$$

where the integration is made over the volume occupied by the charges.

In our case we will take $V(\mathbf{r})$ due to the electrons; $\rho(\mathbf{r})$ in this case is the distribution of nuclear charge, and the integral is taken over the nuclear volume. The potential $V(\mathbf{r})$ may be expanded in a Taylor series around the origin:

$$V(\mathbf{r}) = V(0) + \sum_i x_i \left[\frac{\partial V}{\partial x_i}\right]_0 + \frac{1}{2}\sum_i \sum_j x_i x_j \left[\frac{\partial^2 V}{\partial x_i\, \partial x_j}\right]_0 + \cdots \tag{6.6}$$

where the sums are made over the components 1, 2, 3 (i.e., x, y, z). Alternative approaches use an expansion in spherical harmonics (e.g., Abragam 1961), or in tesseral harmonics (Arif et al. 1975).

Summing and subtracting the term

$$\frac{1}{6}\sum_i \sum_j r^2 \delta_{ij} \left[\frac{\partial^2 V}{\partial x_i\, \partial x_j}\right]_0 = \frac{1}{6}\sum_i r^2 \left[\frac{\partial^2 V}{\partial x_i^2}\right]_0 \tag{6.7}$$

where δ_{ij} is the Kronecker delta, we obtain

$$V(\mathbf{r}) = V(0) + \sum_i x_i \left[\frac{\partial V}{\partial x_i}\right]_0 + \frac{1}{6}\sum_i r^2 \left[\frac{\partial^2 V}{\partial x_i^2}\right]_0 + \frac{1}{6}\sum_i \sum_j (3x_i x_j - r^2\delta_{ij})\left[\frac{\partial^2 V}{\partial x_i\, \partial x_j}\right]_0 + \cdots \tag{6.8}$$

thus, substituting in Eq. (6.5):

$$W = V(0)\int \rho(\mathbf{r})dv + \sum_i \left[\frac{\partial V}{\partial x_i}\right]_0 \int x_i \rho(\mathbf{r})dv + \frac{1}{6}\sum_i \left[\frac{\partial^2 V}{\partial x_i^2}\right]_0 \int r^2 \rho(\mathbf{r})dv +$$

$$+ \frac{1}{6}\sum_i \sum_j \left[\frac{\partial^2 V}{\partial x_i\, \partial x_j}\right]_0 \int \rho(\mathbf{r})(3x_i x_j - r^2 \delta_{ij})dv + \cdots \tag{6.9}$$

The first term of W is the electrostatic energy of the nucleus taken as a point charge (Coulomb term). In the second term, the integral is the electric dipolar term of the nucleus; this is zero, since the center of mass and the center of charge of the nucleus coincide. This may be proved by starting from the fact that the nuclei have well-defined parity [i.e., $\Psi(r) = \pm\Psi(-r)$, therefore $|\Psi(r)|^2 = |\Psi(-r)|^2$]. The third term only gives a displacement in the energy; we will come back to it shortly.

Introducing the notation V_{ij} for the second derivative of the potential, and using the fact that this derivative is equal to the first derivative of the electric field components (with negative sign), we have

$$V_{ij} = \frac{\partial^2 V}{\partial x_i \partial x_j} = -\frac{\partial E_j}{\partial x_i} \tag{6.10}$$

and we speak of a gradient of the electric field, in analogy with the gradient ∇A, where A is a scalar. The integral of the fourth term in Eq. (6.9) is a component of the electric quadrupole moment tensor of the nucleus, Q_{ij}

$$Q_{ij} = \int \rho(\mathbf{r})(3x_i x_j - r^2 \delta_{ij}) dv \tag{6.11}$$

The corresponding term in the expression of the energy therefore remains:

$$-\frac{1}{6}\sum_i \sum_j \frac{\partial E_j}{\partial x_i} Q_{ij} \tag{6.12}$$

that is, it contains the product of the electric field gradient by the electric quadrupolar moment of the nucleus. The electric field gradient is a tensor with components $\partial E_j/\partial x_i$.

To obtain the expression of the electric quadrupole interaction in quantum mechanics, we initially have to substitute the charge density $\rho(\mathbf{r})$ by the operator $\rho^{op}(\mathbf{r})$

$$\rho^{op}(\mathbf{r}) = e \sum_k \delta(\mathbf{r} - \mathbf{r}_k) \tag{6.13}$$

where the sum extends over the Z protons, of coordinates x_{ik}, at the positions $\mathbf{r}_k$. The quadrupole moment tensor operator becomes

$$Q_{ij}^{op} = e \sum_k \int (3x_i x_j - r^2 \delta_{ij})\delta(\mathbf{r} - \mathbf{r}_k) dv \tag{6.14}$$

$$Q_{ij}^{op} = e \sum_k (3x_{ik} x_{jk} - r_k^2 \delta_{ij}) \tag{6.15}$$

and the hamiltonian of the quadrupole interaction results:

$$\mathcal{H}_Q = \frac{1}{6}\sum_i \sum_j V_{ij} Q_{ij}^{op} \tag{6.16}$$

This hamiltonian may be written in simple form, as a function of the operators of the angular momentum of the nucleus, using the Wigner–Eckart theorem, which states that the matrix elements of any vector operator in the space of eigenstates of I^2 and I_z are proportional to the matrix elements of I. This gives, for the matrix elements of the operator Q_{ij} (Slichter 1990, Chapter 10):

$$(Im\zeta|Q_{ij}^{op}|Im'\zeta) = C(Im|\tfrac{3}{2}(I_iI_j + I_jI_i) - \delta_{ij}I^2|Im') \qquad (6.17)$$

where C is a constant and ζ represents the other quantum numbers besides I and m. The quadrupolar hamiltonian therefore remains

$$\mathcal{H}_Q = \frac{eQ}{6I(2I-1)} \sum_{i,j} V_{ij}\left[\frac{3}{2}(I_iI_j + I_jI_i) - \delta_{ij}I^2\right] \qquad (6.18)$$

where Q is a number called the *electric quadrupole moment*, defined as

$$eQ = (II\zeta|e\sum_k (3x_{ik}x_{jk} - r_k^2\delta_{ij})|II\zeta) \qquad (6.19)$$

Taking the axes x, y, and z coincident with the principal axes of the electric field gradient (EFG) tensor V_{ij}, the components of V_{ij} with $i \neq j$ are zero, and (6.18) becomes

$$\mathcal{H}_Q = \frac{e^2qQ}{4I(2I-1)}\left[3I_z^2 - I^2 + \eta(I_x^2 - I_y^2)\right] \qquad (6.20)$$

where we have used Laplace's equation ($\nabla^2 V = 0$). We have introduced $eq \equiv V_{zz}$ and the asymmetry parameter of the electric field gradient $\eta = (V_{xx} - V_{yy})/V_{zz}$. The quantity eq is measured in volts per square meter (SI). The axes are chosen in such way that the components of the EFG tensor obey

$$|V_{zz}| \geq |V_{yy}| \geq |V_{xx}| \qquad (6.21)$$

The quantity η varies between 0 and 1, and measures how much the EFG tensor deviates from axial symmetry.

In solids there are contributions to the EFG from the atom where the nucleus is located and from distant atoms (see Section 6.6); the EFG vanishes at the nuclei of pure S states in cubic symmetry. However, even for free atoms there is a certain amount of intermediate coupling that leads to a mixed ground state. In Gd^{3+}, the ground state becomes a mixture of $^8S_{7/2}$ and $^6P_{7/2}$ (Abragam and Bleaney 1970). For $I = \tfrac{3}{2}$, the eigenvalues of $\mathcal{H}_Q$ are given by

$$E_Q = \frac{e^2qQ}{4I(2I-1)}\left[3m^2 - I(I+1)\right]\left(1 + \tfrac{1}{3}\eta^2\right)^{1/2} \qquad (6.22)$$

In the case of a gradient with axial symmetry, $\eta = 0$, and the hamiltonian (6.20) takes the form

$$\mathcal{H}_Q = \frac{e^2 qQ}{4I(2I-1)} [3I_z^2 - I^2] \tag{6.23}$$

The third integral in the classical expression of W [Eq. (6.9)] involves the laplacian of V ($\nabla^2 V \equiv \sum_i \partial^2 V/\partial x_i^2$). From Poisson's equation, the laplacian is related to the charge present at the point $\mathbf{r}$ (the electronic charge, in this case):

$$\nabla^2 V = -\frac{\rho_e}{\epsilon_0} \tag{6.24}$$

where ρ_e is the electronic charge density and ϵ_0 is the vacuum permittivity [$\epsilon_0 = 1/(\mu_0 c^2)$]; the third integral, in the nuclear volume, is zero for the majority of the electrons. In the case of s electrons (and $p_{1/2}$ electrons) that have nonzero density in the region of the nucleus, the third term of the expansion of the energy [Eq. (6.9)] is not zero. It becomes, using Eq. (6.24) and the fact that the integral of $r^2 \rho(\mathbf{r})$ is equal to $Ze\langle r^2 \rangle$

$$W = -\frac{1}{6\epsilon_0} \rho_e Ze \langle r^2 \rangle = \frac{1}{6\epsilon_0} Ze^2 |\Psi(0)|^2 \langle r^2 \rangle \tag{6.25}$$

where $\langle r^2 \rangle$ is the nuclear mean quadratic radius, and we take into account that the nuclear charge is Ze. We have taken the density ρ_e equal to the electronic density at the origin:

$$\rho_e = -e|\Psi(0)|^2 \tag{6.26}$$

This term in the energy expansion gives rise to the isomer shift, in Mössbauer spectroscopy. In the Mössbauer effect a gamma ray is emitted without recoil by a nucleus in the excited state (at the source) and absorbed resonantly by another nucleus, in the absorber. Since, in principle, the nuclear mean quadratic radii in the ground state and in the excited state are different, and also, the values of $|\Psi(0)|^2$ are in general different in the matrix of the source and in the absorber, the change in W can be measured. The energy displacement (the isomer shift) is proportional to the difference in mean square radius in the ground state (subscript 1) and excited state (subscript 2), and to the difference in electronic density, at the nucleus, between source (subscript s) and absorber (subscript a):

$$\Delta E = \frac{1}{6\epsilon_0} Ze^2 \left(|\Psi(0)|_a^2 - |\Psi(0)|_s^2 \right) \left(<r^2>_2 - <r^2>_1 \right) \tag{6.27}$$

Instead of expressing the isomer shift in terms of the mean square radius, one often uses the nuclear radius R, related through $R^2 = \frac{5}{3} \langle r^2 \rangle$.

The electric field gradients that act on the nuclei in solids arise from the ionic charges from the electrons of the parent atom and also from the conduction electrons, in metals and semiconductors.

For a free ion of total angular momentum **J**, it can be shown that the interaction between the quadrupolar electric moment of the nucleus and the electric field gradient due to the electrons is (Bleaney 1967)

$$\mathcal{H} = B\left[\frac{3(\mathbf{J}\cdot\mathbf{I})^2 + 3/2(\mathbf{J}\cdot\mathbf{I}) - J(J+1)I(I+1)}{2J(2J-1)I(2I-1)}\right] \quad (6.28)$$

where

$$B = -e^2 qQ <r^{-3}><J\|\alpha\|J>J(2J-1) \quad (6.29)$$

with eq the electric field gradient $(\partial^2 V/\partial z^2)$ and Q the electric quadrupole moment of the nucleus; $\langle J\|\alpha\|J\rangle$ is a number tabulated for each ion [for the rare earths, see Elliott and Stevens (1953)].

The electric quadrupole hyperfine interaction in magnetic materials is typically one order of magnitude smaller than the magnetic dipolar hyperfine interaction.

6.3 MAGNETIC DIPOLAR INTERACTIONS

The dominant term in the expansion of magnetic interactions of electrons and nuclei, given by the general expression (6.1), is the interaction with the nuclear magnetic dipole moment. This term arises from the effect of the spin and orbital magnetic moments of the electrons, acting on the dipolar magnetic moments of the nuclei. The magnetic dipolar hyperfine interaction may be written as an interaction of the nuclear magnetic dipolar moment with a magnetic field due to the electrons—the hyperfine field:

$$\mathcal{H}_{\text{hf}} = -\boldsymbol{\mu}_I \cdot \mathbf{B}_{\text{hf}} \quad (6.30)$$

This can be shown from the general expression of the interaction between an electronic current density and the nuclear magnetism. We may also show the different contributions to the hyperfine field; these contributions are due to the orbital momentum of the electrons, the distribution of the spins of the electrons outside the nucleus, and the spin density of the s electrons in the region of the nucleus.

We will initially study the vector potential at **r** associated with an arbitrary distribution of currents, of density **J** at point $\mathbf{r}'$ (Fig. 6.1). The value of **B** at each

166 HYPERFINE INTERACTIONS

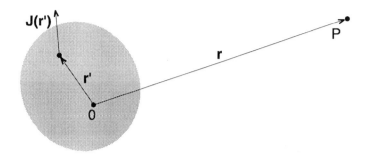

Figure 6.1 The current density J(r') at the point r', in a limited space region, giving rise to a vector potential at a point P, of coordinate r.

point of space P can be obtained from the vector potential $\mathbf{A}(\mathbf{r})$:

$$\mathbf{B}(\mathbf{r}) = \nabla \times \mathbf{A}(\mathbf{r}) \tag{6.31}$$

where the potential $\mathbf{A}(\mathbf{r})$ at the point P, of position $\mathbf{r}$, due to a current density $\mathbf{J}$ at the point $\mathbf{r}'$, is given by the general expression

$$\mathbf{A}(\mathbf{r}) = \frac{\mu_0}{4\pi} \int_V \frac{\mathbf{J}(\mathbf{r}')}{|\mathbf{r} - \mathbf{r}'|} dv' \tag{6.32}$$

where μ_0 is the free space permeability. The integration is performed on a volume contained in a region of finite radius R.

We will study the form of $\mathbf{A}(\mathbf{r})$ for an arbitrary distribution of currents. For this purpose, we will expand the denominator of $\mathbf{A}(\mathbf{r})$ in powers of $\mathbf{r}'$. This is useful for the case of a distribution of currents of small dimensions, compared to $|\mathbf{r}|$. The expansion gives

$$\frac{1}{|\mathbf{r} - \mathbf{r}'|} = \frac{1}{|\mathbf{r}|} + \frac{\mathbf{r} \cdot \mathbf{r}'}{|\mathbf{r}|^3} + \cdots \tag{6.33}$$

The ith component of $\mathbf{A}(\mathbf{r})$ becomes

$$A_i(\mathbf{r}) = \frac{\mu_0}{4\pi|\mathbf{r}|} \left[\int_V J_i(\mathbf{r}')dv' + \frac{\mathbf{r}}{|\mathbf{r}|^2} \cdot \int_V \mathbf{r}' J_i(\mathbf{r}')dv' + \cdots \right] \tag{6.34}$$

Because $\mathbf{J}(\mathbf{r}')$ is spatially limited and has zero divergence, it follows for the first integral of this equation (Jackson 1975, Section V.6):

$$\int_V J_i(\mathbf{r}')dv' = 0 \tag{6.35}$$

Therefore, the term corresponding to the electric monopole in the expansion of the electrostatic interaction (the Coulomb term) is canceled in the magnetic case. The integral of the second term becomes

$$\mathbf{r} \cdot \int_V \mathbf{r}' J_i(\mathbf{r}') dv' \equiv \sum_j x_j \int_V x'_j J_i(\mathbf{r}') dv' = -\frac{1}{2} \sum_j x_j \int_V (x'_i J_j - x'_j J_i) dv' =$$

$$= -\frac{1}{2} \sum_{j,k} \varepsilon_{ijk} x_j \int_V (\mathbf{r}' \times \mathbf{J})_k dv' = -\frac{1}{2} \left[\mathbf{r} \times \int_V (\mathbf{r}' \times \mathbf{J}) dv' \right]_i \quad (6.36)$$

where ε_{ijk} is the Levi–Civita symbol, equal to zero for repeated indices, and $+1$ for circular permutations of i, j and k, and -1 otherwise.

Here we have used (Jackson 1975)

$$\int_V (x'_i J_j + x'_j J_i) dv' = 0 \quad (6.37)$$

The magnetic moment associated with a current density $\mathbf{J}(\mathbf{r}')$ is defined in a general way as

$$\mathbf{m} = \frac{1}{2} \int_V (\mathbf{r}' \times \mathbf{J}(\mathbf{r}')) dv' \quad (6.38)$$

integrated over the region of space (volume) where the currents are circumscribed.

The second term in the expansion of $\mathbf{A}(\mathbf{r})$ can then be expressed in terms of the magnetic moment $\mathbf{m}$, using Eq. (6.38):

$$\mathbf{A}(\mathbf{r}) = \frac{\mu_0}{4\pi} \frac{\mathbf{m} \times \mathbf{r}}{|\mathbf{r}|^3} \quad (6.39)$$

which is the expression of the vector potential at a point $\mathbf{r}$, due to a magnetic dipole at the origin.

The magnetic field $\mathbf{B}$ associated with the vector potential $\mathbf{A}(\mathbf{r})$ is

$$\mathbf{B}(\mathbf{r}) = \frac{\mu_0}{4\pi} \frac{1}{|\mathbf{r}|^5} [3(\mathbf{r} \cdot \mathbf{m})\mathbf{r} - r^2 \mathbf{m}] \quad (6.40)$$

which is the magnetic field of a dipole $\mathbf{m}$. [*Conclusion*: The non-zero term in the expansion up to first order in $\mathbf{r}'$ of the field produced by an arbitrary current density $\mathbf{J}(\mathbf{r}')$ is the field $\mathbf{B}(\mathbf{r})$ due to a magnetic dipole. The magnetic field far from an arbitrary distribution of currents is identical to the field of a dipole.]

We may compute the contributions to the hyperfine field that originate from the magnetic dipolar moment of the electrons. The magnetic dipolar moment of the electron in the atom has a contribution of the spin angular momentum and another of the orbital momentum. We will initially study the spin term.

6.3.1 Contribution of the Electronic Spin to the Magnetic Hyperfine Field

The conduction electron states in a crystal may be described by Bloch functions:

$$\Psi_k(\mathbf{r}) = u_k(\mathbf{r})e^{i\mathbf{k}\cdot\mathbf{r}} \tag{6.41}$$

where $\mathbf{k}$ is the wavevector and $u_k(\mathbf{r})$ is a function that has the periodicity of the crystal lattice; $\Psi_k(\mathbf{r})$ is a plane wave $[\exp(i\mathbf{k}\cdot\mathbf{r})]$ modulated by $u_k(\mathbf{r})$. The spinup electronic density due to electron i at point $\mathbf{r}$ is $\rho_i^\uparrow(\mathbf{r})$; it is given by the probability density of finding the i electron of spinup at the point $\mathbf{r}$, that is, $|u(\mathbf{r}, \uparrow)|^2 = |\Psi(\mathbf{r}, \uparrow)|^2$, or

$$\rho_i^\uparrow(\mathbf{r}) = |u(\mathbf{r}, \uparrow)|^2 \tag{6.42}$$

The magnetization at the point $\mathbf{r}$ due to the electron i is related to the difference in electron density $\Delta\rho_i(\mathbf{r}) = \rho_i^\uparrow - \rho_i^\downarrow$ and has the expression

$$\mathbf{M}(\mathbf{r}) = -g\mu_B \sum_i \mathbf{s}_i [\rho_i^\uparrow(\mathbf{r}) - \rho_i^\downarrow(\mathbf{r})] = -g\mu_B \sum_i \mathbf{s}_i \Delta\rho_i(\mathbf{r}) \tag{6.43}$$

where $\mathbf{s}_i$ is the spin angular momentum of the electron i.

The interaction energy of the magnetization $\mathbf{M}(\mathbf{r})$ at the point $\mathbf{r}$ with the magnetic dipole moment of a nucleus located at the origin, per unit volume, may, in principle, be written in the form of an interaction with a magnetic dipole field:

$$w_s = -\frac{\mu_0}{4\pi} \frac{1}{|\mathbf{r}|^5} \boldsymbol{\mu}_I \cdot \left[3(\mathbf{r}\cdot\mathbf{M})\mathbf{r} - r^2\mathbf{M}\right] \tag{6.44}$$

which is valid only, for $r \neq 0$, of course, and may be written

$$w_s = -\boldsymbol{\mu}_I \cdot \frac{\mathbf{B}_s}{V} \tag{6.45}$$

where $\mathbf{B}_s$ is the spin magnetic dipolar field and V is the volume.

Integrating over the volume of the atom, it follows for the dipolar field $\mathbf{B}_s$ due to the spin momentum of the electrons, that

$$\mathbf{B}_s = \frac{\mu_0}{4\pi} g\mu_B \sum_i [3(\mathbf{s}_i \cdot \mathbf{e}_r)\mathbf{e}_r - \mathbf{s}_i]\langle r_s^{-3}\rangle_i \tag{6.46}$$

with $\mathbf{e}_r$ the unit vector of the direction $\mathbf{r}$ and

$$\langle r_s^{-3}\rangle_i = \int \frac{\Delta\rho_i(\mathbf{r})}{r^3}\,dv \tag{6.47}$$

the mean cubic radius of the electrons with spin $\mathbf{s}$. When the spin density has spherical symmetry, the value of (6.46) is zero; this arises because the magnetic dipole field inside a spherical shell is zero, since it involves the integral

$$\int_0^\pi (1 - 3\cos^2\theta)\sin\theta\,d\theta = 0 \tag{6.48}$$

where θ is the angle between $\mathbf{s}$ and $\mathbf{e}_r$ in Eq. (6.46).

For the electrons that have a nonzero density at $r = 0$, as the s electrons (and the $p_{1/2}$ electrons in the heavy atoms), there is also another term in the hyperfine field, the Fermi contact term, which will be derived below.

The contribution of the magnetization to the magnetic induction inside a sphere with uniform magnetization $\mathbf{M}(0)$ is (Jackson 1975, Section V.10):

$$\mathbf{B} = \frac{\mu_0}{4\pi}\frac{8\pi}{3}\mathbf{M}(0) \tag{6.49}$$

The magnetization due to a single s electron is

$$\mathbf{M}(0) = -g\mu_B \mathbf{s}\rho(0) = -g\mu_B \mathbf{s}|\Psi(0)|^2 \tag{6.50}$$

where $\rho(0)$ is the electronic density at the origin.

Substituting the expression of $\mathbf{M}(0)$ into $\mathbf{B}$ [Eq. (6.49)], we obtain for the contribution to $\mathbf{B}$ of the electron spin density $s\rho(0)$ at the nucleus (the Fermi contact term)

$$\mathbf{B}_c = -\frac{\mu_0}{4\pi}\frac{8\pi}{3}g\mu_B\rho(0)\mathbf{s} \tag{6.51}$$

with the corresponding interaction energy

$$W_c = \frac{\mu_0}{4\pi}\frac{8\pi}{3}g\mu_B\frac{\mu_I}{I}\rho(0)\mathbf{I}\cdot\mathbf{s} \tag{6.52}$$

Using the fact that $\rho(0)$ has the dimension of r^{-3}, we may introduce the expression of the mean value of r^{-3} ($\langle r_c^{-3}\rangle_i$) for electrons that contribute to the contact interaction (mostly s electrons), absorbing into r_c the factor $8\pi/3$ and the ratio $g/2$ (e.g., Bleaney 1967):

$$\mathbf{B}_c = -\frac{\mu_0}{4\pi}2\mu_B\frac{\mu_I}{I}\mathbf{s}\langle r_c^{-3}\rangle_i \tag{6.53}$$

The resulting magnetization is proportional to the difference between the up and down electron spin densities. The spin density due to the superposition of the contributions of all the orbitals at $r = 0$ (each one of quantum number n) is

$$\rho(0)^{\text{tot}} = \sum_n \{|\Psi_{ns}(0,\uparrow)|^2 - |\Psi_{ns}(0,\downarrow)|^2\} \tag{6.54}$$

Since the s electrons have $l = 0$ and their spatial distribution is spherically symmetric, $\mathbf{B}_c$ is their only contribution to the hyperfine field.

The incomplete shells (and also the conduction electrons; see Section 6.5.2) may also give rise to another contribution to the hyperfine field, through the modification of the radial distribution of the closed shells, thus producing a noncompensated spin density at the origin. The expression is the same as in Eq. (6.49), with $\rho(0) = \rho(0)^{\text{tot}}$ and with the sum performed on every shell, both complete and incomplete. This leads to an s magnetization equal to $\mathbf{M}'_s(0)$ at the nucleus, and this term of the hyperfine field, called the *core polarization field*, acts through the contact term and is written

$$\mathbf{B}_{\text{cp}} = \frac{\mu_0}{4\pi} \frac{8\pi}{3} \mathbf{M}'_s(0) \tag{6.55}$$

This term is dominant in the hyperfine field of the S-state rare-earth ions, such as Gd^{3+}, and in the ions of d transition metals, such as Fe. In the free Gd^{3+} ion, the core polarization field has a value of $B_{\text{cp}} = -21$ T; in metallic Fe, it is -27.5 T. In the series of tripositive rare-earth ions, the core polarization field is proportional to the spin component of the total angular momentum J, given, in tesla, approximately by (Netz 1986)

$$B_{\text{cp}} = -6(g-1)J \tag{6.56}$$

In the actinides the core polarization field can be much larger than in the rare-earth elements; in the Am^{2+} ion (S state), for example, the core polarization field is -220 T.

6.3.2 Orbital Contribution to the Magnetic Hyperfine Field

We will now compute the field due to the orbital motion of the electrons. Taking this time the inverse point of view, we will obtain the vector potential at the point $\mathbf{r}$ due to the nuclear magnetic dipole moment $\boldsymbol{\mu}_I$ located at the origin (Fig. 6.2).

This vector potential is given by

$$\mathbf{A}(\mathbf{r}) = \frac{\mu_0}{4\pi} \frac{\boldsymbol{\mu}_I \times \mathbf{r}}{|\mathbf{r}|^3} \tag{6.57}$$

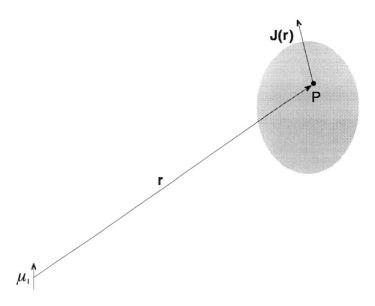

Figure 6.2 The nuclear magnetic moment μ_I creates a vector potential $A(r)$ at point P, of coordinate $\mathbf{r}$, where there is an orbital current of density $J(\mathbf{r})$.

The interaction energy of the nuclear vector potential $\mathbf{A}(\mathbf{r})$ with the electronic current density $\mathbf{J}_e(\mathbf{r})$ is the volume integral[1]

$$W = -\int_V \mathbf{A}(\mathbf{r}) \cdot \mathbf{J}_e(\mathbf{r}) dv = -\frac{\mu_0}{4\pi} \int_V \frac{(\boldsymbol{\mu}_I \times \mathbf{r}) \cdot \mathbf{J}_e(\mathbf{r})}{|\mathbf{r}|^3} dv = -\frac{\mu_0}{4\pi} \int_V \frac{(\mathbf{r} \times \mathbf{J}_e(\mathbf{r})) \cdot \boldsymbol{\mu}_I}{|\mathbf{r}|^3} dv \tag{6.58}$$

where we have used the permutation of the mixed product

$$\mathbf{a} \cdot (\mathbf{b} \times \mathbf{c}) = \mathbf{c} \cdot (\mathbf{a} \times \mathbf{b}) = \mathbf{b} \cdot (\mathbf{c} \times \mathbf{a}) \tag{6.59}$$

We may take the nuclear moment out of the integral; using $\mathbf{J}_e(\mathbf{r})dv = \mathbf{v}\, dq$, where $\mathbf{v}$ is the velocity, dv is a volume element, and dq is an element of charge, it becomes

$$W = -\frac{\mu_0}{4\pi} \boldsymbol{\mu}_I \cdot \int \frac{\mathbf{r} \times \mathbf{v}}{|\mathbf{r}|^3} dq \tag{6.60}$$

---

[1] This expression has to be divided by 2 in the case where the vector potential $\mathbf{A}(\mathbf{r})$ includes contributions of the current density $\mathbf{J}$ (not our present case).

The orbital angular momentum of each electron is $(\mathbf{r} \times m_e \mathbf{v}) = \mathbf{l}\hbar$. Using

$$\int |\mathbf{r}|^{-3} \, dq = -e \langle r_l^{-3} \rangle \tag{6.61}$$

where $\langle r_l^{-3} \rangle$ is the average over the coordinates of the electron with orbital momentum $\mathbf{l}$, it leads to, summing over N electrons:

$$W_L = \frac{\mu_0}{4\pi} \sum_i^N \boldsymbol{\mu}_I \cdot \mathbf{l}_i \frac{e\hbar}{m_e} \langle r_l^{-3} \rangle_i = \frac{\mu_0}{4\pi} 2\mu_B \sum_i^N \boldsymbol{\mu}_I \cdot \mathbf{l}_i \langle r_l^{-3} \rangle_i \tag{6.62}$$

substituting the Bohr magneton $\mu_B = e\hbar/2m_e$.

Finally, the total hyperfine field due to the various electrons, including dipolar spin terms, the contact term, and the orbital term, may be written using Eqs. (6.46), (6.51), and (6.62), and the approximation $g = 2$:

$$\mathbf{B} = -\frac{\mu_0}{4\pi} g\mu_B \sum_i^N \{ [\mathbf{s}_i - 3(\mathbf{s}_i \cdot \mathbf{e}_r)\mathbf{e}_r] \langle r_s^{-3} \rangle_i + \mathbf{s}_i \langle r_c^{-3} \rangle_i - \mathbf{l}_i \langle r_l^{-3} \rangle_i \} \tag{6.63}$$

Neglecting the differences in the effective radii that appear in the spin, contact and orbital hyperfine fields, i.e., making the approximation

$$\langle r_s^{-3} \rangle_i = \langle r_c^{-3} \rangle_i = \langle r_l^{-3} \rangle_i \tag{6.64}$$

we obtain for the expression of the total hyperfine field due to the N electrons

$$\mathbf{B} = -\frac{\mu_0}{4\pi} 2\mu_B \sum_i^N \{ [\mathbf{s}_i - 3(\mathbf{s}_i \cdot \mathbf{e}_r)\mathbf{e}_r] + \mathbf{s}_i - \mathbf{l}_i \} \langle r^{-3} \rangle_i \tag{6.65}$$

For an atom with several electrons and *LS* coupling, the more usual form of coupling of the spin and orbital angular momenta $\mathbf{S}$ and $\mathbf{L}$, the orbital interaction takes the form, assuming that all the electrons in the orbit have the same value of $\langle r_l^{-3} \rangle$:

$$W_L = \left(\frac{\mu_0}{4\pi}\right) 2\mu_B (\boldsymbol{\mu}_I \cdot \mathbf{L}) \langle r_l^{-3} \rangle = \left(\frac{\mu_0}{4\pi}\right) 2\mu_B \left(\frac{\mu_I}{I}\right) (\mathbf{I} \cdot \mathbf{L}) \langle r_l^{-3} \rangle \tag{6.66}$$

From which, if we write

$$W_L = -\boldsymbol{\mu}_I \cdot \mathbf{B}_L \tag{6.67}$$

it follows

$$\mathbf{B}_L = \frac{\mu_0}{4\pi} 2\mu_B \langle r_I^{-3} \rangle \mathbf{L} \qquad (6.68)$$

for the expression of the total orbital hyperfine field.

For a free atom (or molecule) with several electrons, the hyperfine field is not given in terms of the spin and orbital momenta of the individual electrons (Eq. 6.65), but in terms of $\mathbf{S}$ and $\mathbf{L}$, or the total angular momentum $\mathbf{J} = \mathbf{S} + \mathbf{L}$. The hamiltonian of the total magnetic hyperfine interaction can be written

$$\mathcal{H}_{\text{hf}} = A \mathbf{I} \cdot \mathbf{J} \qquad (6.69)$$

where $\mathbf{I}$ and $\mathbf{J}$ are angular momentum operators of the nucleus and of the ion. In the more general case A is the hyperfine tensor; when A is a number, it is called the hyperfine interaction constant. The description of the interaction in terms of the hyperfine field $\mathbf{B}_{\text{hf}}$ in fact applies when A has uniaxial symmetry ($A_z = A \gg A_y, A_x$):

$$\mathcal{H}_{\text{hf}} = A \mathbf{I} \cdot \mathbf{J} = -\boldsymbol{\mu}_I \cdot \mathbf{B}_{\text{hf}} \qquad (6.70)$$

and we may express the hyperfine field operator $\mathbf{B}_{\text{hf}}$ as a function of the hyperfine constant A:

$$\mathbf{B}_{\text{hf}} = -\left(\frac{A}{g_I \mu_N}\right) \mathbf{J} \qquad (6.71)$$

For $T > 0$ K, $\mathbf{J}$ has to be substituted by its thermal average $\langle \mathbf{J} \rangle_T$.

From this definition, it is clear that the hyperfine field represents an effective field, that acting on the nuclear moment, leads to an interaction equal to the total hyperfine interaction. If $\mathbf{B}$ due to the electrons varies from point to point, $\mathbf{B}_{\text{hf}}$ is an average value on the volume of the nucleus. For example, in the derivation of the contact hyperfine field we have used the value of the spin density at the origin $\rho(0)\mathbf{s}$; in fact, the contact field is related to the average of the density in the region occupied by the nucleus. Since nuclei of different isotopes of a given element have different shapes and different average radii, they will in general feel different average spin densities. Thus, the hyperfine fields (or hyperfine constants) will also be different. This effect, called *hyperfine anomaly*, is represented by Δ, defined quantitatively by the relation

$$\frac{A_1}{A_2} = \frac{g_1}{g_2}(1 + \Delta) \qquad (6.72)$$

where A_1, A_2, g_1, and g_2 are the hyperfine constants and nuclear g factors of two isotopes. The values of Δ are normally very small; an exceptionally high value of

−0.5% was observed for the hyperfine anomaly with the isotopes 151 and 153 of Eu in salts of Eu^{2+} (Baker and Williams 1962).

We see from Eq. (6.69) that the hyperfine interaction couples the angular momentum of the nucleus (**I**) and the total angular momentum of the atom (**J**). The total angular momentum (atomic plus nuclear) is

$$\mathbf{F} = \mathbf{I} + \mathbf{J} \tag{6.73}$$

with the corresponding quantum number F, called the *hyperfine quantum number*.

There are several experimental techniques that allow the determination of $\mathbf{B}_{hf}$: nuclear magnetic resonance (NMR), perturbed angular correlation (PAC), Mössbauer spectroscopy (MS), and so on. This is done experimentally from the determination of the eigenstates of $\mathcal{H}_{hf}$. The eigenstates are

$$E_{M,I} = -g_I \mu_N M_I B_{hf}, \quad \text{with} \quad M_I = -I, -I+1, \cdots +I \tag{6.74}$$

Therefore, we may determine B_{hf} from the experimental measurement of the separation between the energies of the hyperfine substates:

$$\Delta E = g_I \mu_N B_{hf} \tag{6.75}$$

The measurement of B_{hf} through NMR (see Chapter 8) consists in the determination of the frequency of the electromagnetic wave (in the radio-frequency or microwave region) that induces transitions between the nuclear hyperfine substates. The frequency ν_0 for which this occurs satisfies

$$h\nu_0 = \Delta E \tag{6.76}$$

Knowing g_I, we may determine the value of the field B_{hf} [from Eq. (6.75)] since μ_N, the nuclear magneton, is a constant. The observed hyperfine splittings ΔE are very small, in the range 10^{-27}–10^{-24} J (10^{-9}–10^{-6} eV); this corresponds to NMR frequencies in the range from a few megahertz to a few gigahertz.

6.4 CONTRIBUTIONS TO B_{hf} IN THE FREE ION

As we have seen previously, a free ion with an incomplete electronic shell presents three contributions to the magnetic hyperfine field: an orbital term, a dipolar term, and a term due to the polarization of the closed shells (McCausland and Mackenzie 1980, McMorrow et al. 1989):[2]

$$\mathbf{B}_{hf} = \mathbf{B}_{orb} + \mathbf{B}_{dip} + \mathbf{B}_{cp} \tag{6.77}$$

---

[2] We have followed these references in discussions in the next sections.

The field $\mathbf{B}_{orb}$ originates from the orbital angular momentum $\mathbf{L}$ of the incomplete shell; it is the most important term for the non-S rare-earth ions. The dipolar field $\mathbf{B}_{dip}$ results from the interaction between the spin of the ion and the nuclear magnetic moment. We have included in this term both contributions of the electronic dipoles: the dipolar term in the strict sense, and the contact term due to the s electrons.

The core polarization field $\mathbf{B}_{cp}$ arises from the deformations of the internal closed shells due to an incomplete shell (e.g., the $4f$ shell). The incomplete shell affects the radial distribution of electrons of spinup (parallel to the spin of the unfilled shell) differently from that of spindown electrons. As a result of the exchange interaction, electrons of the closed shells with spinup are effectively attracted toward the unfilled shell. This leads to different densities of spinup and spindown electrons in the volume occupied by the nucleus. The resulting polarization, or magnetization, interacts with the nuclear magnetic moment through the Fermi contact interaction.

In general, $\mathbf{B}_{orb} > \mathbf{B}_{dip}$ and also $\mathbf{B}_{orb} > \mathbf{B}_{cp}$; if $L = 0$, $\mathbf{B}_{orb} = 0, \mathbf{B}_{dip} = 0$ (by spherical symmetry), and $\mathbf{B}_{cp}$ is dominant. This is the case, for example, with the hyperfine fields measured at nuclei of the ions Eu^{2+} and Gd^{3+} (where $L = 0$).

6.5 HYPERFINE FIELDS IN METALS

The hyperfine interactions of a rare-earth ion located in a metallic matrix will be modified. On one hand, the exchange interactions and the interactions with the crystal field will modify the intraionic interactions previously described by $\mathcal{H}_{hf}$; on the other hand, there will arise interactions with the conduction electrons, and with the magnetic and electrostatic fields due to the neighbor atoms. The total hamiltonian includes in this case intraionic interactions ($\mathcal{H}'$) and extraionic interactions ($\mathcal{H}''$):

$$\mathcal{H}_{hf} = \mathcal{H}' + \mathcal{H}'' \qquad (6.78)$$

Normally, for ions with $L \neq 0$, $\mathcal{H}' \gg \mathcal{H}''$, but for ions with $L = 0$ (and $S \neq 0$), and for nonmagnetic ions ($L = S = 0$), we may have $\mathcal{H}' \approx \mathcal{H}''$.

We will study separately the intraionic interactions and the extraionic interactions of an ion placed in a metallic medium.

6.5.1 Intraionic Interactions in the Metals

We have already seen that there is a hierarchy in the interactions of the free ion of the rare earths:

$$\mathcal{H}_{LS}(\mathbf{L},\mathbf{S}) \gg \mathcal{H}_{el}(\mathbf{J}) \gg \mathcal{H}_{hf}(\mathbf{J},\mathbf{I}) \qquad (6.79)$$

with $\mathcal{H}_{el}(\mathbf{J}) = \mathcal{H}(\text{exchange}) + \mathcal{H}(\text{crystal field})$. Therefore, the hyperfine interaction $\mathcal{H}_{hf}$ is a perturbation in the hamiltonian of the ion, and does not affect much the M_J levels defined by the exchange interaction [i.e., by the magnetic field acting on the ion, or molecular field (see Chapter 3)].

We may, with the purpose of discussing the mechanisms that affect the hyperfine field at the nucleus of an atom in a metallic matrix, separate this field into two parts: one part of the ion in the presence of other ions, and another "extraionic" part, which includes external fields, effects of the conduction electrons, and effects of the neighbor magnetic atoms:

$$\mathbf{B}_t = \mathbf{B}' + \mathbf{B}'' \tag{6.80}$$

The intraionic interaction in a metal is the modified dipolar magnetic interaction (see Section 6.3.2):

$$\mathcal{H}' = \mathbf{a}' \cdot \mathbf{I} \cong A' \mathbf{I} \cdot \mathbf{J} = -\boldsymbol{\mu}_I \cdot \mathbf{B}' \tag{6.81}$$

In the cases where the interaction energy of the ion with the crystal field is much weaker than the Zeeman interaction, the expectation value $\langle \mathbf{J} \rangle$ is the same as that obtained for the free ion, and the intraionic term is equal to the free ion field:

$$\mathbf{B}' = \mathbf{B}_{hf} \tag{6.82}$$

6.5.2 Extraionic Magnetic Interactions

The extraionic magnetic field $\mathbf{B}''$ that acts on the nucleus in a metal is equal to

$$\mathbf{B}'' = \mathbf{B}''_{ext} + \mathbf{B}''_{dip} + \mathbf{B}''_{ce} + \mathbf{B}''_{orb} \tag{6.83}$$

where

$\mathbf{B}''_{ext}$ = applied magnetic field
$\mathbf{B}''_{dip}$ = dipolar field (due to the magnetic dipolar moments in the sample)
$\mathbf{B}''_{ce}$ = field due to the conduction electrons
$\mathbf{B}''_{orb}$ = transferred term induced by the orbital moment.

The dipolar field is given by

$$\mathbf{B}''_{dip} = \sum_j \left(\frac{\mu_0}{4\pi r_j^5} \right) [(3\langle \boldsymbol{\mu}_j \rangle \cdot \mathbf{r}_j)\mathbf{r}_j - r_j^2 \langle \boldsymbol{\mu}_j \rangle] \tag{6.84}$$

where the sum is made over every magnetic dipole μ of the sample, excluding the one from the same atom in which nucleus the field is being measured (parent atom). The dipolar field is usually divided into three terms. To compute them, one thinks of a sphere, with radius much smaller than the dimensions of the sample, but larger than the atomic distances: the Lorentz sphere. The first term is due to the dipoles inside this sphere, and it is zero for cubic crystal lattices. The second term is due to the free magnetic poles on the inside surface of the spherical cavity; it is equal to $\frac{1}{3}\mu_0 \mathbf{M}_l$ and is called the *Lorentz field* ($\mathbf{M}_l$ is the local magnetization in the Lorentz sphere). The last term is the demagnetizing field, arising from the poles at the surface of the sample. This term accounts for the contribution of the dipoles outside the Lorentz sphere.

The demagnetizing field (Section 1.2) is given by

$$\mathbf{B}_d = -\mu_0 N_d \mathbf{M} \tag{6.85}$$

where N_d is the demagnetizing factor, which amounts to $\frac{1}{3}$ for spherical particles in the SI ($4\pi/3$ in the CGS system), and $\mathbf{M}$ is the sample magnetization. Note that for spherical samples the Lorentz field and the demagnetizing field cancel each other (for $\mathbf{M}_l = \mathbf{M}$).

The field at the nucleus due to the conduction electrons has three contributions: one due to the polarization of the electrons by the parent atom ($\mathbf{B}_p''$), another associated with the polarization due to neighbor atoms ($\mathbf{B}_n''$), a third term $K_0 \mathbf{B}_{\text{ext}}$ that arises from the polarization induced by the external magnetic field (the latter is responsible for the Knight shift, observed in NMR measurements in nonmagnetic metals; Section 6.5.4). Therefore

$$\mathbf{B}_{\text{ce}}'' = \mathbf{B}_p'' + \mathbf{B}_n'' + K_0 \mathbf{B}_{\text{ext}} \tag{6.86}$$

$\mathbf{B}_n''$ is usually called the *transferred field*; sometimes under this denomination one includes also the dipolar field inside the Lorentz sphere.

The extraionic hyperfine field due to the conduction electrons is a sum of parent and neighbor contributions:

$$\mathbf{B}_{\text{ce}}'' = K_p \langle \sigma_p \rangle_T + K_n \overline{\sigma} \tag{6.87}$$

with $\langle \overline{\sigma}_p \rangle_T$ representing the thermal average of the parent atom spin at temperature T, and $\overline{\sigma} \cong \langle \overline{\sigma} \rangle_T = (g' - 1) \langle \mathbf{J}' \rangle_T$ the projection of the average spin of the atoms of the matrix. If the matrix is a rare-earth alloy, typical values of the constants in this case will be: $K_0 = 0.005$, $K_p \approx 5$ T, and $K_n \approx -5$ T. In RAl$_2$ intermetallic compounds the measured values are $K_p \approx -5.7$ T and $K_n \approx 0.8$ T (McMorrow et al. 1989).

The contribution of the orbital momentum of the neighbor atoms to the extraionic field is given by

$$\mathbf{B}_{\text{orb}}'' = K_{\text{orb}}(2 - g) \langle \mathbf{J} \rangle_T \tag{6.88}$$

Estimates on rare-earth alloys give $K_{orb} \approx 4.4$ T and in RAl_2 intermetallic compounds $K_{orb} \approx 0.25$ T.

The local extraionic field for a given configuration of neighbor atoms may be taken as proportional to their magnetic moments, or to the projection of their spins $\langle \sigma_j \rangle_T$. We may therefore write,

$$\mathbf{B}_n'' = \sum_j f(\mathbf{r}_j) \langle \sigma_j \rangle_T \tag{6.89}$$

where $f(\mathbf{r}_j)$ is a spatial function, dependent on the crystal structure of the matrix; the sum is made over the relevant neighbors, located at the positions $\mathbf{r}_j$. The fields corresponding to each configuration can be determined with NMR spectroscopy when their lines in the spectrum are resolved. This happens when the linewidth is smaller, or of the order of the difference in field due to a nearest neighbor, and to a distant neighbor of the impurity atom.

In some cases the oscillating character of f as a function of $|\mathbf{r}_j|$ has been demonstrated (Fig. 6.5). The contribution of the neighbor moments located at $\mathbf{r}_j$ may be modified (or amplified) by atoms in sites i that are common neighbors of the probe atom and of the atom j. The moment at $\mathbf{r}_j$ modifies the moment of the atom $\mathbf{r}_i$, and this change affects the hyperfine field at the probe. The perturbation of atom i, and therefore, its amplifying effect, depends on the number $(n_{i,j})$ of neighbors of j in a nonlinear way, following a function $g(n_{i,j})$. In this case we can speak of indirect transferred hyperfine interaction:

$$\mathbf{B}_n''' = \sum_j f^i(\mathbf{r}_j) \langle \sigma_j \rangle_T g(n_{i,j}) \tag{6.90}$$

where the sum includes only atoms j that are neighbors of an atom i neighbor of the probe atom. This type of transferred interaction is observed, for example, in intermetallic compounds of rare earths and iron, in which the iron atoms act as paths for the indirect transferred interactions.

6.5.3 Hyperfine Fields Observed Experimentally

Hyperfine fields have been measured with different experimental techniques; each technique has a characteristic measurement time, typically in the range 10^{-7}–10^{-9} s. Therefore, in systems that present thermal fluctuations in times shorter than these, the observed hyperfine fields may have a null value; this is normally the case for measurements in paramagnetic materials and for magnetic materials above the critical temperature T_C. For this reason, hyperfine fields are usually measured in materials that are magnetically ordered, such as ferromagnets and antiferromagnets below the ordering temperature.

These measurements have been made in many magnetically ordered systems, under different experimental conditions. The conditions that may affect the value

Table 6.1 Values of the hyperfine fields B_{hf} (in tesla) measured at low temperature (4.2 K), computed values of the field of a free ion (B_{fi}) and the core polarization field (B_{cp})

Element or Ion	$B_{fi}(T)$	B_{cp} (T)	B_{hf} (T)
Fe	—	−27.5	−33.9
Co	—	−21.5	−21.5 (bcc)
Ni	—	−7.5	−7.5
Gd	−17	−21	−35
Dy	635	−15	+590
Ho	796	−12	+737.1
Np^{6+}	380	21	—

of the observed hyperfine field include the chemical state of the ion in whose nucleus B_{hf} is measured, the temperature, pressure, the concentration of impurities in the matrix in which the ion is located, and the distance between it and an impurity. Some B_{hf} values are given in Table 6.I, with a preference for metallic matrices; the signs of B_{hf} follow the convention that they are positive when this field is parallel to the atomic magnetic moment.

Table 6.I illustrates how the hyperfine fields in the magnetic systems vary with the atom in which B_{hf} is measured, with the chemical form in which the atom is found, and so on. The free-ion values (B_{fi}) are derived from measurements made in paramagnetic salts, corrected for crystal field effects. The fields listed in Table 6.I vary from several tens of tesla to several hundred teslas. Some

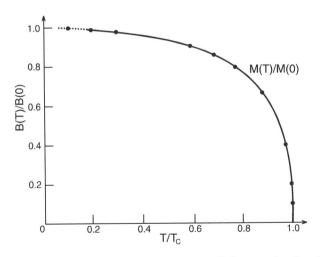

Figure 6.3 Normalized graph of the hyperfine field in metallic iron as a function of temperature, measured at ^{57}Fe nuclei through Mössbauer spectroscopy. The continuous line is the measured magnetization versus temperature, and the dotted line is the variation of NMR frequency with temperature.

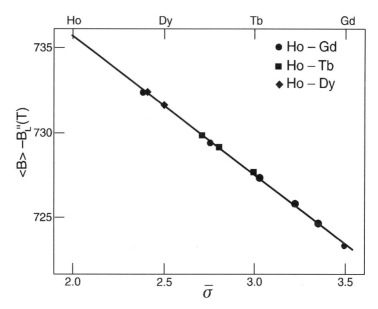

Figure 6.4 Variation of the hyperfine field B_{hf} in ferromagnetic alloys of the rare earths, as a function of the average parameter $\bar{\sigma} = (\bar{g} - 1)\bar{J}$ of the alloy. The measurements were made at 4.2 K through the NMR of ^{165}Ho. [Reprinted from M. A. H. McCausland and I. S. Mackenzie, *Nuclear Magnetic Resonance in Rare Earth Metals*, Taylor & Francis, London, 1980, p. 410.]

observations on the variation of $\mathbf{B}_{hf}$, from atom to atom, can be made from the inspection of Table 6.I: (1) the values of $\mathbf{B}_{hf}$ for the ions of the 3*d* series may reach tens of teslas; (2) the fields for an *S*-state rare-earth ion (e.g., Gd^{3+}) are of the same order of magnitude; and finally (3) $\mathbf{B}_{hf}$ reaches hundreds of teslas at nuclei of non-*S* ions ($L \neq 0$) of rare earths and actinides.

There are countless measurements of $\mathbf{B}_{hf}$, at different nuclei, different ions, different matrices, different conditions of temperature, and pressure, and so forth. As examples of some experimental results we can show the variation of $\mathbf{B}_{hf}$ with some parameters: in Fig. 6.3, the variation of the hyperfine field measured in metallic Fe, as a function of the temperature; in Fig. 6.4, where it is shown that $\mathbf{B}_{hf}$ is a function of the mean value of the quantity $(g-1)J$, in a solid solution of rare earths; and in Fig. 6.5, the last example, which shows the dependence of the hyperfine field at Fe diluted in aluminum, as a function of the distance to another iron impurity.

These two last examples (Figs 6.4 and 6.5) show that, in a metal, the hyperfine field measured on a given site is affected by the atoms that occupy the neighbor sites. In the first example, the field measured at the holmium nucleus varies with the mean value of $(g-1)J$ of the alloy; that is, it is affected by the mean value of the projection of *S* over *J* of the sample. In the last example, the dependence of the effect of the iron moments on $\mathbf{B}_{hf}$ is emphasized. The most remarkable observation in this case is the oscillatory dependence of this influence—the contributions, for example, of the first and second neighbor shells may have

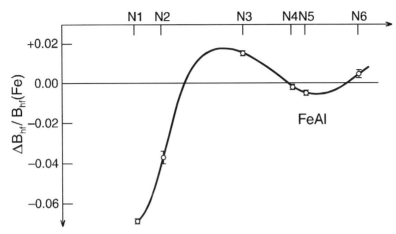

Figure 6.5 Relative change in hyperfine field $\Delta B_{hf}/B_{hf}(Fe)$ at Fe nuclei in a Fe–Al matrix, as a function of the distance of the Fe atom to an Al impurity, measured with Mössbauer spectroscopy of ^{57}Fe. [Reprinted from M. B. Stearns, *Phys. Rev.* **147**, 446 (1966).]

opposite signs. Both examples may be understood if we consider that local magnetic moments induce in a metal a long range polarization. This idea was presented by Vonsovskii (1946) and Zener (1951), and by Ruderman and Kittel (1954); this last model was initially applied to nuclear magnetic moments, and leads to oscillatory polarization of the conduction electrons (Chapter 3).

6.5.4 The Knight Shift

The NMR frequency of a nucleus in a diamagnetic insulating salt is different from that in a metal, for the same value of the applied static magnetic field. This difference arises from the polarization of the conduction electrons in the metal, which contribute an extra magnetic field, and therefore shift the resonance frequency.

The application of an external field **B** polarizes the conduction electrons, creating a magnetization $\mathbf{M}'(0)$, and leading to an additional magnetic field at the nucleus [from Eq. (6.49)]:

$$\Delta \mathbf{B} = \frac{\mu_0}{4\pi} \frac{8\pi}{3} \mathbf{M}'(0) \tag{6.91}$$

The magnetization in this case is given by (Narath 1967)

$$\mathbf{M}'(0) = \langle |u(0)|^2 \rangle_F \chi \mathbf{B} \tag{6.92}$$

with χ the Pauli susceptibility per atom, $\langle |u(0)|^2 \rangle$ the amplitude of the electron wavefunction at the origin, and $\langle \cdots \rangle_F$ indicating an averaging over all electron states at the Fermi level.

The nuclear magnetic resonance of a nucleus with gyromagnetic ratio γ in the total magnetic field $\mathbf{B} + \Delta\mathbf{B}$ will then be observed at the angular frequency

$$\omega = \gamma|(\mathbf{B} + \Delta\mathbf{B})| = \gamma B\left(1 + \frac{\mu_0}{4\pi}\frac{8\pi}{3}\langle|u(0)|^2\rangle_F\chi\right) \quad (6.93)$$

The displacement of the angular frequency due to the electronic polarization is therefore given by

$$\Delta\omega = \gamma B\left(\frac{\mu_0}{4\pi}\right)\frac{8\pi}{3}\langle|u(0)|^2\rangle_F\chi \quad (6.94)$$

This difference is found between the NMR frequency of a nucleus in a nonmagnetic metal and the frequency in a diamagnetic insulating salt.

The shift $\Delta\omega$ is generally positive. The relative (or fractional) shift $\Delta\omega/\omega$ is called the *Knight shift*. It does not depend on the applied magnetic field (or on the frequency, when measured under the form $\Delta B/B$)

$$\frac{\Delta\omega}{\omega} = \left(\frac{\mu_0}{4\pi}\right)\frac{8\pi}{3}\langle|u(0)|^2\rangle_F\chi \quad (6.95)$$

The relative shift $\Delta\omega/\omega$ generally increases with the atomic number Z and is practically independent of the temperature; it is of the order of 0.1–0.3% for many metals; for copper, its value is 0.232% (see Table 6.II).

6.6 ELECTROSTATIC INTERACTIONS IN METALS

The electrostatic hyperfine interactions in condensed matter are modified, in comparison to the situation of the polarized free ion. The electric field gradient (EFG) at the nucleus is affected by contributions of the electric charges at the other atomic sites. This contribution, or lattice EFG, is in turn augmented by deformations induced in the closed atomic shells. It is customary to describe this term and the effect of these deformations by writing the effective extraionic EFG in a solid as

$$eq'' = (1 - \gamma_\infty)eq_{\text{latt}} \quad (6.96)$$

Table 6.II Values of the Knight shift measured at low temperature

Element	Pt	Rh	Zr
$\Delta B/B$ (%)	−3.54	+0.43	+0.33

Source: Reprinted from Landolt-Börnstein, *Magnetic Properties of 3d Elements, New Series III/19a*, Springer-Verlag, New York, 1986. with permission.

where eq_latt is the lattice contribution to the EFG and γ_∞ is an antishielding factor called the *Sternheimer factor*. The value of γ_∞ varies between -10 and -100; thus, the effective field gradient in a solid is multiplied 10–100 times. The value of the Sternheimer factor computed for the rare earths is approximately -75; for Am^{2+}, it is -137.

In a metallic matrix, the effects due to conduction electrons have to be taken into consideration; Eq. (6.96) becomes

$$eq'' = (1 - \gamma_\infty)eq_\text{latt} + (1 - R)eq_\text{ce} \qquad (6.97)$$

with R a core correction factor specific to the probe atom, usually taken as zero, for lack of reliable computed values, and eq_ce the conduction electron contribution.

An experimental correlation was observed between the two terms of this equation (Raghavan et al. 1975), leading to a formulation in terms of the lattice EFG:

$$eq'' = (1 - \gamma_\infty)(1 - K)eq_\text{latt} \qquad (6.98)$$

with the parameter $K \approx 3$ for a number of noncubic metals (Raghavan et al. 1975); K is dependent on the atomic group of the element (e.g., Hagn 1986).

Therefore the electrostatic hyperfine interaction in a solid, taking as principal axes of the EFG tensor the crystal axes abc, is given from Eq. (6.20) as

$$\mathcal{H}''_Q = \frac{e^2 q'' Q}{4I(2I-1)} [3I_c^2 - I^2 + \eta(I_a^2 - I_b^2)] \qquad (6.99)$$

with eq substituted by eq'' and the x, y, z axes substituted by a, b, c.

Some values of the electric field gradient at the nucleus of different elements are given in Table 6.III.

6.7 COMBINED MAGNETIC AND ELECTROSTATIC INTERACTIONS

The total hyperfine hamiltonian for a nucleus subject to magnetic and electrostatic interactions is

$$\mathcal{H}_\text{hf} = \mathcal{H}_\text{mag} + \mathcal{H}_Q \qquad (6.100)$$

In a coordinate system with the axes coincident with the principal axes of the electric field gradient tensor, with a magnetic hyperfine field in the direction (θ, ϕ), we may write the complete hamiltonian

$$\begin{aligned}\mathcal{H}_\text{hf} = &-g\mu_N B[I_z \cos\theta + (I_x \cos\phi + I_y \sin\phi)\sin\theta] \\ &+ \frac{e^2 qQ}{4I(2I-1)}[3I_z^2 - I^2 + \eta(I_x^2 - I_y^2)]\end{aligned} \qquad (6.101)$$

Table 6.III Values of the electric field gradient V_{zz} (in 10^{20} V m^{-2}) at some nuclei in different matrices

Nucleus	^{59}Co	^{67}Zn	^{157}Gd	^{159}Tb	^{237}Np
Matrix	Co	Zn	Gd	Tb	α-Np
Temperature (K)	4.2	4.2	1.6	4.2	4.2
V_{zz} (10^{20} V m^{-2})	2.86×10^{-4}	3.402×10^{-3}	1.62×10^{-3}	4.148×10^{-2}	1.40×10^{3}; 4.45×10^{3}

COMBINED MAGNETIC AND ELECTROSTATIC INTERACTIONS 185

For the case of $\eta = 0$ and $\theta = 0$ (when one of the principal axes of the electric field gradient tensor coincides with the direction of the hyperfine field), $\mathcal{H}_{hf}$ is diagonal, and the eigenvalues can be given in closed form, as a function of the quantum number m:

$$E_m = -g\mu_N Bm + \frac{e^2 qQ}{4I(2I-1)}\left[3m^2 - I(I+1)\right] \qquad (6.102)$$

In the case of $\mathbf{B}_{hf}$ and V_{zz} forming an angle $\theta \neq 0$, for a magnetic interaction much more intense than the electrostatic interaction ($eqQ/\mu H \ll 1$), and $\eta = 0$, the eigenvalues are obtained by perturbation theory as

$$E'_m = E_m + <m|\mathcal{H}_Q|m> \qquad (6.103)$$

where E'_m is the eigenvalue in the new coordinate system, where the z' axis coincides with the direction of $\mathbf{B}$. The nuclear spin operators have to be expressed in this coordinate system; we choose the new axes in such a way that z is in the

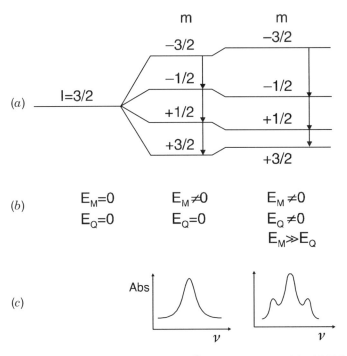

Figure 6.6 (a) Energy levels of a nucleus with $I = \frac{3}{2}$ and (c) positions of the NMR lines of the corresponding spectrum. The diagrams show various situations: (1) zero magnetic and electrostatic interactions; (2) nonzero magnetic interaction with zero electrostatic interaction, and (3) nonzero magnetic interaction with weak electrostatic interaction.

plane $x'0z'$. The operator $[3I_z^2 - I^2]$ becomes, in terms of the components of the operators in the (x',y',z') axes:

$$3(I_z' \cos\theta + I_x' \sin\theta)^2 - I^2 = \tfrac{1}{2}(3\cos^2\theta - 1)(3I_z'^2 - I^2)$$

$$+ \tfrac{3}{2} \sin 2\theta (I_z'I_x' + I_x'I_z') + \tfrac{3}{2} \sin^2\theta (I_x'^2 - I_y'^2) \qquad (6.104)$$

Substituting $\mathcal{H}_Q$ with Eq. (6.104) into Eq. (6.103), we finally obtain for the energy eigenvalues in the case of dominant magnetic interaction ($eqQ/\mu H \ll 1$), $\eta = 0$ and arbitrary θ:

$$E_m = -g\mu_N Bm + \frac{e^2 qQ}{4I(2I-1)}\left[3m^2 - I(I+1)\right]\frac{3\cos^2\theta - 1}{2} \qquad (6.105)$$

To obtain the energy eigenstates for the general case, that is, for an arbitrary ratio of electrostatic and magnetic interactions, and for any value of the angle θ, we need to diagonalize the hamiltonian (6.101). These eigenvalues have been computed and are presented in graphical form (e.g., Parker 1956, Matthias et al. 1962, Kündig 1967).

Figure 6.6 presents in schematic form the sublevels of the hyperfine energy in the case of combined magnetic and electrostatic interactions, for $I = \tfrac{3}{2}$, and the corresponding NMR spectra.

GENERAL READING

Abragam, A., *The Principles of Nuclear Magnetism*, Clarendon Press, Oxford, 1961.
Abragam, A. and B. Bleaney, *Electron Paramagnetic Resonance of Transition Ions*, Clarendon Press, Oxford, 1970.
Allen, P. S., *Contemp. Phys.* **17**, 387 (1976).
Bleaney, B., "Hyperfine Structure and Electron Paramagnetic Resonance," in A. J. Freeman and R. B. Frankel, Eds., *Hyperfine Interactions*, Academic Press, New York, 1967, p. 1.
Bleaney, B., in R. J. Elliott, Ed., *Magnetic Properties of Rare Earth Metals*, Plenum Press, London, 1972.
Goldanskii, V. I. and E. F. Makarov, in V. I. Goldanskii and R. H. Herber, Eds., *Chemical Applications of Mössbauer Spectroscopy*, Academic Press, New York, 1968, p. 1.
Greenwood, N. N. and T. C. Gibb, *Mössbauer Spectroscopy*, Chapman & Hall, London, 1971.
Hagn, E., "Electric Quadrupole Interactions in Non-Magnetic Metals," in N. J. Stone and H. Postma, Eds., *Low-Temperature Nuclear Orientation*, North-Holland, Amsterdam, 1986, p. 527.
Jackson, J. D., *Classical Electrodynamics*, 2nd ed., Wiley, New York, 1975.
Matthias, E., W. Schneider, and R. M. Steffen, *Phys. Rev.* **125**, 261 (1962).
McCausland, M. A. H. and I. S. Mackenzie, *Adv. Phys.* **28**, 305 (1979); also *Nuclear Magnetic Resonance in Rare Earth Metals*, Taylor & Francis, London, 1980.
Netz, G., *Z. Phys. B* **63**, 343 (1986).

Slichter, C. P., *Principles of Magnetic Resonance*, 3rd ed., Springer-Verlag, Berlin, 1990.
Stone, N. J., "Hyperfine Interactions in Magnetic Solids," in N. J. Stone and H. Postma, Eds., *Low-Temperature Nuclear Orientation*, North-Holland, Amsterdam, 1986, p. 351.
Taylor, K. N. R., *Adv. Phys.* **20**, 551 (1971).
Taylor, K. N. R. and M. I. Darby, *Physics of Rare Earth Solids*, Chapman & Hall, London, 1972.

REFERENCES

Abragam, A., *The Principles of Nuclear Magnetism*, Clarendon Press, Oxford, 1961.
Abragam, A. and B. Bleaney, *Electron Paramagnetic Resonance of Transition Ions*, Clarendon Press, Oxford, 1970.
Arif, S. K., D. St. P. Bunbury, G. J. Bowden, and R. K. Day, *J. Phys. F: Metal Phys.* **5**, 1037 (1975).
Baker, J. M. and F. I. B. Williams, *Proc. Roy. Soc.* **A267**, 283 (1962).
Bleaney, B., "Hyperfine Structure and Electron Paramagnetic Resonance," in A. J. Freeman and R. B. Frankel, Eds., *Hyperfine Interactions*, Academic Press, New York, 1967, p. 1.
Elliott, R. J. and K. W. H. Stevens, *Proc. Roy. Soc.* **A218**, 553; **A219**, 387 (1953).
Hagn, E., "Electric Quadrupole Interactions in Non-Magnetic Metals," in N. J. Stone and H. Postma, Eds., *Low-Temperature Nuclear Orientation*, North-Holland, Amsterdam, 1986, p. 527.
Jackson, J. D., *Classical Electrodynamics*, 2nd ed., Wiley, New York, 1975.
Kündig, W., *Nucl. Instrum. Meth.* **48**, 2219 (1967).
Landolt-Börnstein, *Tables of Magnetic Properties of 3d Elements*, Landolt-Börnstein, New Series III/19a, Springer-Verlag, Berlin, 1986.
Matthias, E., W. Schneider, and R. M. Steffen, *Phys. Rev.* **125**, 261 (1962).
McCausland, M. A. H. and I. S. Mackenzie, *Adv. Phys.* **28**, 305 (1979); also *Nuclear Magnetic Resonance in Rare Earth Metals*, Taylor & Francis, London, 1980.
McMorrow, D. F., M. A. H. McCausland, Z. P. Han, and J. S. Abell, *J. Phys.: Condens. Matter* **1**, 10439 (1989).
Narath, A., "Nuclear Magnetic Resonance in Magnetic and Metallic Solids," in A. J. Freeman and R. B. Frankel, Eds., *Hyperfine Interactions*, Academic Press, New York, 1967, p. 287.
Netz, G., *Z. Phys. B* **63**, 343 (1986).
Parker, P. M., *J. Chem. Phys.* **24**, 1096 (1956).
Raghavan, R. S., E. N. Kaufmann, and P. Raghavan, *Phys. Rev. Letters* **34**, 1280 (1975).
Ruderman, M. A. and C. Kittel, *Phys. Rev.* **96**, 99 (1954).
Slichter, C. P., *Principles of Magnetic Resonance*, 3rd ed., Springer-Verlag, Berlin, 1990.
Stearns, M. B., *Phys. Rev.* **147**, 439 (1966).
Vonsovskii, S. V., *Zh. Eksper. Teor. Fiz.* **16**, 981 (1946).
Zener, C., *Phys. Rev.* **81**, 440 (1951).

EXERCISES

6.1 *Nuclear Populations.* Compute the Boltzmann populations of a nucleus with $I = 2$ and magnetic moment 0.3 μ_N in a magnetic induction of 25 T, for the temperatures 300 K (room temperature), 4 K (liquid helium temperature), and 0.01 K. What is the magnetic induction necessary to produce an energy separation ΔE approximately equal to kT at room temperature?

6.2 *Hyperfine Interaction in the Free Atom.* Rubidium possesses two stable isotopes, ^{85}Rb ($I = \frac{5}{2}$) and ^{87}Rb ($I = \frac{3}{2}$). The electronic configuration of rubidium is that of the alkali metals, containing one single electron in an s state. Draw a scheme of the hyperfine structure of these two isotopes, and show the splitting of the energy levels in the presence of a strong applied magnetic field.

6.3 *Hyperfine Interaction in Metals.* Assume that a spin of a conduction electron in a metal feels a magnetic hyperfine field arising from the interaction with the nuclear spins. Let the component z of the field felt by the electron be given by

$$B_i = \left(\frac{a}{N}\right) \sum_{j=1}^{N} I_j^z$$

where I_j^z may be $\pm \frac{1}{2}$. Show that $\langle B_i^2 \rangle = (a/2N)^2 N$ and $\langle B_i^4 \rangle = 3(a/2N)^4 N^2$ for $N \gg 1$.

6.4 *Effects of the Crystalline Field.* For a good number of rare earths that present magnetic order, the hyperfine field is proportional to $\langle J_z \rangle$, the z component of the total angular momentum of the ion. In the presence of a crystalline field, the value of $\langle J_z \rangle$ may be reduced, the so-called 'quenching' of the angular momentum. If $\mathcal{H}_m$ describes the exchange interaction and $\mathcal{H}_{CF}$ the interaction with the crystal field, in the limit in which $\mathcal{H}_{CF} \ll \mathcal{H}_m$ the reduction in $\langle J_z \rangle$ is given by:

$$\langle J_z \rangle = J \left\{ 1 - \left(\frac{C}{X}\right)^2 \right\}$$

where X is a factor relating to the magnitude of the molecular field, and the C values are the nondiagonal elements of the crystal field, given by

$$C^2 = \frac{1}{(g-1)^2 J} \sum_{M=-J}^{J-2} \frac{1}{J-M} |\langle J; M|\mathcal{H}_{CF}|J; J\rangle|^2$$

Assume that $\mathcal{H}_{CF}$ is given by

$$\mathcal{H}_{CF} = \frac{3}{4} B_2^2 J_-^2$$

where J_- is the spin-lowering operator. Compute $\langle J_z \rangle$ for Tb^{3+}. Take $B_2^2/k \approx 1$ K, $g = \frac{3}{2}$, $J = 6$, and $X \approx 100$ K.

7

NUCLEAR MAGNETIC RESONANCE

7.1 THE PHENOMENON OF MAGNETIC RESONANCE

Let us imagine a system of N identical particles with magnetic dipole moment μ and collinear angular momentum $\hbar \mathbf{I}$, sufficiently separated in such a way that the interactions among them are negligible. In the initial configuration the directions of the individual moments are distributed at random, every orientation being equally probable. The projections of the magnetization $\mathbf{M}$ [$\mathbf{M} = (1/V) \sum_i \boldsymbol{\mu}_i$] (where V is the volume) are zero along any direction; the same applies to the total angular momentum.

If a static magnetic field $\mathbf{B}_0$ is applied to this system at the time $t = 0$, the moments will start to precess around the direction of $\mathbf{B}_0$ (we will define the z axis as the direction of $\mathbf{B}_0$) in such a way that a projection of the ith magnetic moment on the z axis will be $\mu \cos \theta_i$. In the classical description, θ may have any value; since, by hypothesis, they are all equally probable, a parallel orientation is as likely as an antiparallel one, and therefore $M_z = (1/V) \sum_i \mu \cos \theta_i = 0$. The magnetic moments precess at the same angular frequency $\omega = g\mu_B B_0/\hbar = \gamma B_0$ and consequently, maintain the phase differences among themselves (g is the g-factor and γ is the gyromagnetic ratio). As a result, the components of $\mathbf{M}$ perpendicular to $\mathbf{B}_0$ remain zero ($M_x = M_y = 0$).

Such a system of isolated spins, therefore, cannot be magnetized (Fig. 7.1). More precisely, an isolated system under the action of an external magnetic field would acquire a magnetization very slowly, as the spins lost the magnetic energy through radiation.

In a more realistic situation, the spins are energetically coupled to a thermal

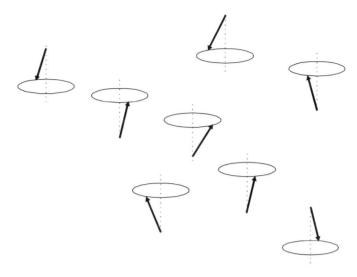

Figure 7.1 Ensemble of isolated magnetic moments precessing under the action of an applied magnetic field.

reservoir, which we will call the *lattice*; the microscopic mechanisms through which this coupling occurs will not be discussed at this point. Under the action of a magnetic field $\mathbf{B}_0$, the spins will exchange their magnetic energies $E_i = -\mu B_0 \cos \theta_i$ with the reservoir, of temperature $T > 0$, and the populations $p(E_i)$ will follow a Boltzmann distribution. Since the states with lower energies E_i are more populated, a magnetization parallel to z will appear. This process of thermalization (or relaxation) occurs in a characteristic time T_1, called the *spin–lattice relaxation time*. The coupling between the spins and the lattice gives rise, under the action of $\mathbf{B}_0$, to a nonzero component M_z.

If the system had a certain transverse magnetization $M_x(0)$ at the moment of application of the field $B_0 \mathbf{k}$, the phase differences among the isolated spins would gradually change, their motion would lose coherence, and the transverse magnetization would decay. One contribution to this relaxation of the perpendicular components of $\mathbf{M}$ is due to interactions among the spins. This spin–spin interaction leads to different precession frequencies, and therefore modifies the relative phases in the precession; each spin feels the random fields due to the other spins and as a result, its precession becomes momentarily slower or faster. This process through which the spin system loses the "memory" of its initial phase relation has a characteristic time T_2, usually called *spin–spin relaxation time*.

Magnetic resonance is observed when a system of spins with magnetic energy levels (called *Zeeman levels*) separated by an interval $\Delta E = E_m - E_{m-1}$ is irradiated with photons of energy $h\nu = \Delta E$ (Figs. 7.2 and 7.3). The radio-frequency field associated with the electromagnetic radiation induces transitions among the states of energy E_m of the system, which absorbs energy.

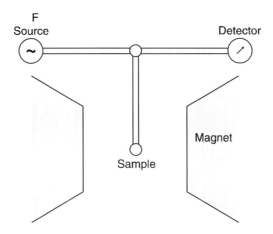

Figure 7.2 Schematic representation of a device for the observation of nuclear magnetic resonance: a radiofrequency source of variable frequency sends photons that impinge on the sample under a static magnetic field. The reflected radiation is detected and measured.

7.2 EQUATIONS OF MOTION: BLOCH EQUATIONS

The magnetic dipoles that are present in the spin system also have associated collinear angular momenta. Assuming that the system is formed of atomic nuclei, of angular momentum $\hbar I$, the magnitudes of the magnetic moments μ and angular momentum $\mathbf{I}$ will be connected through the relation

$$\boldsymbol{\mu} = \gamma \hbar \mathbf{I} \tag{7.1}$$

where γ is the nuclear gyromagnetic ratio (or magnetogyric ratio). The magnetic moment may also be written [Eq. (6.3)]

$$\boldsymbol{\mu} = g\mu_N \mathbf{I} \tag{7.2}$$

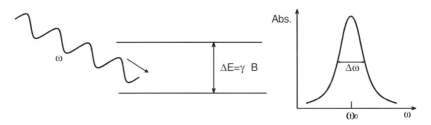

Figure 7.3 Schematic quantum-mechanical description of the phenomenon of magnetic resonance: at resonance, the photons of electromagnetic radiation carry an energy equal to the difference ΔE between the energy levels of the magnetic moments–magnetic field system.

where μ_N is the nuclear magneton and g is the nuclear g factor (or spectroscopic splitting factor) of the nuclear species in question. Analogously to the electronic magneton—the Bohr magneton [Eq. (2.4)]—the nuclear magneton is given by [Eq. (6.4)]:

$$\mu_N = \frac{e\hbar}{2m_p} \tag{7.3}$$

where m_p is the proton mass. Since this mass is 1836 times larger than the electron mass, the nuclear magneton is smaller than the Bohr magneton, by the same factor. The value of the nuclear magneton is 5.0509×10^{-27} J T^{-1}.

The rate of variation of the angular momentum $(d\hbar\mathbf{I}/dt)$ is equal to the torque acting on the elementary dipoles; in an applied magnetic field $\mathbf{B}$ the torque is given by $\boldsymbol{\mu} \times \mathbf{B}$, and we have

$$\frac{d(\hbar\mathbf{I})}{dt} = \frac{1}{\gamma}\frac{d\boldsymbol{\mu}}{dt} = \boldsymbol{\mu} \times \mathbf{B} \tag{7.4}$$

or

$$\dot{\boldsymbol{\mu}} = \gamma\, \boldsymbol{\mu} \times \mathbf{B} \tag{7.5}$$

The magnetization $\mathbf{M} = n\boldsymbol{\mu}$ (n is the number of moments per unit volume) follows the same equation:

$$\dot{\mathbf{M}} = \gamma\, \mathbf{M} \times \mathbf{B} \tag{7.6}$$

Equation (7.5) describes classically the motion of the magnetic moments (or the motion of the magnetization). The quantum-mechanical description of the motion of $\boldsymbol{\mu}$ (taking $\boldsymbol{\mu}$ as the magnetic moment operator $\boldsymbol{\mu} = g\mu_N\mathbf{I}$) is given, in the Heisenberg representation, by the commutator

$$i\hbar\dot{\boldsymbol{\mu}} = [\boldsymbol{\mu}, \mathcal{H}] \tag{7.7}$$

where $\mathcal{H}$ is the hamiltonian of the interaction of the moment with the magnetic field.

In a magnetic resonance experiment, the magnetic moment $\boldsymbol{\mu}$ is in the presence of a static field $\mathbf{B}_0$ and is submitted to an electromagnetic wave (Fig. 7.3). The moment $\boldsymbol{\mu}$ feels the oscillating magnetic field $\mathbf{B}_1(t)$ associated with the electromagnetic wave. The total applied field is therefore

$$\mathbf{B}(t) = \mathbf{B}_0 + \mathbf{B}_1(t) \tag{7.8}$$

and the hamiltonian of the interaction with the magnetic moment is

$$\mathcal{H} = -\boldsymbol{\mu} \cdot \mathbf{B}(t) = -g\mu_N \mathbf{I} \cdot \mathbf{B}(t) = -\gamma \hbar \mathbf{I} \cdot \mathbf{B}(t) \tag{7.9}$$

Expanding the scalar product, we obtain

$$-\boldsymbol{\mu} \cdot \mathbf{B}(t) = -g\mu_N (I_x B_x + I_y B_y + I_z B_z) \tag{7.10}$$

Using the commutation rules for I_x, I_y, and I_z, and inserting into Eq. (7.7), it follows for the x component that

$$i\hbar \dot{\mu}_x = ig^2 \mu_N^2 (I_y B_z - I_z B_y) \tag{7.11}$$

Taking all the components, we obtain

$$i\hbar \dot{\boldsymbol{\mu}} = ig^2 \mu_N^2 \, \mathbf{I} \times \mathbf{B} = ig\mu_N \boldsymbol{\mu} \times \mathbf{B} \tag{7.12}$$

In terms of the gyromagnetic ratio $\gamma = g\mu_N/\hbar$, the equation of motion of the operator $\boldsymbol{\mu}$ is written

$$\dot{\boldsymbol{\mu}} = \gamma \, \boldsymbol{\mu} \times \mathbf{B} \tag{7.13}$$

which is formally identical to the classical description given by Eq. (7.5). The magnetization of this system of moments $\boldsymbol{\mu}$ is given by $\mathbf{M} = n\langle \boldsymbol{\mu} \rangle$, where $\langle \boldsymbol{\mu} \rangle$ is the expectation value of the operator $\boldsymbol{\mu}(t)$ and n is the number of magnetic moments contained in a unit volume. If we include the temperature dependence, the thermal average of the operator $\boldsymbol{\mu}(t)$ is given by the trace of the density matrix in the equilibrium state ρ_0:

$$\langle \boldsymbol{\mu}(t) \rangle = \text{Tr}[\rho_0 \boldsymbol{\mu}] = \frac{\text{Tr}[\boldsymbol{\mu}(t) \exp(-\mathcal{H}_0/kT)]}{\text{Tr}[\exp(-\mathcal{H}_0/kT)]} \tag{7.14}$$

$\mathcal{H}_0$ is the hamiltonian of the interaction with the field; from Eq. (7.13), we obtain the equation of motion for the magnetization:

$$\dot{\mathbf{M}} = \gamma \mathbf{M} \times \mathbf{B} \tag{7.15}$$

which is the same as the classical equation of motion [Eq. (7.6)]. The magnetization of an ensemble of nuclei with spin $I = \frac{1}{2}$ in the presence of a static magnetic field pointing along the z direction is illustrated in Fig. 7.4; note that the magnetization $\mathbf{M}$ also points along z.

We have so far considered the motion of the magnetization in the presence of a magnetic field $\mathbf{B}(t)$, neglecting relaxation processes. From Eq. (7.6) it follows that

$$\frac{d}{dt}M^2 = \frac{d}{dt}(\mathbf{M} \cdot \mathbf{M}) = 2\mathbf{M} \cdot \frac{d\mathbf{M}}{dt} = 2\mathbf{M} \cdot (\gamma \mathbf{M} \times \mathbf{B}) \equiv 0 \quad (7.16)$$

that is, the magnitude of $\mathbf{M}$ is constant in this case, $\mathbf{M}$ only changes direction, as a function of time.

Taking $\mathbf{B} = B\mathbf{k}$, the three components of Eq. (7.15) are

$$\frac{dM_x}{dt} = \gamma M_y B \quad (7.17a)$$

$$\frac{dM_y}{dt} = -\gamma M_x B \quad (7.17b)$$

$$\frac{dM_z}{dt} = 0 \quad (7.17c)$$

From Eq. (7.17c) it immediately follows $M_z(t) = M_z(0)$. The first two equations are coupled and may be written as follows, making $M_\pm = M_x \pm iM_y$:

$$\frac{dM_\pm}{dt} = \frac{dM_x}{dt} \pm i\frac{dM_y}{dt} = \mp i\gamma B M_\pm \quad (7.18)$$

whose solution is

$$M_\pm(t) = M_\pm(0) e^{\mp i\gamma Bt} \quad (7.19)$$

thus

$$M_x(t) = M_x(0)\cos(\gamma Bt) + M_y(0)\sin(\gamma Bt) \quad (7.20a)$$

$$M_y(t) = -M_x(0)\sin(\gamma Bt) + M_y(0)\cos(\gamma Bt) \quad (7.20b)$$

$$M_z(t) = M_z(0) \quad (7.20c)$$

These equations describe the motion of $\mathbf{M}$ as a simple precession around z, with angular frequency $\omega = \gamma B$.

When relaxation processes are active, the motion of $\mathbf{M}$ may be described with an equation derived from Eq. (7.6), assuming that simultaneously with the precession, the deviations of the longitudinal and transverse components of $\mathbf{M}$ in relation to equilibrium decay exponentially, with characteristic times T_1 and T_2, respectively. With this hypothesis, we obtain a phenomenological

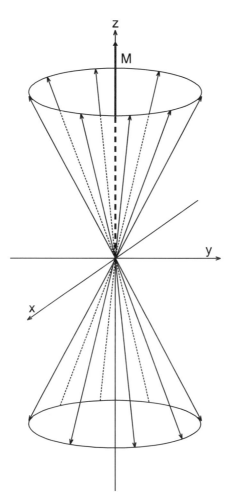

Figure 7.4 Magnetization **M** of an ensemble of magnetic moments of nuclei with spin $I = \frac{1}{2}$, showing the alignment of **M** with the magnetic field **B**, in the z direction.

equation

$$\dot{\mathbf{M}} = \gamma \mathbf{M} \times \mathbf{B} - \frac{M_x \mathbf{i} + M_y \mathbf{j}}{T_2} - \frac{[M_z - M(0)]\mathbf{k}}{T_1} \quad (7.21)$$

which is known as the *Bloch equation* (Bloch 1946).

In Bloch's equation, the same relaxation time (T_2) was assumed for the x and y components of the magnetization. The x and y axes are equivalent because the physical system has axial symmetry along the z direction, which is the direction of the magnetic field. The z component relaxes with a different characteristic time T_1; this difference in behavior is connected to the physical fact that the transverse

196 NUCLEAR MAGNETIC RESONANCE

relaxation (process T_2) conserves energy, while the longitudinal relaxation does not. In fact, T_2 may also involve exchange of energy, in the case of saturation (a condition defined in the next section)—the high rf power causes stimulated emission, and the transverse relaxation time is modified.

Besides the static magnetic field, there may be a rotating magnetic field, given by (in the stationary frame)

$$\mathbf{B}_1(t) = B_1[\mathbf{i}\cos(\omega t + \phi) + \mathbf{j}\sin(\omega t + \phi)] \quad (7.22)$$

We may take the phase $\phi = 0$, which is equivalent to having $\mathbf{B}_1 = B_1\mathbf{i}$ at $t = 0$.

Substituting $\mathbf{B}_0 = B_0\mathbf{k}$ and $\mathbf{B}_1(t)$ into the Bloch equation [Eq. (7.21)], the components will be

$$\frac{dM_x}{dt} = \gamma M_y B_0 - \gamma M_z B_1 \sin \omega t - \frac{M_x}{T_2} \quad (7.23a)$$

$$\frac{dM_y}{dt} = \gamma M_z B_1 \cos \omega t - \gamma M_x B_0 - \frac{M_y}{T_2} \quad (7.23b)$$

$$\frac{dM_z}{dt} = \gamma M_x B_1 \sin \omega t - \gamma M_y B_1 \cos \omega t - \frac{M_z - M(0)}{T_1} \quad (7.23c)$$

The description of the motion of the magnetization is considerably simplified if we adopt a coordinate system with axes x', y', z' that turns around the $z = z'$ axis with angular frequency equal to the Larmor frequency [Eq. (2.10)]. We will show below how we can rewrite the Bloch equations in this new system.

7.3 MAGNETIZATION FROM ROTATING AXES

Let us consider two coordinate systems xyz and $x'y'z'$ with a common origin; we assume that $x'y'z'$ rotates with angular velocity $-\Omega$ relative to xyz. Since both coordinate systems have the same origin, the position vectors of a point P, $\mathbf{r}$ and $\mathbf{r}'$, will be identical, although with different coordinates in the two systems. The time dependence of any vector, however, differs when described from xyz or $x'y'z'$. For example, the time derivative of the unit vectors of the directions x, y, and z, ($\mathbf{i}$, $\mathbf{j}$, and $\mathbf{k}$, respectively) are nonzero if described from $x'y'z'$; since they have unitary lengths, only their directions can change. If we describe the derivative of the vector $\mathbf{i}$ from a coordinate system $x'y'z'$ that rotates with angular velocity $-\Omega$ relative to xyz, we will have

$$\frac{d'\mathbf{i}}{dt} = \Omega \times \mathbf{i} \quad (7.24)$$

and analogously for $\mathbf{j}$ and $\mathbf{k}$. The time derivative of an arbitrary vector

$\mathbf{V} = V_x\mathbf{i} + V_y\mathbf{j} + V_z\mathbf{k}$ in this rotating system will be

$$\frac{d\mathbf{V}}{dt} = \frac{dV_x}{dt}\mathbf{i} + \frac{dV_y}{dt}\mathbf{j} + \frac{dV_z}{dt}\mathbf{k} + V_x\frac{d'\mathbf{i}}{dt} + V_y\frac{d'\mathbf{j}}{dt} + V_z\frac{d'\mathbf{k}}{dt} \quad (7.25)$$

or

$$\frac{d'\mathbf{V}}{dt} = \frac{d\mathbf{V}}{dt} + \mathbf{\Omega} \times \mathbf{V} \quad (7.26)$$

where $d\mathbf{V}/dt$ is the derivative in relation to the stationary system (xyz).

For the magnetization vector evolving under a constant field $\mathbf{B}_0$, we will have

$$\frac{d'\mathbf{M}}{dt} = \frac{d\mathbf{M}}{dt} + \mathbf{\Omega} \times \mathbf{M} = \gamma\mathbf{M} \times \mathbf{B}_0 + \mathbf{\Omega} \times \mathbf{M} \quad (7.27)$$

thus

$$\frac{d'\mathbf{M}}{dt} = \mathbf{M} \times (\gamma\mathbf{B}_0 - \mathbf{\Omega}) = \gamma\mathbf{M} \times \left(\mathbf{B}_0 - \frac{\mathbf{\Omega}}{\gamma}\right) \quad (7.28)$$

The gyromagnetic ratio γ can be positive or negative, but in this chapter we are assuming $\gamma > 0$. We may write this equation for a rotating coordinate system, under the same form that it presents in the stationary system:

$$\frac{d'\mathbf{M}}{dt} = \gamma\mathbf{M} \times \mathbf{B}_{\text{eff}} \quad (7.29)$$

with the effective field given by

$$\mathbf{B}_{\text{eff}} = \mathbf{B}_0 - \frac{\mathbf{\Omega}}{\gamma} \quad (7.30)$$

We may choose a rotating system such that $B_{\text{eff}} = 0$; it suffices to take its angular frequency

$$\Omega = \gamma B_0 = \omega_0 \quad (7.31)$$

that is, it is sufficient to choose a system that rotates around the z axis with angular frequency equal to the Larmor frequency. In this case

$$\frac{d'\mathbf{M}}{dt} = 0 \quad (7.32)$$

thus, the magnetization is stationary in relation to this system.

A time-dependent magnetic field, rotating in the stationary coordinate system, given by Eq. (7.22), will be in the rotating system:

$$\mathbf{B}_1(t) = B_1 \mathbf{i}' \tag{7.33}$$

The effective field in the presence of $\mathbf{B}_1(t)$ becomes

$$\mathbf{B}_{\text{eff}} = \mathbf{B}_0 + \mathbf{B}_1(t) - \frac{\Omega}{\gamma} \tag{7.34}$$

and in the coordinate system (x', y', z') that rotates at the Larmor frequency, the only field acting on the magnetization is $\mathbf{B}_1(t)$, and the equation of motion becomes

$$\frac{d'\mathbf{M}}{dt} = \gamma \mathbf{M} \times \mathbf{B}_1(t) \tag{7.35}$$

Therefore, the rotating system simplifies the description of the motion of spins given by Eq. (7.6).

Thus, the Bloch equations [Eq. (7.21)] in the system rotating with angular frequency $-\Omega$ take the form

$$\dot{\mathbf{M}}' = \gamma \mathbf{M} \times \left(\mathbf{B}_0 - \frac{\Omega}{\gamma} + \mathbf{B}_1 \right) - \frac{M'_x \mathbf{i} + M'_y \mathbf{j}}{T_2} - \frac{[M'_z - M(0)]\mathbf{k}}{T_1} \tag{7.36}$$

The field B_{eff} is given by:

$$\mathbf{B}_{\text{eff}} = \left(B_0 - \frac{\Omega}{\gamma} \right) \mathbf{k}' + B_1 \mathbf{i}' \tag{7.37}$$

In a rotating coordinate system turning with angular frequency $\Omega = \omega$, with natural frequency of precession (Larmor frequency) $\omega_0 = \gamma B_0$ [Eq. (7.31)], the magnetic field can be written

$$\mathbf{B}_{\text{eff}} = \frac{1}{\gamma}(\omega_0 - \omega)\mathbf{k}' + B_1 \mathbf{i}' \tag{7.38}$$

And the equations of motion in the rotating frame become

$$\frac{dM'_x}{dt} = (\omega_0 - \omega)M'_y - \frac{M'_x}{T_2} \tag{7.39a}$$

$$\frac{dM'_y}{dt} = -(\omega_0 - \omega)M'_x + \gamma M'_z B_1 - \frac{M'_y}{T_2} \tag{7.39b}$$

$$\frac{dM'_z}{dt} = -\gamma M'_y B_1 - \frac{M'_z - M(0)}{T_1} \tag{7.39c}$$

The Bloch equations under this form show the evolution of the components of the magnetization in the rotating system, as a function of the angular frequency of precession ω in this system; ω_0 is the Larmor frequency.

There are two experimental forms of studying magnetic resonance. In the first and more traditional way, called the *continuous-wave* (CW) technique, the behavior of the resonant system is studied in the stationary regime, with the rf (oscillating field) applied continuously. The resonance curve is obtained by recording the magnetization (or the absorbed power, see discussion below) as a function of the frequency ω. In the *pulsed-resonance method*, on the other hand, the spins are submitted to short (compared to T_1 and T_2) pulses, and the evolution of the magnetization is observed as a function of time.

The continuous-wave technique can also be employed keeping the frequency ω fixed and sweeping the applied field; the shape of the lines is the same as that obtained sweeping in frequency, with $\gamma(B - B_0)$ in place of $(\omega - \omega_0)$. To make sure that the system remains in the steady state, the sweep has to be slow in comparison with T_1 and T_2 (slow passage); the condition is $dB/dt \ll B_1/(T_1 T_2)^{1/2}$. We will consider both techniques:

Continuous-Wave Technique The solution of the equations of motion in the stationary regime is obtained making

$$\frac{dM_x}{dt} = \frac{dM_y}{dt} = \frac{dM_z}{dt} = 0$$

The system of equations (7.39) can be solved, with the following result (in the rotating axes):

$$M'_x = \frac{\gamma B_1 (\omega_0 - \omega) T_2^2}{1 + (\omega_0 - \omega)^2 T_2^2 + \gamma^2 B_1^2 T_1 T_2} M(0) \tag{7.40a}$$

$$M'_y = \frac{\gamma B_1 T_2}{1 + (\omega_0 - \omega)^2 T_2^2 + \gamma^2 B_1^2 T_1 T_2} M(0) \tag{7.40b}$$

$$M'_z = \frac{1 + (\omega_0 - \omega)^2 T_2^2}{1 + (\omega_0 - \omega)^2 T_2^2 + \gamma^2 B_1^2 T_1 T_2} M(0) \tag{7.40c}$$

The curves M'_x and M'_y are known as curves of *dispersion* (out-of-phase magnetization) and *absorption* (in-phase magnetization). In the cases of low rf field intensity, $\gamma B_1 \ll T_1, T_2$ (or $\gamma^2 B_1^2 T_1 T_2 \ll 1$), the magnetizations are

$$M'_x = \frac{\gamma B_1 (\omega_0 - \omega) T_2^2}{1 + (\omega_0 - \omega)^2 T_2^2} M(0) \tag{7.41a}$$

$$M'_y = \frac{\gamma B_1 T_2}{1 + (\omega_0 - \omega)^2 T_2^2} M(0) \tag{7.41b}$$

$$M'_z = M(0) \tag{7.41c}$$

The graphs of M'_x and M'_y (or the corresponding susceptibilities) versus $(\omega_0 - \omega) T_2$ are shown in Fig. 7.5; the lineshape of these curves is called lorentzian. The full width at half maximum (fwhm) of the lorentzian is $\Delta \omega_{1/2} = 2/T_2$ (in radians per second) (Fig. 7.5); in frequency units, the width is $2/\pi T_2$ Hz. In the resonance curve obtained with fixed frequency, varying the magnetic field, the width is $\Delta B_{1/2} = 2/\gamma T_2$, in units of **B** (teslas).

The three components of the magnetization are proportional to $M(0)$, the equilibrium magnetization of the spin system. In the case of magnetic resonance of nuclei, $M(0)$ is the nuclear magnetization, related to the static nuclear

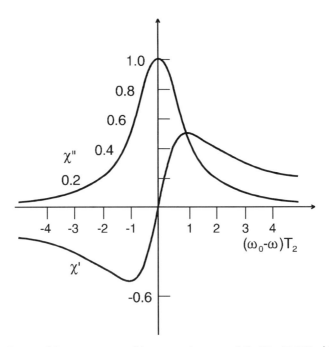

Figure 7.5 Curves of the components of the magnetic susceptibility [Eq. (7.49)]: χ' (dispersion), and χ'' (absorption), versus $(\omega_0 - \omega) T_2$. The lineshape of these curves is usually called lorentzian [Eqs. (7.41a) and (7.41b)].

susceptibility χ_n:

$$M(0) = \chi_n H = \chi_n \frac{B_0}{\mu_0} \tag{7.42}$$

where H is the magnetic field intensity. The nuclear susceptibility χ_n is given as a function of the angular momentum I of the nucleus (for $kT \gg \mu_0\gamma\hbar IH$):

$$\chi_n = \frac{\mu_0 n \gamma^2 \hbar^2 I(I+1)}{3kT} \tag{7.43}$$

From this expression we can see that the nuclear magnetization, and therefore the intensity of NMR signal, are inversely proportional to T, following the Curie law [analogous to Eq. (2.63)].

The nuclear magnetization under an applied field B_0, for n nuclei per unit volume is given by Eq. (7.42), or $M(0) = nB_0\gamma^2\hbar^2 I(I+1)/3kT$. This quantity is a measure of the intensity of the NMR signal; the NMR relative sensitivity for a given nuclide is equal to $M(0)/M(0)_H$, where $M(0)_H$ is the magnetization of the same number of hydrogen nuclei. The relative sensitivity multiplied by the natural isotopic abundance gives the absolute NMR sensitivity; the values of these quantities are listed in Appendix A for the different nuclear species.

In an NMR experiment, the applied time-dependent field (rf field) is linearly polarized, and can be described in the laboratory as a superposition of two rotating fields, one turning clockwise, and the other counterclockwise:

$$\mathbf{B}_1(t) = B_1(\mathbf{i}\cos\omega t - \mathbf{j}\sin\omega t) \tag{7.44a}$$

and

$$\mathbf{B}_1(t) = B_1(\mathbf{i}\cos\omega t + \mathbf{j}\sin\omega t) \tag{7.44b}$$

Adding up, one obtains the oscillating field along the x axis:

$$\mathbf{B}_1(t) = 2B_1\mathbf{i}\cos\omega t \tag{7.45}$$

We may write the two rotating fields in terms of the projection ω_z of the angular velocity vector $\boldsymbol{\omega}$, with the equation

$$\mathbf{B}_1(t) = B_1(\mathbf{i}\cos\omega_z t + \mathbf{j}\sin\omega_z t) \tag{7.46}$$

for ω_z positive and negative.

The components of the magnetization in the laboratory M_x and M_y may be obtained from the components in the rotating system:

$$M_x = M'_x \cos \omega t + M'_y \sin \omega t \tag{7.47a}$$

$$M_y = -M'_x \sin \omega t + M'_y \cos \omega t \tag{7.47b}$$

From (7.47a) one can define the susceptibilities χ' and χ'' of the spin system submitted to the linear field in the laboratory, of amplitude $2B_1$:

$$M_x = (\chi' \cos \omega t + \chi'' \sin \omega t) 2B_1 \tag{7.48}$$

The magnetizations M'_x and M'_y are then directly proportional to the susceptibilities χ' and χ''; χ' and $-\chi''$ respectively may be considered the real and imaginary components of a complex susceptibility χ:

$$\chi = \chi' - i\chi'' \tag{7.49}$$

The component χ' is also called the *dispersive part* and χ'', the *absorptive part* of the susceptibility.

The x component of $\mathbf{B}_1$ in the laboratory can be written, under complex form, as

$$B_{1x}(t) = 2B_1 e^{i\omega t} \tag{7.50}$$

and the magnetization becomes

$$M_x(t) = \mathcal{R}e(2\chi B_1 e^{i\omega t}) \tag{7.51}$$

which is equivalent to Eq. (7.47a).

The average power absorbed by the system of spins is given, in terms of the magnetic energy E by

$$P = \overline{\frac{dE}{dt}} = -\overline{\mathbf{M} \cdot \frac{d\mathbf{B}}{dt}} \tag{7.52}$$

where this expression has to be averaged over a full period (from $t = 0$ to $t = 2\pi/\omega$). Since we are considering the stationary regime, the average of $d\mathbf{B}/dt$ can be identified to the instantaneous value:

$$\overline{\frac{d\mathbf{B}}{dt}} = \frac{d\mathbf{B}}{dt} = -\omega \times B_1 \mathbf{i}' = -\omega B_1 \mathbf{j}' \tag{7.53}$$

and Eq. (7.52) becomes

$$P = \omega M'_y B_1 \tag{7.54}$$

The absorbed power is therefore proportional to χ'', or M'_y. The expression for the average power $\mathcal{P}$ is then

$$\mathcal{P} = \frac{\omega \gamma B_1^2 T_2}{1 + (\omega_0 - \omega)^2 T_2^2} M(0) \tag{7.55}$$

computed here for a situation far from saturation (i.e., for $\gamma^2 B_1 T_1 T_2 \ll 1$).

Since M'_y is proportional to χ'', we can re-write the expression of the power:

$$\mathcal{P} = \omega \chi'' B_1^2 \tag{7.56}$$

Therefore, the power absorbed by the spin system is proportional to the square of the amplitude of the rf field B_1^2.

Pulsed Resonance Technique In this technique, more widely used, the radiofrequency field $B_1(t)$ is applied in pulses of short duration. At resonance, with $B_1(t)$ rotating at the Larmor frequency, this field is stationary in the system of rotating axes. By varying the duration of the pulse, we can make the magnetization turn any angle we want, for example, π radians (or 180°); in this case the magnetization is turned to a position opposite to that of thermodynamic equilibrium. Equations (7.39) describe the evolution of the components of the magnetization as a function of time.

In general, the values of B_1 employed in pulsed resonance are larger, corresponding to the situation of saturation described above. Pulsed magnetic resonance is discussed in more detail in Section 7.6.

7.4 RELAXATION

The longitudinal relaxation, of characteristic time T_1, and the transverse relaxation (T_2) are caused by fluctuations in the magnetic field and in the electric field gradient felt by the nuclei. These fluctuations originate in the random motions associated with the thermal vibrations. Their effects depend on the timescale in which they occur, or, which is equivalent, depend on their Fourier spectrum. Thus, the fluctuations in the interactions that have a characteristic time much shorter than the inverse of the Larmor frequency of the nuclei, that is, below about 10^{-7} s, do not contribute to the relaxation times. The motion of the electrons and the molecular vibrations are included in this category.

In this section we will consider the contributions to the relaxation of the fluctuations of the magnetic fields. The fluctuations $\delta \mathbf{B}$ measure the extent of deviation of the instantaneous field from its average value:

$$\delta \mathbf{B} = \mathbf{B} - \overline{\mathbf{B}} \tag{7.57}$$

where $\overline{\mathbf{B}}$ is the average value of the magnetic field.

If the spectrum of transverse fluctuations $\delta B_\pm(\omega)$ has components in the Larmor frequency ($\omega_0 = \gamma \overline{B}$), these will be able to induce transitions between the states of different magnetic energy of the nucleus. This is equivalent to the fact that in pulsed NMR the rf field $B_1(\omega)$ is effective for the inversion of the spins when it oscillates with angular frequency equal to the Larmor frequency, that is, $\omega = \omega_0$. Through transitions between the states with $\Delta m = \pm 1$, the nuclear magnetization reaches its equilibrium value in a time T_1. This relaxation induces (or is equivalent to) a broadening of the resonance line, which is called *secular broadening*.

The longitudinal fluctuations δB_z have the effect of increasing or reducing the angular velocity of precession, and therefore, affect the relative phase of the rotating spins, contributing in this way to T_2. If, at the instant $t_0 = 0$, the spins are precessing in phase in the xy plane, after a time of the order of T_2 they will lose coherence and the transverse magnetization $M_\pm$ will decay to zero.

In general we can assume that the fluctuations of the components $B_i(t)$ of the field (with $i = x, y, z$) are isotropic. To study the time variation of the fluctuations of B it is useful to define the correlation function (or self-correlation) between the value of the field fluctuations $\delta B_i(t)$ and $\delta B_i(t + \tau)$ at two instants t and $t + \tau$ as

$$F = \overline{\delta B_i(t) \delta B_i(t + \tau)} \tag{7.58}$$

We assume that F varies exponentially with the interval τ, with the form:

$$F = \delta B_i^2 \exp(-|\tau|/\tau_0) \tag{7.59}$$

where τ_0 is the correlation time, a measure of the average time during which the spin feels the magnetic field fluctuation of magnitude δB. If the correlation function has the above form, and if τ_0 is the same for all the components i, we can conclude that the relaxation rate $1/T_2$ is given by (Slichter 1990, Chapter 5)

$$\frac{1}{T_2} = \frac{1}{T_2'} + \frac{1}{2T_1} \tag{7.60}$$

where $1/T_2'$ is the term due to the time fluctuations of B_z (secular broadening) and $1/2T_1$ is the broadening due to the lifetime of the state (nonsecular broadening).

We can obtain an approximate expression that relates the fluctuations δB to the relaxation time. Let us assume that a fluctuation δB_z, acting during a time t on a nucleus, produces a significant variation of phase of the precessing spins, of the order of one radian. This dephasing will be given by

$$\delta \theta = \gamma \, \delta B_z t \approx 1 \tag{7.61}$$

Variations of the phase of the individual spins of this order will make the transverse magnetization of the ensemble of spins decay to zero. We may, therefore, consider this time $t \approx T_2$. It follows that

$$T_2 \approx \frac{1}{\gamma \, \delta B_z} \tag{7.62}$$

A more realistic estimate should consider that this loss of phase memory will come about after n increments δB_z in the instantaneous magnetic field, each value of the field remaining constant for a time τ_0 (correlation time). At the end, a time $t = n\tau_0$ will have elapsed, and the mean square dephasing will be

$$\overline{\delta\theta^2} = n \, \overline{\delta\theta^2} = n\gamma^2 \, \overline{\delta B_z^2}\tau_0^2 = \gamma^2 \, \overline{\delta B_z^2} t \tau_0 \tag{7.63}$$

And finally, it follows, identifying this time t for the dephasing of one radian to T_2, that

$$\frac{1}{T_2} = \gamma^2 \, \overline{\delta B_z^2}\tau_0 \tag{7.64}$$

where we can note that $1/T_2$ depends on τ_0. From this relation we may conclude that for shorter correlation times τ_0, the rate $1/T_2$ is reduced, a phenomenon known as *motional narrowing*. In other words, very rapid variations in the field B_z have on the average no effect on the transverse relaxation; this may occur, for example, for a nucleus that diffuses through a liquid, feeling for very short times the local magnetic dipolar fields due to the different atoms.

A more careful derivation of the expressions for the rates $1/T_1$ and $1/T_2$, still assuming isotropy in the fluctuations of B, leads to the following results (see Carrington and McLachlan 1967):

$$\frac{1}{T_1} = \gamma^2 (\overline{\delta B_x^2} + \overline{\delta B_y^2}) \frac{\tau_0}{1 + \omega_0^2 \tau_0^2} \tag{7.65}$$

$$\frac{1}{T_2} = \gamma^2 \left[\overline{\delta B_z^2}\,\tau_0 + \frac{1}{2}(\overline{\delta B_x^2} + \overline{\delta B_y^2}) \frac{\tau_0}{1 + \omega_0^2 \tau_0^2} \right] \tag{7.66}$$

which contain the spectral density function $J(\omega)$:

$$J(\omega) = \frac{\tau_0}{1 + \omega^2 \tau_0^2} \tag{7.67}$$

for $\omega = \omega_0$ (the Larmor frequency). The spectral density function $J(\omega)$ is in fact the Fourier transform of the correlation function, and is related to the average power of the fluctuations at the frequency ω.

One can note in Eqs. (7.65)–(7.66) that the longitudinal and the transverse relaxations depend in a different way on the correlation time τ_0: T_1 goes through a minimum for $\omega_0 \tau_0 \approx 1$, and the same does not occur with T_2. The minimum of

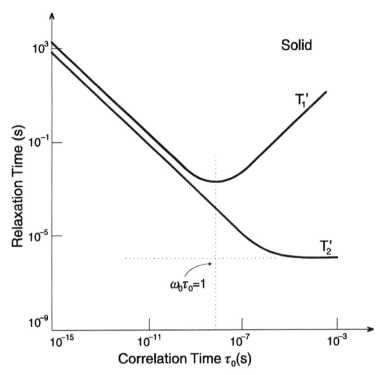

Figure 7.6 Variation of the secular broadening rate $(1/T_2')$ and nonsecular broadening $(1/2T_1) \equiv (1/T_1')$ as a function of the correlation time τ_0. Note the maximum of $1/T_1'$ for $\tau_0 \cong 1/\omega_0$ (ω_0 is the Larmor frequency).

T_1 occurs for correlation times comparable to the inverse of the Larmor frequency ω_0, namely, for fluctuations of the magnetic fields with larger Fourier intensity at this frequency.

From the expressions (7.65) and (7.66) we can see that the fluctuations of all the components of the magnetic field contribute to T_2, but only the fluctuations of the transverse components (x and y) contributes to T_1. We can also note that in the case $\omega_0 \tau_0 \ll 1$, the equality $T_1 = T_2$ (for isotropic fluctuations) follows from Eqs. (7.65) and (7.66).

The behavior of the relaxation times versus correlation time can be seen in Fig. 7.6. One can imagine the figure as representing the relaxation processes in a liquid that becomes gradually more viscous. With the increase in viscosity, the correlation time τ_0 increases and the local field fluctuations at the resonant frequency ω_0 decrease, becoming zero for $\omega_0 \tau_0 \gg 1$. For short correlation times, on the other hand, one can see that T_1 increases with $1/\tau_0$; for long correlation times, it increases with τ_0 [from Eq. (7.65)].

In general, the correlation time τ_0 is a function of the temperature of the

Table 7.I Nuclear relaxation times T_1 and T_2 measured in some metallic systems (in ms)

Nucleus	Matrix	T_1		T_2	
		4.2 K	$\cong$ 300 K	4.2 K	$\cong$ 300 K
^{57}Fe	Fe	10–500	0.9–6.5	10–500	0.9–6.5
^{59}Co	Co	0.2–17	0.1–0.5	0.088	0.025

Source: I. D. Weisman, L. J. Swartzendruber, and L. H. Bennett, in *Techniques of Metal Research*, E. Passaglia, Ed., Vol. VI, Part 2. Copyright © 1973, Wiley-Interscience, New York. Reprinted with permission of John Wiley & Sons, Inc.

sample. A common dependence of τ_0 is of the type

$$\tau_0 = \tau_0^0 \exp\left(\frac{E_a}{kT}\right) \quad (7.68)$$

which corresponds to a thermally activated process, with activation energy E_a. Table 7.I presents some values of T_1 and T_2 observed experimentally.

7.4.1 Longitudinal Relaxation

Let us consider a system of spins in the presence of a magnetic field, and in contact with a thermal reservoir (or lattice), that is at a temperature T (Fig. 7.7). The characteristic time after which the system, if disturbed, returns to the thermal equilibrium configuration is T_1, which is the inverse of the relaxation rate of the longitudinal magnetization $(1/T_1)$ in Bloch equations (Section 7.2). When the spin subsystem is in equilibrium with the lattice, the occupation of its energy states is a function of T given by the Boltzmann distribution (Fig. 7.8).

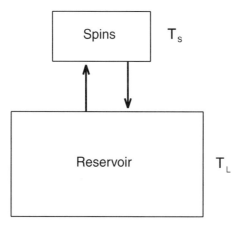

Figure 7.7 Coupling between the reservoir (lattice) at temperature T_L and the spin system at temperature T_s. The characteristic time in which the spins reach the lattice temperature is T_1.

208 NUCLEAR MAGNETIC RESONANCE

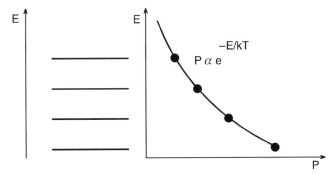

Figure 7.8 Representation of the populations of the energy states of an ensemble of spins at temperature T, following a Boltzmann distribution.

In situations where the interaction among the spins is much stronger than that between the spins and the lattice, the population of the levels of an ensemble of spins can be described by a temperature T_s; this is called the *spin temperature*, and it may, in principle, be different from the lattice temperature.

Let T_1 be the time interval in which the spin subsystem, with N energy levels E_i and probability of occupation $p(E_i)$, will reach the lattice temperature T_L. The computation of T_1 is done deriving an equation that describes the way in which the temperature of the spins evolves with time to reach the value T_L (Slichter 1990).

The average energy of the N_s spins, if the probability of occupation of the levels is $p(E_i) = p_i$, is

$$\langle E \rangle_T = \sum_i^N p_i E_i \qquad (7.69)$$

and the rate of variation of this energy is

$$\frac{d\langle E \rangle_T}{dt} = \frac{d\langle E \rangle_T}{d\beta} \frac{d\beta}{dt} \qquad (7.70)$$

where $\beta = 1/kT$. But $d\langle E \rangle_T / dt = \sum_i E_i \, dp_i/dt$, where $dp_i/dt = \sum_j (p_j W_{ji} - p_i W_{ij})$, and W_{ij} is the transition rate from the state i to the state j. Thus

$$\frac{d\langle E \rangle_T}{dt} = \sum_{i,j}^N E_i (p_j W_{ji} - p_i W_{ij}) \qquad (7.71)$$

or, changing the indices

$$\frac{d\langle E \rangle_T}{dt} = \sum_{i,j}^N E_j (p_i W_{ij} - p_j W_{ji}) \qquad (7.72)$$

Summing Eqs. (7.71) and (7.72) and dividing by 2, we obtain

$$\frac{d\langle E\rangle_T}{dt} = \frac{1}{2}\sum_{i,j}^{N}(p_j W_{ji} - p_i W_{ij})(E_i - E_j) \tag{7.73}$$

Assuming a Boltzmann distribution, the probabilities p_i will be given by

$$p_i = \frac{\exp(-\beta E_i)}{\sum_i^N \exp(-\beta E_i)} \tag{7.74}$$

We can expand the exponential:

$$\exp(-\beta E_i) = 1 - \beta E_i + \tfrac{1}{2!}\beta^2 E_i^2 + \cdots \tag{7.75}$$

Therefore, in the limit of high temperature, $p_i \approx (1 - \beta E_i)/N$. In this limit we will have

$$\frac{d\langle E\rangle_T}{d\beta} = \frac{d}{d\beta}\sum_i^N p_i E_i = \frac{1}{N}\frac{d}{d\beta}\sum_i^N (1 - \beta E_i)E_i \tag{7.76}$$

or

$$\frac{d\langle E\rangle_T}{d\beta} = -\frac{1}{N}\sum_i^N E_i^2 \tag{7.77}$$

Using Eq. (7.70), it follows that

$$\frac{d\langle E\rangle_T}{dt} = -\frac{1}{N}\sum_i^N E_i^2 \frac{d\beta}{dt} \tag{7.78}$$

When the spin system is in equilibrium with the lattice (the thermal reservoir), $d\langle E\rangle_T/dt = 0$, and it comes out of Eq. (7.71) that

$$p_j^L W_{ji} = p_i^L W_{ij} \tag{7.79}$$

where p_j^L and p_i^L are the equilibrium probabilities of occupation of the i and j energy levels, that is, the probabilities of occupation of the levels at the lattice temperature.

This is the so-called principle of detailed balance—in equilibrium, the product of the probability of occupation of a level by the probability of transition of this level to another level is equal to the probability of occupation of this other level times the probability of transition in the inverse sense. Then

$$W_{ji} = W_{ij}\frac{p_i^L}{p_j^L} = W_{ij}\exp[-(E_i - E_j)\beta_L] \qquad (7.80)$$

Substituting the expression of W_{ji} given by this equation into Eq. (7.71), we obtain, after expanding the exponential

$$\frac{d\langle E\rangle_T}{dt} = \frac{1}{2N}\sum_{i,j}^{N} W_{ij}(E_i - E_j)^2(\beta - \beta_L) \qquad (7.81)$$

From Eq. (7.78), it follows that

$$\frac{d\beta}{dt} = \frac{1}{2}\frac{\sum_{i,j}^{N} W_{ij}(E_i - E_j)^2}{\sum_i E_i^2}(\beta_L - \beta) \qquad (7.82)$$

This is the rate of change of the spin temperature toward the lattice temperature T_L; this defines the longitudinal relaxation rate $1/T_1$:

$$\frac{d\beta}{dt} \simeq \frac{\beta_L - \beta}{T_1} \qquad (7.83)$$

The time T_1 is thus the time constant of the exponential evolution of the spin temperature. The expression of the relaxation rate $\delta = 1/T_1$ is given by

$$\delta = \frac{1}{T_1} = \frac{1}{2}\frac{\sum_{i,j}^{N} W_{ij}(E_i - E_j)^2}{\sum_i E_i^2} \qquad (7.84)$$

The time T_1 is the characteristic time elapsed until the system of spins establishes thermal equilibrium with the reservoir (e.g., the crystal lattice or the electron gas). In a time of the order of T_1, the occupation probabilities p_i reach the equilibrium values p_i^L.

7.4.2 Transverse Relaxation

The relaxation time T_2, introduced in a phenomenological way into the Bloch equations [Eq. (7.21)], measures the time interval in which the transverse magnetization tends to zero. In the expression of the component $M_y(\omega)$ of the magnetization, T_2 is directly related to the inverse of the width of the resonance line ($T_2 = 2/\Delta\omega_{1/2}$).

The contribution to T_2 due to the fluctuation in the dipolar field of a neighboring nucleus can be estimated, taking as a starting point the fact that this field is approximately given by [from Eq. (6.40)]:

$$\delta B_z^d \cong \frac{\mu_0}{4\pi} \frac{\mu}{r^3} \qquad (7.85)$$

where μ is the nuclear magnetic moment and r, the separation between the nuclei. From Eq. (7.62)

$$T_2 \cong \frac{1}{\gamma \delta B_z^d} = \frac{4\pi}{\mu_0} \frac{r^3}{\gamma \mu} \qquad (7.86)$$

For protons, we have $\gamma = 2.675 \times 10^8$ radians s^{-1} T^{-1} and $\mu = 1.41 \times 10^{-26}$ J T^{-1}; using $r = 2$ Å, we obtain $T_2 \approx 10^{-4}$ s for the relaxation time due to the dipolar contribution of the other nuclei.

The magnitude of the fluctuation in the field δB_z^d obtained in this example is $\approx 10^{-4}$ T, or 1 G. This is therefore the magnetic resonance linewidth expected in a solid, from nuclear dipole–dipole interactions; these interactions are usually the most important source of line broadening in solids. In solids with paramagnetic ions, the dipolar fields due to the atomic moments dominate, and the linewidths are much larger (or T_2 much shorter).

7.4.3 Nuclear Magnetic Relaxation Mechanisms

Longitudinal Relaxation The simplest way to study the mechanisms that give rise to nuclear magnetic relaxation in a solid is to consider the nuclei submitted to an oscillating magnetic field due to the lattice vibrations. The longitudinal component $B_z(t)$ of this field contributes to the relaxation rate $1/T_2$, and the transverse component contributes to $1/T_2$ and $1/T_1$, as seen in Section 7.4.

Taking Eq. (7.84) as a starting point, we may compute the relaxation time T_1 for a given spin system, provided the rate W_{ij} is known; this quantity will depend on the microscopic mechanism that couples the spins to the thermal reservoir (interaction with the electron gas, with phonons, magnons, etc). In nonmagnetic insulators, phonons are the dominant mechanism; in nonmagnetic metals the coupling is done mainly through the conduction electrons. In magnetic matrices the relaxation mechanism may involve magnons.

If we assume that the thermalization of the nuclei occurs through their interaction with the electron gas, the magnetic relaxation rate $1/T_1$ may be calculated from the expression of the transition probability W_{ij}. In this case, W_{ij} is a function of the densities of states of occupied and nonoccupied electronic states:

$$W_{ij} = \sum_{k\sigma_{\text{occ.}}} \sum_{k'\sigma'_{\text{non-occ.}}} W_{jk\sigma, ik'\sigma'} = \sum_{kk'\sigma\sigma'} W_{jk\sigma, ik'\sigma'} f(k\sigma)[1 - f(k'\sigma')] \qquad (7.87)$$

where $f(k\sigma)$ is the Fermi–Durac distribution, k is the electron wavevector, and σ

labels the spin state (up or down). Assuming that the interaction between the nuclei and the electrons is the Fermi contact interaction (see Section 6.3.1)

$$V = \frac{8\pi}{3} \gamma_e \gamma_n \hbar^2 \mathbf{I} \cdot \mathbf{S} \delta(\mathbf{r}) \tag{7.88}$$

where γ_e and γ_m are the electronic and nuclear gyromagnetic ratios, and $W_{jk\sigma,ik'\sigma'}$ is given by

$$W_{jk\sigma,ik'\sigma'} = \frac{2\pi}{\hbar} \frac{64\pi^2}{9} \gamma_e^2 \gamma_n^2 \hbar^4 \sum_{\alpha,\alpha'=x,x'} (i|I_\alpha|j)(j|I_{\alpha'}|i)(\sigma|S_\alpha|\sigma')(\sigma'|S_{\alpha'}|\sigma)$$

$$\times |u_k(0)|^2 |u_{k'}(0)|^2 \delta(E_j - E_i + E_{k\sigma} - E_{k'\sigma'}) \tag{7.89}$$

Substituting in W_{ij} and making the calculations for the case of one electron, we obtain (Slichter 1990):

$$\frac{1}{T_1} = \frac{64\pi^3}{9} \gamma_e^2 \gamma_n^2 \hbar^3 < |u_k(0)|^2 >_{E_F} n^2(E_F) kT \tag{7.90}$$

where $\langle |u_k(0)|^2 \rangle_{E_F}$ is the density of electrons at the Fermi level at $r = 0$, and $n(E_F)$ is the electronic density of states, also at the Fermi level. This gives approximately

$$\frac{1}{T_1} = \frac{\Delta B}{B} \frac{\gamma_n^2}{\gamma_e^2} \frac{4\pi k}{\hbar} T \tag{7.91}$$

This equality is the *Korringa relation*. $\Delta B/B$ is the *Knight shift* (see Section 6.5.4). Note that the preceding result points to a relaxation rate due to conduction electrons proportional to the temperature. This can be simply understood from the fact that W_{ij} is proportional to the function $f(E)[1 - f(E)]$, which, near the Fermi level, is a narrow function, of width proportional to kT.

In ferromagnetic metals, another magnetic relaxation mechanism is the Weger process (see McCausland and Mackenzie 1980, Bobek et al. 1993), which consists in the nuclear relaxation via emission of magnons, that exchange energy with the conduction electrons. The resulting relaxation rate is also proportional to T.

The longitudinal relaxation times in magnetic materials are usually longer in domains than in domain walls; for this reason the measured relaxation times are in general dependent on the rf power used in an NMR experiment (see Chapter 8). The relaxation rates measured at high power in metals (exciting preferentially nuclei in domains) usually present a linear dependence on temperature. This suggests that the relaxation in the domains has strong participation of the conduction electrons.

In those magnetic systems where the spin–spin interaction is weak compared to the spin–lattice interaction (e.g., [57]Fe in domain walls in metallic Fe, [61]Ni in metallic Ni above 77 K), this last interaction dominates the transverse decay, and $1/T_1$ is of the order of $1/T_2$ (Weisman et al. 1973).

Transverse Relaxation As shown at the beginning of Section 7.4, the transverse relaxation rate $1/T_2$ depends on the time fluctuations of both the z component and the transverse components of the magnetic fields acting on the nucleus. The spin–lattice relaxation is a process that contributes to these fluctuations, as can be seen from the relation connecting $1/T_2$ and $1/T_1$ [Eq. (7.60)]. More interesting is the study of the contributions of the spin–spin interaction to $1/T_2$.

In metals, the Ruderman–Kittel interaction, which consists in the coupling of a pair of magnetic moments through the electron gas, is one of the interaction mechanisms: one electron is scattered by a nuclear magnetic moment, then by another, and the information on the spin state of the first nucleus is thus transmitted to the second one. The resulting coupling of the nuclear moments can be described under the form

$$\mathcal{H} = \mathcal{J}_{RK}(r_{ij}) \mathbf{I}_i \cdot \mathbf{I}_j \qquad (7.92)$$

where $\mathcal{J}_{RK}$ is the effective coupling constant. This coupling does not depend on the temperature, since the conduction electrons neither lose nor gain energy, and therefore, do not depend on the existence of empty states to which they can be promoted. This interaction, relevant in the coupling between atomic magnetic moments [known in this case as Ruderman–Kittel–Kasuya–Yosida (RKKY) interaction] is discussed in Section 3.3.

Another mechanism of spin–spin interaction, relevant in the magnetically ordered materials, is the Suhl–Nakamura (SN) interaction. In this interaction, two nuclear magnetic moments are coupled through the hyperfine interaction of two atomic moments, the latter connected through the exchange interaction. We may describe this interaction as a coupling mediated by the emission and absorption of magnons. The transverse components of the angular momentum **I** are more effective in this process. The hamiltonian is written

$$\mathcal{H} = \mathcal{J}_{SN}(r_{ij}) I_i^+ I_j^- \qquad (7.93)$$

It is important to remark the difference between these two interactions; in the Ruderman–Kittel interaction, only the relative orientation of the moments matter. In the Suhl–Nakamura coupling, only the components of **I** transverse to the quantization axis, in this case the direction of magnetization, participate.

The Suhl–Nakamura interaction is independent of temperature, has long range, and is the main mechanism of transverse relaxation in concentrated magnetic systems. For example, in metallic cobalt, the relaxation of ^{59}Co (100% abundant) has an important contribution from the Suhl–Nakamura mechanism.

7.5 DIFFUSION

In pulsed NMR the loss of memory of the transverse magnetization in liquid samples is enhanced by the diffusion of the nuclei toward regions of different magnetic field: this effect is called *spatial diffusion*. This effect gives rise to an attenuation factor that affects the NMR signal (specifically, the spin echo intensity, see next section), proportional to τ^3 (τ is the separation between the two pulses), to the diffusion coefficient D, and to the magnetic field gradient (Carr and Purcell 1954). The spin echo intensity is given in this case by:

$$E(2\tau) = M_0 \exp\left(\frac{-2\tau}{T_2}\right)\left[\exp\left(-D\left(\gamma\frac{\partial B}{\partial z}\right)^2 \frac{2\tau^3}{3}\right)\right] \quad (7.94)$$

We may regard the spatial diffusion as another channel of transverse relaxation, which leads to shorter effective values of T_2.

An analogous phenomenon is observed in solids with inhomogeneously broadened NMR lines. The nuclei excited in a frequency interval $\delta\omega$ of the resonance line, interact with other nuclei, and distribute among these nuclei their magnetic energy. The loss of magnetization of the originally excited nuclei appears as a contribution to the relaxation, and this is referred to as *spectral diffusion*, or *frequency diffusion*.

In NMR experiments in magnetic materials, where the nuclei in domain walls are preferentially excited, this excitation in the domain walls may be transferred to nuclei in adjacent domains, in a special form of spatial diffusion.

7.6 PULSED MAGNETIC RESONANCE

The NMR technique requires the application, on the nuclear magnetic moments, of a static magnetic field and also of a magnetic field that varies sinusoidally with time. The latter may be produced, for example, through the incidence of microwaves. In the pulsed-resonance method, the radiofrequency (rf) is applied during time intervals that are short compared to the characteristic times of the nuclear spin system (the relaxation times T_1 and T_2 of the system). An rf pulse submits the nuclei to a magnetic field of intensity B_1 stationary in the rotating reference system (e.g., parallel to y' and perpendicular to B_0). Such a pulse, with duration t_a, will make the nuclear magnetization precess by an angle θ given by

$$\theta = \gamma B_1 t_a \quad (7.95)$$

where γ is the nuclear gyromagnetic ratio.

The application of a $\pi/2$ pulse (i.e., of such duration that $\theta = \pi/2$) will take the magnetization to the plane xy (Fig. 7.9). In the laboratory reference system the magnetization will perform a motion of nutation; this is a precessional

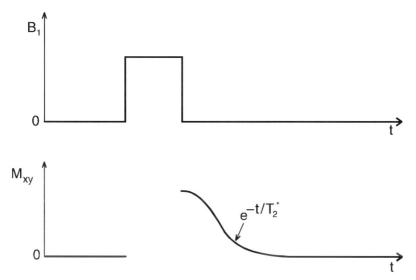

Figure 7.9 Time dependence of the transverse nuclear magnetization after a $\pi/2$ pulse. This is the free induction decay (FID); the magnetization decays exponentially with a time constant T_2^*.

motion with a variable angle with the axis of precession. In a reference system rotating at the Larmor frequency, the magnetization will precess around $\mathbf{B}_1$ (e.g., assumed to be parallel to y'). After a $\pi/2$ pulse, the transverse magnetization is maximum; the magnetization that exists after this pulse is over, is called *free induction*. We may also use this expression to refer to the signal induced by the transverse magnetization on a coil.

The transverse magnetization remaining after the application of the pulse, or free induction, decays with time, since the motions of the individual magnetic moments lose coherence under the action of two factors: (1) each moment feels a magnetic field that varies randomly with time (due to fluctuations in $\mathbf{B}$, especially arising from the other moments), and (2) the moments may feel different magnetic fields due to spatial field inhomogeneity. This free induction decay (FID) has a characteristic time T_2^* given by

$$\frac{1}{T_2^*} = \frac{1}{T_2} + \gamma \, \Delta B \tag{7.96}$$

where T_2 is the spin–spin relaxation time strictly considered (the term due to fluctuations in the field) and ΔB is the field inhomogeneity.

The term T_2 gives a measure of the transverse relaxation inside each set of magnetic moments that precess with the same angular frequency ω_n (these sets are called *isochrones* or *isochromats*). The term $1/\gamma \Delta B$ is due to the field inhomogeneity; ΔB is the width of the distribution of values of B. All the relaxation processes (spin–spin, but also spin–lattice; see Section 7.4) that affect

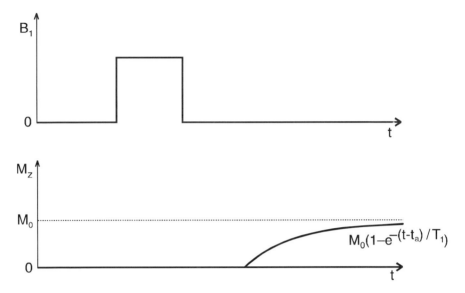

Figure 7.10 Time evolution of the longitudinal nuclear magnetization after a pulse of $\pi/2$, showing the exponential recovery of the magnetization. In a time T_1 after the pulse, the magnetization differs by M_0/e from the equilibrium value M_0.

the lifetime of the nuclear Zeeman levels, contributing an energy uncertainty $\Delta E \approx \hbar/T_2$, are included in T_2; in this way, T_2 contains the contributions of the so-called homogeneous broadening. The term $1/T_2$ is the thermodynamically irreversible contribution to $1/T_2^*$, and $\gamma \Delta B$ is a reversible term (see discussion below).

In a long timescale (such that $t \gg T_2$ and $t \geq T_1$), the longitudinal component M_z of the magnetization returns to its equilibrium value M_0 (Fig. 7.10). In the rotating system, after a pulse of $\pi/2$, the magnetization, for $t > t_a$ and $\mathbf{B}_1 = B_1 \mathbf{i}'$, will be

$$\mathbf{M}'(t) = M_0 \left[e^{-(t-t_a)/T_2^*} \mathbf{j}' + (1 - e^{-(t-t_a)/T_1}) \mathbf{k}' \right] \tag{7.97}$$

In the laboratory, we will have

$$\mathbf{M}(t) = M_0 \left[(1 - e^{-(t-t_a)/T_1}) \mathbf{k} + \sin \omega t \, e^{-(t-t_a)/T_2^*} \mathbf{j} + \cos \omega t \, e^{-(t-t_a)/T_2^*} \mathbf{i} \right] \tag{7.98}$$

In the absence of magnetic field inhomogeneity, the transverse nuclear magnetization decays with the characteristic time T_2; in inhomogeneous magnetic fields, when $1/\gamma \Delta B \ll T_2$, it decays with $T_2^* = 1/\gamma \Delta B$. In zero-field NMR in magnetic materials (see Chapter 8), inhomogeneous broadening is important, and the last situation is the most usual.

The total magnetic induction, due to the ensemble of nuclear moments, is

$E(t)$. It results, at each instant, from the precession of the sum of the projections, on the rotating y' axis, of the magnetizations of the isochrones.

The expression of the decay of the free induction is obtained from the contribution to the transverse magnetization due to one isochrone, or packet of spins (e.g., Borovik-Romanov et al. 1984), with $\Delta\omega = \omega_n - \omega$ (difference between the frequency of the spin packet and the frequency of the applied rf). The magnetic moment of one isochrone is

$$m(\Delta\omega) = m_0 F(\Delta\omega)\delta\Delta\omega \qquad (7.99)$$

where $F(\Delta\omega)$ is the NMR spectrum shape, and $\delta\Delta\omega$ is the frequency width of the spin packet.

Immediately after application of an rf pulse of duration t_a along the x' axis, the transverse magnetization due to one isochrone will be

$$m_y(\Delta\omega) = m(\Delta\omega) \sin(\gamma B_1 t_a) \qquad (7.100)$$

The magnetic moments begin to dephase, and there appears an x component of the magnetization. The perpendicular magnetization will be given in complex form by

$$m_\perp = m_y - im_x \qquad (7.101)$$

Including the time evolution, each isochrone contributes to the perpendicular magnetization

$$m_\perp(\Delta\omega, t) = m(\Delta\omega) \sin(\gamma B_1 t_a) \exp(-i\Delta\omega t) \qquad (7.102)$$

where we have substituted

$$\exp(-i\Delta\omega t) = \cos(\Delta\omega t) - i\sin(\Delta\omega t) \qquad (7.103)$$

To obtain the total free induction, it is necessary to sum over all the spin packets, that is, to sum over a frequency interval $\Delta\omega$ where all the isochromats that have been excited are included. When this range of excitation is so large that it encompasses the whole NMR spectrum, the sum is made over the spectrum shape $F(\Delta\omega)$, and one can integrate from $-\infty$ to $+\infty$:

$$m(t) = m_0 \sin(\gamma B_1 t_a) \int_{-\infty}^{\infty} F(\Delta\omega) \exp(-i\Delta\omega t) d\Delta\omega \qquad (7.104)$$

The integral in this equation is equal to the Fourier transform $G(t)$ of the lineshape, thus, the time dependence of the magnetization that gives the free induction can be written

$$m(t) = m_0 \sin(\gamma B_1 t_a) G(t) \tag{7.105}$$

If the lineshape is a lorentzian, with half-width $1/T_2^*$, the function $F(\Delta\omega)$ has the form

$$F(\Delta\omega) = \frac{1}{\pi} \frac{1/T_2^*}{(\Delta\omega - \Delta\omega_0)^2 + (1/T_2^*)^2} \tag{7.106}$$

In this case the Fourier transform is $G(t) = \exp(-t/T_2^*)$, and we will have for the free induction decay

$$m(t) = m_0 \sin(\gamma B_1 t_a) \exp(-t/T_2^*) \tag{7.107}$$

which is the equation of an exponential decay with characteristic time T_2^*; an exponential decay with characteristic time corresponding to the inverse of the half-width of the NMR line is strictly correct only for the conditions of the present derivation.

The application of a sequence of two consecutive pulses leads to a new effect—the spin echo—discovered by E. Hahn in 1950. Let us exemplify with a pulse sequence $(\pi/2, \pi)$ applied along the same axis y', separated by a time interval τ (Fig. 7.11). Once the time τ has elapsed after the first pulse ($\tau \gg T_2^*$), the transverse magnetization vanishes completely. The inversion of the spin "pancake," due to the action of the second pulse, places the spins that precess more rapidly behind the slower spins. Since the sense of precession does not

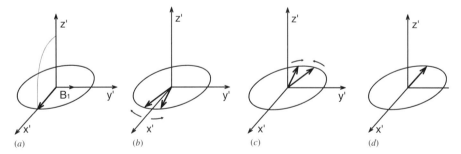

Figure 7.11 Formation of the spin echo after a sequence of a $\pi/2$ pulse and a π pulse, separated by a time interval τ. After the $\pi/2$ pulse is applied in the y' direction (a), the nuclear magnetization aligns itself with the x' axis. Because of the inhomogeneity of the static field B_0 and the spin–spin interaction, the isochrones precess with different angular velocities, and the magnetization in the $x'y'$ plane decays to zero, with a characteristic time T_2^* (b). In succession, a π pulse is applied (c), and the spin packets refocus at the time 2τ (d). The maximum in the transverse magnetization is the spin echo.

change with this inversion—because it depends only on the direction of $\mathbf{B}_0$—after a second interval of duration τ all the spins refocus, producing a maximum in the transverse magnetization. This ensuing magnetization maximum constitutes the spin echo.

The transverse magnetization that contributes to the spin echo at the frequency ω, due to a spin packet that precesses at the frequency ω_n ($\Delta\omega = \omega_n - \omega$), is given by (Hahn 1950):

$$m_\perp(\Delta\omega, t) = m(\Delta\omega) \sin(\gamma B_1 t_a) \sin^2 \frac{\gamma B_1 t_b}{2} \exp[-i\Delta\omega(t - 2\tau)] \quad (7.108)$$

From this expression one can see that the echo is a nonlinear function of the angles of rotation $\theta_a = \gamma B_1 t_a$ and $\theta_b = \gamma B_1 t_b$ due to the two pulses. One can also note that the echo occurs at $t = 2\tau$. The maximum echo is obtained for $\theta_a = \pi/2$ and $\theta_b = \pi$. For two equal pulses of duration t_a (i.e., $t_b = t_a$), we see that the maximum echo will arise when $\theta = 2\pi/3$. For small values of θ, the echo is proportional to B_1^3.

To obtain the total transverse magnetization, we have to sum over all the isochromats, that is, over the NMR spectrum shape [see remark on the derivation of the FID, before Eq. (7.104)]:

$$m(t) = m_0 \sin(\gamma B_1 t_a) \sin^2(\gamma B_1 t_b/2) \int_{-\infty}^{+\infty} F(\Delta\omega) \exp[-i\Delta\omega(t - 2\tau)] d\Delta\omega \quad (7.109)$$

The result is the transverse magnetization at resonance after the second pulse (i.e., for $t \geq \tau$), in the rotating system (Hahn 1950):

$$m(t) = m_0 \sin(\omega_1 t_a) \sin^2 \frac{\omega_1 t_a}{2} \exp\left[\frac{-(t - 2\tau)^2}{2T_2^{*2}}\right]$$

$$- \cos^2 \frac{\omega_1 t_a}{2} \exp\left(\frac{-t^2}{2T_2^{*2}}\right) \exp\left(\frac{-kt^3}{3} - \frac{t}{T_2}\right) + m'(t) \quad (7.110)$$

with $\omega_1 = \gamma B_1$, and k a parameter proportional to the diffusion coefficient D (see Section 7.5). One can see from the first term that the echo is formed at the instant $t = 2\tau$, with a gaussian shape and a width at half-height of $\approx 2T_2^*$. The term $m'(t)$ describes the free induction decay:

$$m'(t) = -m_0 \sin(\omega_1 t_a) \left[-\cos^2 \frac{\omega_1 t_a}{2} \exp\left(\frac{-t^2}{2T_2^{*2}}\right) \exp\left(\frac{-kt^3}{3} - \frac{t}{T_2}\right) \right]$$

$$- M_z(\tau) \sin(\omega_1 t_a) \exp\left[-\frac{1}{2}\left(\frac{t-\tau}{T_2^*}\right)^2 - \frac{t-\tau}{T_2} - \frac{k(t-\tau)^3}{3} \right] \quad (7.111)$$

From this expression we can see (in the first term) the decay of the free induction (FID) with characteristic time T_2^*.

The results shown above can be obtained more directly by solving the Bloch equations with a matrix method (Jaynes 1955, Bloom 1955).

As the echo is formed, the refocusing of the transverse magnetization is not complete, however; only the loss of memory due to the inhomogeneity, not the loss due to the spin–spin interaction, is recovered. This interaction produces random magnetic field fluctuations, and therefore the resulting decay (with characteristic time T_2) is irreversible.

The spin echo is observed experimentally as the electromotive force (or the voltage) induced in a coil wound around the sample where the resonance is being observed; its magnitude is proportional to the time derivative of the magnetic flux due to the precessing moments. The echo signal is, therefore

$$E(t) = c\ \omega m(t) \tag{7.112}$$

where c is the constant that takes into account parameters such as the quality factor Q and the filling factor of the coil. In a magnetic sample, the signal due to the precession of the magnetization $m(t)$ is multiplied by the factor η, the enhancement factor (to be discussed in Section 8.3). This factor η amplifies the rf field B_1 felt by the nucleus, as well as the magnetic flux through the coil, due to the precession of the nuclear magnetic moments.

A remarkable aspect of the spin echo technique is the possibility of studying physical processes with characteristic time T_2, even in the presence of a much faster relaxation, with rate $1/T_2^* \approx \gamma \Delta B$. In the continuous-wave (CW) technique, a T_2 process leads to a line broadening that is masked, in the case of a magnetic sample, by a much larger broadening due to the inhomogeneity (with which one associates the rate $\gamma \Delta B$). The pulsed technique allows the direct study of the relevant relaxation rates, instead of inferring them from the widths of static measurements, which may lead to errors due to the presence of the inhomogeneous broadening. The half width derived from a CW experiment would be equal to $1/T_2^*$ [Eq. (7.106)], containing, therefore, the sum of a homogeneous broadening (corresponding to the rate $1/T_2$) with an inhomogeneous broadening (rate $\gamma \Delta B$).

The spin echo technique also presents an advantage of practical character in relation to other pulse techniques; it involves measurement of signals after time intervals sufficiently long to allow the decay of the instrumental perturbation caused by the excitation pulses.

The large potential of the pulsed NMR technique comes from the fact that the signal in the time domain (either the free induction decay or the spin echo) can be easily Fourier-transformed to give the frequency spectrum. This is the basis of operation of Fourier transform NMR. The spin echoes or the FIDs contain contributions from a wide range of nuclear precessing frequencies because the excitation pulses have a broad Fourier frequency spectrum. For a pulse of width t_a and frequency ν_0, this spectrum has a half-width of the order of $2\pi/t_a$ around

ν_0. If the magnetic field inhomogeneity is not larger than this width, the full frequency spectrum can be obtained by Fourier transforming.

Besides the spin echo described here, known as the *Hahn echo*, other echoes have been observed; for example, a pulse applied at a time T after the pair of pulses, will produce an echo at time $T + \tau$, known as the *stimulated echo*.

7.7 QUADRUPOLE OSCILLATIONS

In NMR experiments where electric quadrupole interactions are combined with the dominant magnetic interactions (Section 6.7), the NMR spectrum has $2 \times I$ lines, where I is the spin of the nucleus. The lines are separated by a frequency interval $\Delta\nu = 2a$, with a, the quadrupole interaction parameter, given by

$$a = \frac{3e^2qQ}{4I(2I-1)} \frac{1}{2}(3\cos^2\theta - 1) \qquad (7.113)$$

which depends on the electric field gradient eq ($\equiv V_{zz}$), the nuclear quadrupole moment Q, and the angle θ between the major axis of the electric field gradient and the hyperfine field; an axial EFG was assumed.

In pulsed NMR in systems that present only magnetic interactions, a spin echo is observed at $t = 2\tau$ (Section 7.6); when there are also quadrupole interactions, other echoes may appear, depending on the nuclear spin I and the degree of homogeneity of the magnetic interaction (Butterworth 1965).

When the linewidth is such that the $2I$ lines cannot be resolved, there appear oscillations in the echo amplitude as a function of the separation τ between the pulses (Abe et al. 1966). These oscillations may be interpreted as beats between the transition frequencies between unequally spaced nuclear Zeeman levels, an effect of the quadrupole interactions. The frequency of the oscillations is given by multiples of $\nu_0 = a/\pi$, where a is the quadrupole interaction parameter.

Abe and collaborators have computed the amplitudes of the quadrupole oscillations for the case $I = \frac{5}{2}$, using a perturbative method, and starting from the hamiltonian

$$\mathcal{H} = \mathcal{H}_0 + \mathcal{H}_{\text{int}} \qquad (7.114)$$

formed by a static part and a time-dependent part. The static part, with $\omega = \gamma B$ (see Section 6.7), is given by a magnetic term and an electrostatic term

$$\mathcal{H}_0 = (\omega_0 - \omega)I_z + aI_z^2 \qquad (7.115)$$

and the time-dependent part takes a different form in the following time regions of the NMR experiment (Fig. 7.12): (1) when the first rf pulse is on (region I), (2) between the two pulses (region II), (3) during the second pulse (III), and

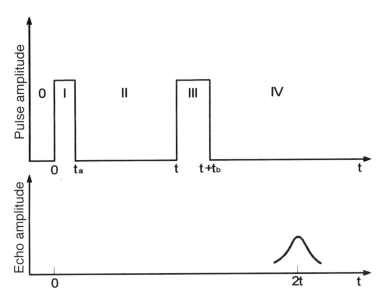

Figure 7.12 Time regions in a spin echo NMR experiment. Region I, application of the first rf pulse; region II, interval between the pulses; region III, application of the second pulse, and finally, region IV, evolution of the magnetization with the formation of the spin echo.

(4) after the complete sequence (region IV). The time dependent part is written

$$\mathcal{H}_{\text{int}} = \omega_1 I_z \quad \text{in regions I and III} \tag{7.116a}$$

$$\mathcal{H}_{\text{int}} = 0 \quad \text{in regions II and IV} \tag{7.116b}$$

The results show multiple echoes: one echo at time 2τ (Hahn echo); other echoes at 3τ, 4τ, 5τ, and 6τ; and oscillating amplitudes for the different echoes. Leaving aside the exponential decay with constant T_2, one obtains for the first echo an amplitude, in the case $I = \frac{5}{2}$ given by (Abe 1966):

$$E^{(1)}(2\tau) = C_0^{(1)} + C_1^{(1)} \cos(2a\tau + \delta_1) + C_2^{(1)} \cos(4a\tau + \delta_2)$$
$$+ C_3^{(1)} \cos(6a\tau + \delta_3) + C_4^{(1)} \cos(8a\tau + \delta_4) \tag{7.117}$$

where the δ_i are phase angles. The second echo is given by

$$E^{(2)}(3\tau) = C_1^{(2)} \cos(2a\tau + \delta_1) + C_3^{(2)} \cos(6a\tau + \delta_3)$$
$$+ C_5^{(2)} \cos(10a\tau + \delta_5) + C_7^{(2)} \cos(14a\tau + \delta_7) \tag{7.118}$$

where a is the electric quadrupole interaction parameter [Eq. (7.113)], the C

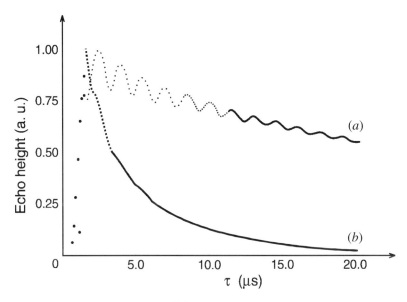

Figure 7.13 Quadrupole oscillations in ^{59}Co spin echo height, versus separation τ between pulses, in the compound GdCo$_2$, at (a) 4.2 K, and (b) 38 K. [Reprinted from A. C. Barata and A. P. Guimarães, *Physica* **130B**, 485 (1985), with permission from Elsevier North-Holland, NY.]

values are coefficients that depend on the matrix elements of the interaction matrix. From the preceding expressions one can see that the echoes present oscillations as a function of the time separation τ between the pulses, and the value of the quadrupole interaction may be extracted from the spectrum of these oscillations (see also Fig. 7.13). This is a useful feature of these oscillations, since the quadrupole interaction may be measured in this way, even in cases where the quadrupole lines in the NMR spectrum are not resolved.

GENERAL READING

Abragam, A., *Principles of Nuclear Magnetism*, Oxford Univ. Press, Oxford, 1961.

Ailion, D. C. and W. D. Ohlsen, "Magnetic Resonance Methods for Studying Defect Structure in Solids," in J. N. Mundy, S. J. Rothman, and L. C. Smedskjaer, Eds., *Methods of Experimental Physics*, Vol. 21, Solid State: Nuclear Methods, Academic Press, Orlando, 1983.

Aleksandrov, I. V., *The Theory of Nuclear Magnetic Resonance*, Academic Press, New York, 1966.

Carrington, A. and A. D. McLachlan, *Introduction to Magnetic Resonance*, Harper, New York, 1967.

Farrar, C. and E. D. Becker, *Pulse and Fourier Transform NMR*, Academic Press, New York, 1971.

Fukushima, E. and S. B. W. Roeder, *Experimental Pulse NMR: A Nuts and Bolts Approach*, Addison-Wesley, Reading, MA, 1981.

Gerstein, B. C. and C. R. Dybowski, *Transient Techniques in NMR of Solids*, Academic Press, Orlando, 1985.
Mims, W. B., "Electron Spin Echoes," in S. Geschwind, Ed., *Electron Paramagnetic Resonance*, Plenum Press, New York, 1972.
Poole, C. P. and H. A. Farach, *Relaxation in Magnetic Resonance*, Academic Press, New York, 1971.
Slichter, C. P., *Principles of Magnetic Resonance*, 3rd ed., Springer-Verlag, Berlin, 1990.

REFERENCES

Abe, H., H. Yasuoka, and A. Hirai, *J. Phys. Soc. Japan* **21**, 77 (1966).
Barata, A. C. and A. P. Guimarães, *Physica* **130B**, 484 (1985).
Bloch, F., *Phys. Rev.* **70**, 460 (1946).
Bloom, A. L., *Phys. Rev.* **98**, 1105 (1955).
Bobek, C., R. Dullenbacher, and E. Klein, *Hyp. Int.* **77**, 327 (1993).
Borovik-Romanov, A. S., Yu. M. Bun'kov, B. S. Dumesh, M. I. Kurkin, M. P. Petrov, and V. P. Chekmarev, *Sov. Phys. Usp.* **27**, 235 (1984).
Butterworth, J., *Proc. Phys. Soc.* **86**, 297 (1965).
Carr, H. Y. and E. M. Purcell, *Phys. Rev.* **94**, 630 (1954).
Carrington, A. and A. D. McLachlan, *Introduction to Magnetic Resonance*, Harper, New York, 1967.
Hahn, E. L., *Phys. Rev.* **80**, 580 (1950).
Jaynes, E. T., *Phys. Rev.* **98**, 1099 (1955).
McCausland, M. A. H. and I. S. Mackenzie, *Adv. Phys.* **28**, 305 (1979), also in *Nuclear Magnetic Resonance in Rare Earth Metals*, Taylor & Francis, London, 1980.
Slichter, C. P., *Principles of Magnetic Resonance*, 3rd ed., Springer-Verlag, Berlin, 1990.
Weisman, I. D., L. J. Swartzendruber, and L. H. Bennett, "Nuclear Magnetic Resonance in Ferromagnetic Materials (FNR)," in E. Passaglia, Ed., *Techniques of Metal Research*, Vol. VI, Part 2, Interscience, New York, 1973, p. 336.

EXERCISES

7.1 *NMR Spectrum of ^{157}Gd.* Consider the nucleus of the isotope ^{157}Gd that possesses spin $I = \frac{3}{2}$ and $\mu = -0.3398\ \mu_N$. Let a hyperfine field $B_{hf} = 10$ T act on the nucleus. Compute the NMR frequency (in MHz) from the condition $\omega_0 = \gamma B_{hf}$. Assume that, in addition to the magnetic interaction, there is an electric quadrupole interaction given by Eq. (6.23). Make a sketch showing how the energy levels will be modified by a quadrupole interaction that corresponds to 1% of the magnetic interaction, and also the NMR spectrum in the cases in which (a) the linewidth is smaller than the quadrupole splitting and (b) the linewidth is larger than the quadrupole splitting.

7.2 *Rotating Reference System.* Take the vector $\mathbf{A}(t) = A_x(t)\hat{\mathbf{x}} + A_y(t)\hat{\mathbf{y}} + A_z(t)\hat{\mathbf{z}}$. Assume that the coordinate system $\hat{\mathbf{x}}$, $\hat{\mathbf{y}}$, $\hat{\mathbf{z}}$ rotates with angular velocity Ω.

(a) Show that $d\hat{x}/dt = \Omega_y\hat{z} - \Omega_z\hat{y}$, and so on.

(b) Show that $d\mathbf{A}/dt = (d\mathbf{A}/dt)_G + \times \mathbf{A}$, where $(d\mathbf{A}/dt)_G$ is the derivative of $\mathbf{F}$ seen from the rotating system G.

(c) If $\mathbf{\Omega} = -\gamma B_0\hat{z}$, where B_0 is a magnetic field applied in the laboratory system, in the rotating system there will be no static field. Assume that a field B_1 is applied along the $\hat{x}$ direction during a time interval t. Assuming that the magnetization points initially along $\hat{z}$, find an expression for the time t at the end of which $\mathbf{M}$ will point along $-\hat{z}$ (neglect relaxation effects).

7.3 *Ferromagnetic Resonance.* Consider a spherical sample of a ferromagnet with anisotropy energy of the form $U_K = -K\sin^2\theta$ where θ is the angle between the magnetization M_s and the axis z. Assume that K is positive. Show that in the presence of an external field $B_0\hat{z}$ the system will have only one resonance frequency given by $\omega_0 = \gamma(B_0 + B_a)$ where $B_a = 2K/M_s$.

7.4 *RF Saturation.* Let a system with two energy levels of a spin in a magnetic field $B_0\hat{k}$ be in equilibrium at the temperature T. Let N_1 and N_2 be the respective populations of the two levels and W_{12}, W_{21} the transition rates $1 \to 2$ and $2 \to 1$. An rf signal is applied in such way as to induce a transition rate W_{rf} between the two levels.

(a) Derive an equation for dM_z/dt and show that in the stationary state

$$M_z = \frac{M_0}{1 + 2W_{rf}T_1}$$

where $1/T_1 = W_{12} + W_{21}$. Note that as $2W_{rf}T_1 \ll 1$, the rf does not appreciably modify the populations of the two levels.

(b) Define $N = N_1 + N_2$, $n = N_1 - N_2$ and $n_0 = N(W_{21} - W_{12})/(W_{21} + W_{12})$, and from the expression of n write the rate of energy absorption from the rf field. What happens when W_{rf} approaches $1/2T_1$? This effect is called *saturation* and may be used to measure T_1.

8

MAGNETIC RESONANCE IN MAGNETIC MATERIALS

8.1 NUCLEAR MAGNETIC RESONANCE

The nuclear magnetic resonance (NMR) in magnetically ordered materials—also called *ferromagnetic nuclear resonance* (FNR)—differs from the magnetic resonance observed in diamagnetic and paramagnetic materials in several aspects. These differences arise from the fact that in ordered materials there are two magnetic species in interaction: atomic nuclei and magnetic ions. NMR in magnetic materials is, therefore, more complex, being essentially a resonance of two coupled spin systems. It may be described through a pair of coupled Bloch equations: one equation for the nuclear magnetic moments, and another for the atomic moments (see Section 7.2).

The phenomenology of this type of nuclear magnetic resonance presents the following differences in relation to the usual NMR:

1. *Hyperfine Fields.* In the magnetically ordered materials (ferromagnets and antiferromagnets), the atomic nuclei are under the action of local static magnetic fields, roughly proportional to the spontaneous magnetization of the sublattice where they are located. These fields arise essentially from the hyperfine interactions (see Chapter 6), and allow the observation of nuclear magnetic resonance without requiring external applied magnetic fields.

2. *Enhancement (or Amplification) of the Radiofrequency Field.* In magnetically ordered materials, the time-dependent transverse magnetic field felt by the nuclei is much larger than the oscillating field $B_1(t)$ applied to the sample. This effect can be explained as follows. Inside the domains, the field $B_1(t)$ tilts the

magnetization M from its equilibrium direction, and the transverse component of the hyperfine field ($\propto M$) is many (10–100) times larger than the intensity of $B_1(t)$. Inside the domain walls B_1 is also amplified by a factor of 10^3–10^5; this arises from the displacement of the walls by the rf field, and the consequent change in the direction of the hyperfine fields that act on the nuclei in this region, also generating the appearance of large transverse oscillating components of the hf field (see Section 8.3).

3. *Linewidth.* The NMR linewidths in magnetic materials are usually some orders of magnitude larger than those found in diamagnetic matrices. This is due to the distribution of hyperfine fields (and demagnetizing fields) naturally found in magnetic samples; this effect is known as *inhomogeneous broadening*. Electric quadrupole interactions also contribute to the linewidth.

4. *Effects of Spin Waves.* Contrary to what occurs in nonmagnetically ordered matrices, where under an external magnetic field one observes the

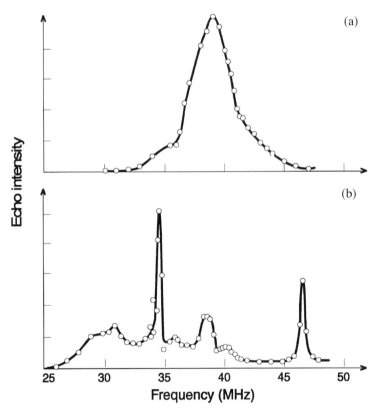

Figure 8.1 NMR spectra of the amorphous alloy $Fe_{86}B_{14}$ at 4.2 K: (*a*) as-quenched, and (*b*) annealed at 420°C (for 40 min). The spectra show ^{11}B resonances; a ^{57}Fe resonance appears near 47 MHz in the spectrum of the annealed alloy. [Reprinted from Y. D. Zhang, J. I. Budnick, J. C. Ford, and W. A. Hines, *J. Mag. Mag. Mat.* **100**, 31 (1991), with permission from Elsevier North-Holland, NY.]

precession in phase of the atomic magnetic moments (uniform mode), in magnetic materials we have spatially nonuniform oscillations: the spin waves.

Two nuclei, coupled to the magnetic moments of the corresponding atoms via the hyperfine interaction, may couple to one another through the spin waves. In the language of quantum mechanics, this indirect interaction is due to the virtual emission and absorption of magnons. This indirect process represents an additional contribution to the nuclear magnetic relaxation, and is observed only in magnetically ordered materials (the Suhl–Nakamura effect). Other effects, including a shift in the nuclear resonance frequency (dynamic frequency shift, or frequency pulling), can be explained in terms of spin wave interactions.

8.1.1 NMR Studies of Magnetically Ordered Solids

The NMR technique has been applied to the study of many magnetically ordered matrices. The large number of NMR nuclides and the possibility of probing the atomic environment in an atomic scale has stimulated many investigations in magnetism. This is illustrated in the zero-field NMR spectra of Figs. 8.1–8.3; in every case graphs show intensity of the NMR signal versus frequency. Figure 8.1 shows the NMR spectrum of an amorphous alloy of FeB. Figure 8.2 depicts a ^{147}Sm NMR spectrum of Sm_2Fe_{17} at 4.2 K showing the electric quadrupole split septet, since for ^{147}Sm $I = \frac{7}{2}$. Finally, Fig. 8.3 shows a ^{59}Co spectrum of Co/Cu multilayers; there are seven lines identified in the spectrum, each one corresponding to a specific environment of the Co atoms. The main line appears at a frequency close to that of bulk fcc Co (217 MHz); the small shift is due to strain in the film.

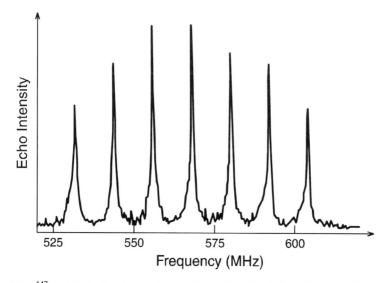

Figure 8.2 ^{147}Sm NMR spin echo spectrum at 4.2 K of the rhombohedral intermetallic compound Sm_2Fe_{17}. [Reprinted with permission from Cz. Kapusta, J. S. Lord, G. K. Tomka, P. C. Riedl, and K. H. J. Buschow, *J. Appl. Phys.* **79**, 4599 (1996). Copyright © 1996, American Institute of Physics.]

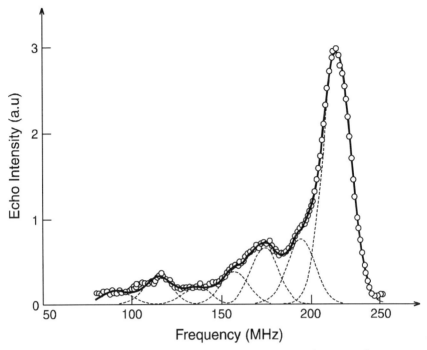

Figure 8.3 ^{59}Co NMR spectrum of a Co/Cu multilayer [40 ×(12.3 Å Co + 42 Å Cu)] at 1.4 K showing the lines corresponding to the different atomic environments of the Co atoms. The more intense line comes from fcc Co in the interior of the film; its frequency is slightly shifted from that of bulk Co (217 MHz) as a result of strain. [Reprinted from H. A. M. de Gronckel et al., *Phys. Rev.* **B44**, 911 (1991).]

Some values of magnetic hyperfine fields measured by NMR in magnetically ordered metals and intermetallic compounds are given in Table 8.I.

8.2 RESONANCE IN A COUPLED TWO-SPIN SYSTEM

Under the action of a radiofrequency field, magnetically ordered matrices may exhibit the phenomenon of magnetic resonance, with the participation of two magnetic species: the magnetic ions and the atomic nuclei. These species interact through the hyperfine interaction. NMR in magnetically ordered matrices can thus be described in a simple fashion by a system of coupled Bloch equations; each equation describes the motion of one type of magnetic moment. If the magnetizations of these two magnetic species, nuclear and ionic are, respectively, **m** and **M**, we will have, in the laboratory system

$$\frac{d\mathbf{m}}{dt} = \gamma_n \mathbf{m} \times \mathbf{b} + \mathbf{r} \qquad (8.1a)$$

Table 8.1 NMR frequencies and hyperfine fields B_{hf} of some nuclides in different magnetically ordered matrices, measured at low temperature

Nuclide	Matrix	Frequency (MHz)	B_{hf} (T)	Temperature (K)
$^{11}B^a$	Fe$_2$B	40.2	2.94	4.2
^{55}Mn^b	YMn$_2$	118.0, 130.0	11.24; 12.38	4.2
^{55}Mn^b	GdMn$_2$	13.6	12.9	4.2
^{57}Fe^c	Fe metal	46.67	33.93	4.2
^{57}Fe^b	YFe$_2$	29.36	21.34	4.2
^{57}Fe^d	GdFe$_2$	31.55, 33.46, 35.50	22.94; 24.33; 25.81	4.2
^{59}Co^c	fcc Co	—	21.73	0
^{59}Co^c	hcp Co	—	22.80	0
^{59}Co^d	GdCo$_2$	61.6	6.13	4.2
^{61}Ni^c	Ni metal	28.46	7.491	4.2
^{61}Ni^c	NiFe$_3$	63.5	16.7	1.4
^{89}Y^b	YFe$_2$	45.94	22.02	4.2
^{89}Y^b	YCo$_5$	—	10.17	4.2
^{143}Nd^b	NdAl$_2$	786	339	1.4
^{147}Sm^b	SmCo$_2$	—	318.8	1.4
^{147}Sm^b	SmFe$_2$	—	304.2	1.4
^{147}Sm^e	Sm$_2$Fe$_{17}$	568.3	323.3	4.2
^{155}Gd^c	Gd metal	30.7	23.5	4.2
^{155}Gd^d	GdFe$_2$	56.64	43.33	4.2
^{159}Tb^f	Tb metal	3120	307.9	4.2
^{163}Dy^c	Dy metal	1163	572.9	1.4
^{163}Dy^b	DyAl$_2$	1183.5	583.00	1.4
^{165}Ho^f	Ho metal	6467	726.5	4.2
^{165}Ho^g	HoFe$_2$	6933	778.9	1.4
^{165}Ho^g	Ho$_1$Gd$_{99}$Fe$_2$	7015	788.1	1.4
^{167}Er^f	Er metal	913	748	4.2

General note: The values of B_{hf} have been calculated from the resonance frequencies, using the factors $\gamma/2\pi$ from Dormann (1991).

^{a}Zhang et al. (1991);
^{b}Dormann (1991);
^{c}Weisman et al. (1973);
^{d}Tribuzy and Guimarães (1977);
^{e}Kapusta (1996);
^{f}McCausland and Mackenzie (1979);
^{g}Guimarães (1971).

$$\frac{d\mathbf{M}}{dt} = \gamma_e \mathbf{M} \times \mathbf{B} + \mathbf{R} \qquad (8.1b)$$

where **r** and **R** are the corresponding relaxation terms, **b** and **B** are the magnetic fields acting on each of the two species, and γ_n and γ_e are the respective gyromagnetic ratios.

Solving this system of differential equations, we may obtain the transverse magnetization, and from it the transverse susceptibility measured in a magnetic resonance experiment. We may also show that the coupling of the two resonant systems leads to an enhancement, or amplification, of the applied radiofrequency field (e.g., Turov and Petrov 1972).

For this purpose we will assume that in equilibrium we have $\mathbf{m}(t) = \mathbf{m}_0 e^{i\omega t}$ and $\mathbf{M}(t) = \mathbf{M}_0 e^{i\omega t}$, and also consider the magnetic fields acting on the nuclei and ions given within the molecular field approximation (see Section 2.6) by

$$\mathbf{b} = \mathbf{B}_0 + \mathbf{B}_1 e^{i\omega t} + \lambda_m \mathbf{M} \tag{8.2a}$$

$$\mathbf{B} = \mathbf{B}_0 + \mathbf{B}_a + \mathbf{B}_1 e^{i\omega t} + \lambda_m \mathbf{m} \tag{8.2b}$$

where

$\mathbf{B}_a$ = anisotropy field acting on the ions (in the z direction)
$\mathbf{B}_0$ = external field (in the z direction)
λ_m = molecular field parameter
$\mathbf{B}_1$ = circularly polarized rf field.

The molecular field felt by the nuclei is identified to the magnetic hyperfine field:

$$\mathbf{B}_{\mathrm{hf}} = \lambda_m \mathbf{M} \tag{8.3}$$

We are not considering demagnetizing fields that would contribute to $\mathbf{B}$. The magnetizations in the x–y plane can be written

$$\mathbf{M}_\pm = M_x \pm i M_y \tag{8.4a}$$

$$\mathbf{m}_\pm = m_x \pm i m_y \tag{8.4b}$$

The components of the magnetic field are

$$\mathbf{B}_\pm = \mathbf{B}_{1\pm} + \lambda_m \mathbf{m}_\pm \tag{8.5a}$$

$$\mathbf{b}_\pm = \mathbf{B}_{1\pm} + \lambda_m \mathbf{M}_\pm \tag{8.5b}$$

$$B_z = B_0 + B_a + \lambda_m m \tag{8.5c}$$

$$b_z = B_0 + \lambda_m M \tag{8.5d}$$

where we have taken $m_z \approx m$ and $M_z \approx M$ (the equilibrium values of the magnetizations). Although the nuclear magnetization is normally far from saturation, this approximation can be justified (see de Gennes 1963). In the stationary regime, $\mathbf{M}_\pm = \mathbf{M}_\pm(0) e^{i\omega t}$ and $\mathbf{m}_\pm = \mathbf{m}_\pm(0) e^{i\omega t}$; taking this into the system of equations (8.1), neglecting the relaxation terms, and the terms in $B_{1\pm}$, we get:

$$-\gamma_n \lambda_m m M_\pm + [\pm\omega + \gamma_n(B_0 + B_{hf})]m_\pm = 0 \tag{8.6a}$$

$$[\pm\omega + \gamma_e(B_0 + B_a + \lambda_m m)]M_\pm - \gamma_e B_{hf} m_\pm = 0 \tag{8.6b}$$

The solutions (the normal modes ω_i) of this system of equations are given by equating the determinant of the coefficients to zero:

$$\pm\omega^2 \pm \omega[\gamma_n(B_0 + B_{hf}) + \gamma_e(B_0 + B_a + \lambda_m m)]$$
$$+ \gamma_n\gamma_e(B_0 + B_a + \lambda_m m)(B_0 + B_{hf}) - \gamma_n\gamma_e\lambda_m m B_{hf} = 0 \tag{8.7}$$

The roots are given using the approximation $(a^2 - \epsilon)^{1/2} \approx a - \epsilon/2a$, where we have taken the term $\epsilon = 4\gamma_n\gamma_e\lambda_m m B_{hf}$ smaller than the other contributions. We also take into account that $\omega_e = \gamma_e(B_0 + B_a) \gg \omega_n = \gamma_n(B_0 + B_{hf})$.

The first solution is the nuclear resonance frequency:

$$\Omega_n = |\omega_1| = \left|-\gamma_n\left[B_0 + B_{hf}\left(1 - \eta\frac{m}{M}\right)\right]\right| \tag{8.8}$$

where the quantity η, known as the *enhancement factor*, or *amplification factor*, is given by

$$\eta = \frac{B_{hf}}{B_0 + B_a} \tag{8.9}$$

This result shows that the nuclear resonance frequency is shifted from the value $\omega_n = \gamma_n(B_0 + B_{hf})$ by a term proportional to the ratio of the nuclear magnetization to the atomic magnetization (in equilibrium); this is normally a very small contribution.

The shift in NMR frequency is given by (for $B_0 = 0$)

$$\delta\omega = \omega_n \eta \frac{m}{M} \tag{8.10}$$

and this effect is known as the *dynamic frequency shift* or "frequency pulling" (see de Gennes 1963), and in the extreme cases where it is significant, the NMR frequency is not a direct measure of the hyperfine field. This occurs at very low temperatures, since this shift is proportional to the nuclear magnetization m, which is proportional to $1/T$; also, it is most relevant for 100% abundant nuclei (large m). Since m is dependent on rf power, the nuclear resonance frequency Ω_n in the presence of dynamic shift will also show a small power dependence. As an example of large dynamic shift, the NMR in a Mn^{2+} ion shows $\delta\omega/\omega \cong 3 \times 10^{-3}/T$.

The other root of Eq. (8.7) is the electronic resonance frequency:

$$\Omega_e = |\omega_2| = \left|-\gamma_e(B_0 + B_a)\left(1 + \eta\frac{m}{M}\right)\right| \quad (8.11)$$

This is the frequency of magnetic resonance of the atomic moments in a ferromagnetically ordered material; it is the ferromagnetic resonance (FMR) frequency. It also shows a small displacement from $\omega_e = \gamma_e(B_0 + B_a)$, proportional to the ratio m/M of the magnetizations.

We now proceed to obtain the expression of the transverse magnetization. We rewrite the equations for $M_\pm$ and $m_\pm$, this time including the rf field $\mathbf{B}_{1\pm}$:

$$-\gamma_n\lambda_m m M_\pm + [\pm\omega + \gamma_n(B_0 + B_{hf})]m_\pm - \gamma_n m B_{1\pm} = 0 \quad (8.12a)$$

$$[\pm\omega + \gamma_e(B_0 + B_a + \lambda_m m)]M_\pm - \gamma_e B_{hf} m_\pm - \gamma_e M B_{1\pm} = 0 \quad (8.12b)$$

The expression for $M_\pm$ [from Eq. (8.12b)] is

$$M_\pm = \frac{\gamma_e(B_{hf}m_\pm + B_{1\pm}M)}{\pm\omega + \gamma_e(B_0 + B_a + \lambda m)} \approx \frac{B_{hf}m_\pm + B_{1\pm}M}{B_0 + B_a} \quad (8.13)$$

where we have neglected $\lambda_m m$ in comparison with $B_0 + B_a$, and ω in comparison with $\omega_e = \gamma_e(B_0 + B_a)$; the latter approximation is justified since we are looking for the transverse magnetization near the nuclear resonance frequency ω_n, which is much smaller than ω_e.

Substituting into Eq. (8.12a), we obtain

$$m_\pm = \frac{\gamma_n\gamma_e m B_{1\pm}(B_0 + B_a + B_{hf})}{\gamma_n[\pm\omega + \gamma_n(B_0 + B_{hf})][\pm\omega + \gamma_e(B_0 + B_a)] - \gamma_n\gamma_e\lambda_m m B_{hf}} \quad (8.14)$$

or

$$m_\pm = (1 + \eta)\frac{\gamma_n B_{1\pm} m}{\pm\omega + \Omega_n} \quad (8.15)$$

with Ω_n given by Eq. (8.8). Taking the negative sign in $\pm\omega$, we obtain

$$m_\pm = -(1 + \eta)\frac{\gamma_n B_{1\pm} m}{\omega - \Omega_n} \quad (8.16)$$

The transverse susceptibility of the nuclei becomes

$$\chi_m^\pm = \mu_0\frac{m_\pm}{B_{1\pm}} = -\mu_0(1 + \eta)\frac{\gamma_n m}{(\omega - \Omega_n)} \quad (8.17)$$

where μ_0 is the vacuum permeability. With the static nuclear susceptibility defined by [see Eq. (7.43)]:

$$\chi_n = \mu_0 \frac{m}{B_0} \qquad (8.18)$$

and using $\omega_0 = \gamma_n B_0$, the nuclear resonance frequency in the applied magnetic field, we finally obtain

$$\chi_m^\pm = -(1+\eta)\frac{\omega_0}{(\omega - \Omega_n)}\chi_n \qquad (8.19)$$

We can see from this equation that the transverse susceptibility of the nuclei in a ferromagnet appears augmented by a factor $(1+\eta)$, compared to its expression in a nonmagnetic matrix. This susceptibility presents a maximum for the frequency $\omega = \Omega_n$.

Substituting $m_\pm$ [Eq. (8.14)] into the expression of $M_\pm$ [Eq. (8.13)], one obtains:

$$M_\pm = \frac{\gamma_e M B_{1\pm}}{[\pm\omega + \gamma_e(B_0 + B_a + \lambda_m m)]} + \frac{\gamma_n m B_{hf}(B_0 + B_a + B_{hf})}{[\pm\omega + \gamma_e(B_0 + B_a + \lambda_m m)](B_0 + B_a)} \qquad (8.20)$$

Dividing by $B_{1\pm}$, and making some simplifications, one obtains

$$\chi_M^\pm = \chi_e - \eta(1+\eta)\frac{\omega_0}{\omega - \Omega_n}\chi_n \qquad (8.21)$$

where $\chi_e = \mu_0 M/(B_0 + B_a)$ is the electronic static susceptibility, a term that is independent of frequency.

The total transverse susceptibility for the system at frequency ω is the sum of the nuclear term and the electronic (or ionic) term:

$$\chi(\omega) = \chi_m^\pm + \chi_M^\pm \qquad (8.22)$$

and is given by

$$\chi(\omega) = \chi_e - (1+\eta)^2\frac{\omega_0}{(\omega - \Omega_n)}\chi_n \qquad (8.23)$$

From this equation one may see that the total response (electronic plus nuclear) to the transverse field contains a multiplicative factor $(1+\eta)^2$. It is interesting to note that this total susceptibility is $(1+\eta) \approx \eta$ times larger than the enhanced nuclear susceptibility [Eq. (8.19)] (apart from a constant term). This means that the largest contribution to the total transverse susceptibility arises from the ions, even for a frequency near the nuclear resonance frequency ($\omega \approx \Omega_n$).

The absorbed power may be calculated writing the total susceptibility under complex form [Eq. (7.49)]:

$$\chi = \chi' - i\chi'' \tag{8.24}$$

To identify χ' and χ'' in this case, we go back to the coupled Bloch equations [Eqs. (8.1)] and substitute **m** and **M** by $\mathbf{m}(t) = \mathbf{m}_0 e^{i\omega t}$ and $\mathbf{M}(t) = \mathbf{M}_0 e^{i\omega t}$, this time including in ω an imaginary part, which leads to exponential decay, or exponential relaxation of the magnetizations. This is equivalent to the substitution of the nuclear magnetic frequency in the total field (Ω_n) by a complex expression:

$$\Omega_n \to \Omega_n + i\Gamma_n \tag{8.25}$$

where Γ_n is the nuclear relaxation term, equal to the half width of the line (in the frequency spectrum). Substituting into Eq. (8.23), it follows that

$$\chi'(\omega) = \chi_e + (1+\eta)^2 \chi'_n(\omega) \tag{8.26}$$

$$\chi''(\omega) = (1+\eta)^2 \chi''_n(\omega) \tag{8.27}$$

with

$$\chi'_n(\omega) = -\frac{\omega_0(\omega - \Omega_n)}{(\omega - \Omega_n)^2 + \Gamma_n^2} \chi_n \tag{8.28}$$

and

$$\chi''_n(\omega) = \frac{\omega_0 \Gamma_n}{(\omega - \Omega_n)^2 + \Gamma_n^2} \chi_n \tag{8.29}$$

The power absorbed by the spin system is given as a function of the imaginary part of the susceptibility $\chi''(\omega)$ by Eq. (7.56) (see Section 7.3):

$$\mathcal{P} = \omega \chi''(\omega) B_1^2 \tag{8.30}$$

Substituting $\chi''(\omega)$, we have

$$\mathcal{P} = \omega \chi''_n(\omega)(1+\eta)^2 B_1^2 \cong \omega \chi''_n(\omega)(\eta B_1)^2 \tag{8.31}$$

In conclusion, the absorbed power is proportional to the (nonenhanced) nuclear susceptibility, to the frequency, and to the square of ηB_1; this last quantity is in fact, the effective rf field B_2 acting on the nucleus in the ferromagnet:

$$B_2 = \eta B_1 \tag{8.32}$$

And the absorbed power is then given by

$$\mathcal{P} \cong \omega \chi_n''(\omega) B_2^2 \qquad (8.33)$$

In this derivation the relaxation term of the atomic moment (Γ_e) was not taken into account. Its inclusion leads to the appearance in the absorbed power of an additional contribution proportional to the electronic term $\chi_M''(\omega)$, and to the nuclear dispersive term $\chi_n'(\omega)$ (e.g., Narath 1967).

8.3 NMR ENHANCEMENT FACTOR : DOMAINS AND DOMAIN WALLS

As we have shown, when an rf field of intensity $\mathbf{B}_1$ is applied to a sample of magnetic material, the nuclei feel a field augmented by η, a quantity known as the NMR enhancement factor, or amplification factor.

The expression for the enhancement factor given in the preceding section [Eq. (8.9)] is applicable to domains. This quantity has different values in domains and in the domain walls; in domains it measures from 1 to 100, and in the domain walls it is of the order of $10^3 - 10^5$. Table 8.II shows values of some enhancement factors η observed experimentally.

It is easy to obtain the expression of the domain enhancement factor [Eq. (8.9)] from geometric arguments (Fig. 8.4). Assuming that inside the domains the atomic magnetic moments feel an anisotropy field B_a along the z direction, a perpendicular rf field B_1 displaces the magnetization from its equilibrium position. The appearance of a perpendicular component of the atomic moment leads to a hyperfine field component in the same direction of B_1, since the hyperfine field is approximately proportional to $\mathbf{M}$. From Fig. 8.4, one can see that

$$\frac{B_{\text{hf}}^\perp}{B_1} = \frac{B_{\text{hf}}^\parallel}{B_a} \approx \frac{B_{\text{hf}}}{B_a} \qquad (8.34)$$

Which is the expression of the enhancement factor in the domains

$$\eta_d = \frac{B_{\text{hf}}}{B_a} \qquad (8.35)$$

Table 8.II Values of the enhancement factor η measured in some metallic matrices

Nucleus	Matrix	Location	Temperature (K)	η
^{57}Fe	Fe	Wall center	4.2	6 100 (300)
^{57}Fe	Fe	Wall center	295	25 000 (2000)
^{61}Ni	Ni	Wall center	1.3	4 000 (500)
^{61}Ni	Ni	Crystal average	$\cong 300$	1 600–43 000

Source: Reprinted from Landolt-Börnstein, *Magnetic Properties of 3d Elements, New Series III/19a*, Springer-Verlag, New York, 1986, with permission.

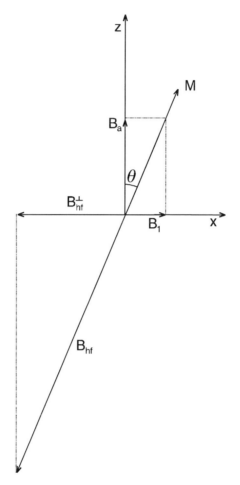

Figure 8.4 Amplification mechanism of the rf field B_1 inside a domain; the magnetic moment is turned by an angle θ, and the transverse component of the applied field becomes $B_{hf}^\perp$, much larger than B_1.

In the presence of an external field $\mathbf{B}_0$, also parallel to z, the enhancement factor is reduced, and measures [Eq. (8.9)]:

$$\eta_d = \frac{B_{hf}}{B_a + B_0} \qquad (8.36)$$

The nuclei at the domain wall edge (DWE) normally have enhancement factors larger than in the domains.

Inside a domain wall, the field B_1 is amplified by the factor η_w, usually much larger than η_d. This amplification effect can be understood as follows. The field

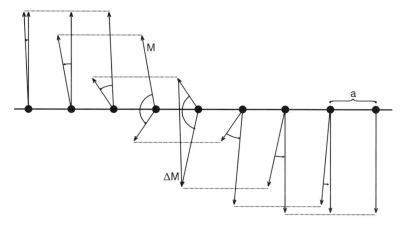

Figure 8.5 Enhancement in the domain walls: the rf field B_1 displaces the wall, and the magnetic moments **M** of the atoms turn, leading to the appearance of large transverse components of the hyperfine fields ($\propto \Delta M$) acting on the nuclei.

B_1 displaces the wall, favoring the growth of the domains with direction of magnetization close to the direction of B_1 (see Sections 5.4 and 5.5). This displacement induces a rotation of the magnetization inside the wall, which leads to the appearance of components of the hyperfine field along B_1, which add to this rf field—this is the mechanism of enhancement in the walls (Fig. 8.5). This enhancement depends on the position x inside the domain wall; for the nuclei at the domain wall center (DWC) (Fig. 8.6) the factor η_w reaches a maximum. The enhancement factor η_w is proportional to the displacement δx of the wall, for small values of δx.

The domain wall enhancement factor for a domain wall of thickness δ, inside a particle of diameter D, demagnetizing factor N_d, saturation magnetization M_s, and hyperfine field B_{hf} is (Portis and Gossard 1960):

$$\eta_w = \frac{\pi D B_{hf}}{\mu_0 N_d \delta M_s} \qquad (8.37)$$

In a pulsed NMR experiment, the nuclear magnetization is turned from the equilibrium direction (the z direction) by the application of the radiofrequency field B_1 in the xy plane. The angle of rotation of the nuclear magnetization after an rf pulse of duration t_a is given by (see Section 7.6)

$$\theta = \gamma \eta t_a B_1 \qquad (8.38)$$

where γ is the nuclear gyromagnetic ratio, B_1 is the radiofrequency field, and η is the enhancement factor. The power applied to the sample is related to the field B_1

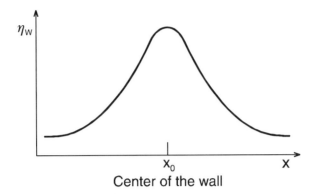

Figure 8.6 Dependence of the domain wall enhancement factor η_w on x, inside a Bloch wall perpendicular to this axis [η_w is given by an expression of the form of Eq. (8.42)].

through the equation [Eq. (8.31)]:

$$\mathcal{P} = cB_1^2 \tag{8.39}$$

The amplitude of the free induction decay (FID) after a single pulse, and the amplitude of the spin echo after a sequence of two pulses (to simplify, of equal duration), depend on the angle θ. The amplitude of the free induction decay presents an oscillatory dependence with θ in a nonmagnetic sample. For the same pulse duration, a periodic variation versus the rf field intensity B_1 is expected. The amplitude of the echo is also a periodic function of θ (or B_1) in these samples.

The case we want to discuss here, however, is that of ferromagnetic samples; in these, several factors contribute to make the results more complex. In the simplest hypothesis, assuming a constant enhancement factor, two pulses of equal width t_a and rf field B_1 perpendicular to the static field B_0, the amplitude of the echo is given by (Bloom 1955)

$$E(t_a, B_1) = C \sin(\gamma \eta B_1 t_a) \sin^2 \frac{\gamma \eta B_1 t_a}{2} \tag{8.40}$$

which is essentially the result obtained previously [Eq. (7.108)], with the addition of the enhancement factor.

The analysis of the problem of formation of spin echoes in magnetic materials has been extended to include domain wall enhancement in multidomain samples (Stearns 1967). The following factors had to be taken into account in this treatment: (1) the distribution of angles θ between B_1 and the directions of magnetization of the different domains; (2) the spatial variation of η inside the Bloch walls; (3) the oscillatory motion of these walls (assumed to be of circular shape), like drum membranes; and (4) the distribution of the areas of these walls. The resulting function derived for the echo amplitude at the resonance frequency

ω_0 is:

$$E(\omega_0, B_1, \tau) = \frac{1}{2} m_0 \eta_0 \int_0^\infty \int_0^1 \sin^2\left(\frac{\alpha_0 z \operatorname{sech}(x)}{2}\right) \times \sin[\alpha_0 z \operatorname{sech}(x)] z \operatorname{sech}(x) \, p(z) dz \, dx \quad (8.41)$$

where $z = (1-r^2) h_m \cos\theta$, $p(z) = \frac{1}{2}\ln^2(1/z)$ for $Ap(A) = \text{const.}$
A = area of a wall
$p(A)$ = probability of finding walls with a given value of the area A
h_m = displacement of center of wall, normalized to the maximum displacement, which is that of the wall of largest radius
m_0 = nuclear magnetization
α_0 = maximum angle of rotation m_0 after excitation by rf pulses.

The variation of the wall enhancement factor η_w with the position of the nucleus inside the wall is described by an even function, with maximum at the center of the wall ($x = 0$); the function

$$\eta(x) = \eta_0 \operatorname{sech}(x)(1-r^2) h_m \quad (8.42)$$

was postulated, with r representing the normalized distance of the nucleus from the axis of the (circular) wall, varying between 0 and 1.

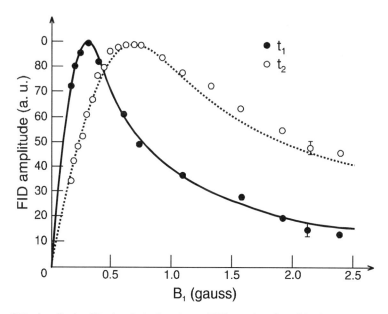

Figure 8.7 Amplitude of the free induction decay (FID) as a function of the intensity of the rf field (B_1) in metallic iron, for two different pulse lengths ($t_1 = 3$ μs, and $t_2 = 1.2$ μs); the curves are computer fits using the model described in the text. [Reprinted from M. B. Stearns, *Phys. Rev.* **162**, 496 (1967).]

The experimental data for the echo amplitudes as a function of rf field, obtained with metallic Fe and metallic Ni, fitted well the function (8.41); Fig. 8.7 shows the data for the FID amplitudes.

In the case of NMR in metallic matrices, the radiofrequency field B_1 is attenuated, and its intensity inside the sample decays exponentially with the depth (the skin effect; see Section 5.6). This phenomenon gives rise to a spatial inhomogeneity in the rf field, and as a consequence, a reduction of the effective volume of the sample in a magnetic resonance experiment. To minimize this problem, metallic samples are usually studied in the form of powders or thin foils.

8.4 FERROMAGNETIC RESONANCE

We have shown in Section 8.2 that the coupled system of atomic magnetic moments and nuclear magnetic moments found in a ferromagnet has two resonant frequencies, the NMR frequency, and the ferromagnetic resonance (FMR) frequency. We will derive below the ferromagnetic resonance frequency, and discuss the result in more detail, neglecting the influence of the nuclear magnetic moments.

Ferromagnetic resonance (FMR) is observed when a sample of ferromagnetic material is submitted to an rf field of frequency equal to the precession frequency of the atomic moments (or the magnetization $\mathbf{M}$). With the precession of $\mathbf{M}$, the demagnetizing fields $H_d = -N_d \mathbf{M}$ along different directions have to be taken into account, and since the demagnetizing factors depend on the shape, the form of the sample is important. In the case of an ellipsoidal sample, with principal axes coinciding with the coordinate axes, N_d is a diagonal tensor given by

$$\mathbf{N}_d = \begin{pmatrix} N_d^x & 0 & 0 \\ 0 & N_d^y & 0 \\ 0 & 0 & N_d^z \end{pmatrix} \tag{8.43}$$

and

$$\mathbf{B}_d = -\mu_0 \mathbf{N}_d \mathbf{M} = -\mu_0 \begin{pmatrix} N_d^x & 0 & 0 \\ 0 & N_d^y & 0 \\ 0 & 0 & N_d^z \end{pmatrix} \begin{pmatrix} M_x \\ M_y \\ M_z \end{pmatrix}$$

$$= -\mu_0 (N_d^x M_x \mathbf{i} + N_d^y M_y \mathbf{j} + N_d^z M_z \mathbf{k}) \tag{8.44}$$

The total field acting on $\mathbf{M}$ is

$$\mathbf{B} = \mathbf{B}_0 + \mathbf{B}(t) + \mathbf{B}_d \tag{8.45}$$

with $\mathbf{B}_0 = B_0 \mathbf{k}$, and $\mathbf{B}(t) = \mathbf{B}_1 e^{i\omega t}$.

Neglecting relaxation effects, the motion of **M** is given by

$$\frac{d\mathbf{M}}{dt} = \gamma \mathbf{M} \times (\mathbf{B}_0 + \mathbf{B}(t) + \mathbf{B}_d) \tag{8.46}$$

One can assume that the magnetization does not deviate much from the equilibrium value (the saturation magnetization), that is, $\mathbf{M} \approx \mathbf{M}_s$. In the stationary regime, $\mathbf{M}_x = \mathbf{M}_x(0)e^{i\omega t}$, and $\mathbf{M}_y = \mathbf{M}_y(0)e^{i\omega t}$. Inserting these expressions into Eq. (8.46), we obtain

$$i\omega M_x = -\gamma[\mu_0 M_y M_z (N_d^z - N_d^y) + M_y B_0 + M_z B_1] \tag{8.47a}$$

$$i\omega M_y = -\gamma[\mu_0 M_x M_z (N_d^x - N_d^z) + M_x B_0 + M_z B_1] \tag{8.47b}$$

The condition for the existence of solutions for this system of equations is that the determinant of the coefficients be equal to zero; for $\mathbf{B}(t) = B_1 = 0$, the root of the resulting equation is the frequency

$$\omega_0 = \gamma\{[B_0 + \mu_0(N_d^y - N_d^z)M_s][B_0 + \mu_0(N_d^x - N_d^z)M_s]\}^{1/2} \tag{8.48}$$

where $M_z \approx M_s$.

This is the frequency of precession of the magnetization; the individual magnetic moments precess in phase, and therefore ω_0 is called the *uniform mode precession* frequency. For a spherical sample, the demagnetizing factors are $N_d^x = N_d^y = N_d^z = \frac{1}{3}$, and the resonance frequency simplifies to

$$\omega_0 = \gamma B_0 \tag{8.49}$$

For a sample in the form of a thin film, in the plane xy, $N_d^x = N_d^y = 0$ and $N_d^z = 1$, and we have

$$\omega_0 = \gamma(B_0 - \mu_0 M_s) \tag{8.50}$$

In real magnetic materials one has in general to consider other sources of magnetic anisotropy, especially crystalline anisotropy. Magnetic anisotropy can be described in the simplest approximation through the use of an anisotropy field $B_a = \mu_0 H_a$; for axial anisotropy of crystalline origin, $B_a = 2K_1/M$ (Section 5.2.2). Introducing this field applied along the z direction, one has

$$\mathbf{B} = \mathbf{B}_0 + \mathbf{B}(t) + \mathbf{B}_d + \mathbf{B}_a \tag{8.51}$$

In the case of a spherical sample, and when $B_1 = 0$, it follows:

$$\omega_0 = \gamma(B_0 + B_a) \tag{8.52}$$

From this result one can conclude that ferromagnetic resonance can be observed even in the absence of an external magnetic field (i.e., with $B_0 = 0$). Equation (8.52) shows that the FMR frequency in this case is proportional to the magnitude of the anisotropy field B_a:

$$\omega_0 = \gamma B_a \tag{8.53}$$

For the more general case of crystalline anisotropy, not necessarily axial, it is convenient to write the anisotropy field as $\mathbf{B}_a = N_a \mathbf{M}$, with N_a the anisotropy tensor. For example, for a spherical sample of a cubic crystal, with $\mathbf{B}_0$ along the [111] direction (see Chapter 5), it can be shown that

$$\mathbf{B}_a = \mathbf{B}_K = -\mu_0 \frac{(12K_1 + 4K_2)}{9M_s} \tag{8.54}$$

and from $\mathbf{B}_a$, one obtains

$$\omega_0 = \gamma \left[B_0 - \mu_0 \frac{(12K_1 + 4K_2)}{9M_s} \right] \tag{8.55}$$

The magnetic susceptibility of a ferromagnet under an applied rf field can be obtained from $\mathbf{M}(t) = \chi \, \mathbf{B}(t)/\mu_0$, where χ is the magnetic susceptibility tensor. To obtain χ, we have to solve Eqs. (8.47) for M_x and M_y.

For an isotropic sample (i.e., $\mathbf{B}_a = 0$), the susceptibility is given by

$$\chi = \begin{pmatrix} \chi_{11} & \chi_{12} & 0 \\ \chi_{21} & \chi_{22} & 0 \\ 0 & 0 & 0 \end{pmatrix} \tag{8.56}$$

with components

$$\chi_{11} = \frac{\gamma^2 M_s (B_0 - \mu_0 N_d^z M_s + \mu_0 N_d^y M_s)}{\omega_0^2 - \omega^2} \tag{8.57a}$$

$$\chi_{12} = -\chi_{21} = -\frac{i\gamma \omega M_s}{\omega_0^2 - \omega^2} \tag{8.57b}$$

$$\chi_{22} = \frac{\gamma^2 M_s (B_0 - \mu_0 N_d^z M_s + \mu_0 N_d^x M_s)}{\omega_0^2 - \omega^2} \tag{8.57c}$$

where ω_0 is as given by Eq. (8.55). Note that χ is the extrinsic susceptibility, so called since it measures the response of the magnetic sample to the external rf field $\mathbf{B}(t)$; if one had used the time-dependent field inside the sample, it would have led to the intrinsic susceptibility.

The magnetic response of a sample to the rf field $\mathbf{B}(t)$ can alternatively be described through the relative permeability tensor $\boldsymbol{\mu}$ (we have dropped the subscript r to simplify) which connects $\mathbf{B}(t)$ and $\mathbf{H}(t)$: $\mathbf{B}(t) = \mu_0 \boldsymbol{\mu}\, \mathbf{H}(t)$. This is done using the relation $\boldsymbol{\mu} = \mathbf{I} + \boldsymbol{\chi}$, where $\mathbf{I}$ is the identity matrix, or the equivalent relations:

$$\mu_{11} = 1 + \chi_{11} \qquad (8.58a)$$

$$\mu_{12} = \chi_{12} = -\mu_{21} \qquad (8.58b)$$

$$\mu_{22} = 1 + \chi_{22} \qquad (8.58c)$$

The motion of the magnetization in a ferromagnet, and the phenomenon of ferromagnetic resonance in the presence of magnetic relaxation can be described with a phenomenological equation due to Gilbert:

$$\frac{d\mathbf{M}}{dt} = \gamma \mathbf{M} \times \mathbf{B} + \alpha \frac{\mathbf{M}}{|M|} \times \frac{d\mathbf{M}}{dt} \qquad (8.59)$$

The solutions of Gilbert's equation for isotropic media and spherical samples can be shown to be (e.g., Morrish 1965)

$$\chi_{11} = \chi_{22} = \frac{\gamma M_s(\gamma B_0 + i\alpha\omega)}{(\omega_0 + i\alpha\omega)^2 - \omega^2} \qquad (8.60a)$$

$$\chi_{12} = -\chi_{21} = -\frac{i\gamma\omega M_s}{(\omega_0 + i\alpha\omega)^2 - \omega^2} \qquad (8.60b)$$

Since in this case there is relaxation, or damping, all the matrix elements of the magnetic susceptibility tensor are complex. Another description of FMR is due to Landau and Lifshitz; the two forms are equivalent. Landau and Lifshitz's equation is

$$\frac{d\mathbf{M}}{dt} = \gamma \mathbf{M} \times \mathbf{B} - \frac{\lambda}{M^2} \mathbf{M} \times (\mathbf{M} \times \mathbf{B}) \qquad (8.61)$$

Since ferromagnetic resonance phenomena occur in strongly coupled magnetic systems, the relaxation or damping term has an important contribution arising from spin waves. In conducting media, eddy currents are also important.

In the preceding discussion we have assumed that the external magnetic field is sufficient to saturate the ferromagnetic sample; if this condition is not satisfied, the samples are multidomain and the description of FMR becomes more complex.

Magnetic resonance is also observed in antiferromagnetic and ferrimagnetic materials. In antiferromagnets the resonance is called *antiferromagnetic resonance* (AFMR). It can easily be described along the same lines of the FMR phenomenon, assuming two sublattices with magnetizations $\mathbf{M}_1$ and $\mathbf{M}_2$, with $M = |\mathbf{M}_1| = |\mathbf{M}_2|$. The exchange coupling between the sublattices is described through a molecular field of modulus $B_m = \lambda M$, acting on either sublattice. The AFMR frequency is equal to

$$\omega_0 = \gamma[B_a(B_a + 2B_m)]^{1/2} \tag{8.62}$$

In the usual situation where the mean field B_m is much larger than the anisotropy field B_a, one has

$$\omega_0 = \gamma(2B_aB_m)^{1/2} \tag{8.63}$$

GENERAL READING

NMR in Magnetic Materials

Ailion, D. C. and W. D. Ohlsen, "Magnetic Resonance Methods for Studying Defect Structure in Solids," in J. N. Mundy, S. J. Rothman, and L. C. Smedskjaer, Eds., *Methods of Experimental Physics*, Vol. 21, Solid State: Nuclear Methods, Academic Press, Orlando, 1983.

Dormann, E., "NMR in Intermetallic Compounds," in K. A. Gschneidner, Jr., and L. Eyring, Eds., *Handbook on the Physics and Chemistry of Rare Earths*, Vol. 14, Elsevier, 1991, p. 63.

Figiel, H., *Mag. Res. Rev.* **16**, 101 (1991).

McCausland, M. A. H. and I. S. Mackenzie, *Nuclear Magnetic Resonance in Rare Earth Metals*, Taylor & Francis, London, 1980.

Narath, A., "Nuclear Magnetic Resonance in Magnetic and Metallic Solids," in A. J. Freeman and R. B. Frankel, Eds., *Hyperfine Interactions*, Academic Press, New York, 1967, p. 287.

Portis, A. M. and R. H. Lindquist, "Nuclear Resonance in Ferromagnetic Materials," in G. T. Rado and H. Suhl, Eds., *Magnetism*, Vol. 2A, Academic Press, New York, 1965, p. 357.

Turov, E. A. and M. P. Petrov, *Nuclear Magnetic Resonance in Ferro and Antiferromagnets*, Halsted Press, New York, 1972.

Weisman, I. D., L. J. Swartzendruber, and L. H. Bennett, "Nuclear Magnetic Resonance in Ferromagnetic Materials (FNR)," in E. Passaglia, Ed., *Techniques of Metal Research*, Vol. VI, Part 2, Interscience, New York, 1973, p. 336.

Zinn, W., *At. Energy Rev.* **12**, 709 (1974).

Ferromagnetic Resonance

Craik, D., *Magnetism, Principles and Applications*, Wiley, Chichester, 1995.

Foner, S., "Antiferromagnetic and Ferrimagnetic Resonance," in G. T. Rado and H. Suhl, Eds., *Magnetism*, Vol. 1, Academic Press, New York, 1963, p. 384.

Morrish, A. H., *Physical Principles of Magnetism*, Wiley, New York, 1965.

Sparks, M., *Ferromagnetic Relaxation Theory*, McGraw-Hill, New York, 1964.
Stancil, D. D., "Microwave and Optical Magnetics," in R. Gerber, C. D. Wright and G. Asti, Eds., *Applied Magnetism*, Kluwer, Dordrecht, 1994, p. 457.
Vonsovskii, S. V., Ed., *Ferromagnetic Resonance*, Pergamon Press, Oxford, 1966.

REFERENCES

Bloom, A. L., *Phys. Rev.* **98**, 1105 (1955).
de Gennes, P. G., P. A. Pincus, F. Hartmann-Boutron, and J. M. Winter, *Phys. Rev.* **129**, 1105 (1963).
de Gronckel, H. A. M., K. Kopinga, W. J. M. de Jonge, P. Panissod, J. P. Schillé, and F. J. A. den Broeder, *Phys. Rev. B* **44**, 9100 (1991).
Dormann, E., "NMR in Intermetallic Compounds," in K. A. Gschneidner, Jr. and L. Eyring, Eds., *Handbook on the Physics and Chemistry of Rare Earths*, Vol. 14, Elsevier, 1991, p. 63.
Guimarães, A. P., unpublished thesis, Univ. Manchester, 1971.
Kapusta, Cz., J. S. Lord, G. K. Tomka, P. C. Riedi, and K. H. J. Buschow, *J. Appl. Phys.* **79**, 4599 (1996); Cz. Kapusta, private communication (1997).
Landolt-Börnstein, *Magnetic Properties of Metals*, Landolt-Börnstein Tables, New Series III/19a, New York, 1986.
McCausland, M. A. H. and I. S. Mackenzie, *Adv. Phys.* **28**, 305 (1979).
Morrish, A. M., *Physical Principles of Magnetism*, Wiley, New York, 1965.
Narath, A., "Nuclear Magnetic Resonance in Magnetic and Metallic Solids," in A. J. Freeman and R. B. Frankel, Eds., *Hyperfine Interactions*, Academic Press, New York, 1967, p 287.
Portis, A. M. and A. C. Gossard, *J. Appl. Phys.* **31**, 205S (1960).
Stearns, M. B., *Phys. Rev.* **162**, 496 (1967).
Tribuzy, C. V. B. and A. P. Guimarães, unpublished (1997).
Turov, E. A. and M. P. Petrov, *Nuclear Magnetic Resonance in Ferro and Antiferromagnets*, Halsted Press, New York, 1972.
Weisman, I. D., L. J. Swartzendruker, and L. H. Bennett, "Nuclear Magnetic Resonance in Ferromagnetic Materials (FNR)," in E. Passaglia, Ed., *Techniques of Metal Research*, Vol. VI, Part 2, Interscience, New York, 1973, p. 336.
Zhang, Y. D., J. I. Budnick, J. C. Ford, and W. A. Hines, *J. Mag. Mag. Mat.* **100**, 13 (1991).

EXERCISES

8.1 *Fourier Spectrum of a Rectangular Pulse.* Let a pulse $B_1(t)$ be given by

$$B_1(t) = \begin{cases} B_1 & \text{if } |t| < T/2 \\ 0 & \text{if } |t| > T/2 \end{cases}$$

Compute the Fourier spectrum of the pulse, and show that the power is concentrated in the frequency range $-1/T$ and $1/T$. Make an estimate of this frequency band for $T = 10$ μs.

8.2 *Enhancement Factor inside a Bloch Wall.* Consider a grain of a ferromagnet of volume V composed of two ferromagnetic domains, separated by a 180° Bloch wall of width W and area A. Let M_s be the saturation magnetization in each domain. Assume that an rf field $\mathbf{B}_1$ is applied along the plane of the wall in the direction parallel to the direction of magnetization of the domains.

(a) Show that the increase in the magnetization of the grain caused by the rf field is:

$$\delta M_1 = \frac{2AM_s\,\delta x}{V}$$

where δx is the instantaneous displacement of the wall in the direction perpendicular to the field $\mathbf{B}_1$.

(b) Given the electronic susceptibility to the rf field defined by

$$\delta M_1 = \frac{\chi_e}{\mu_0} B_1$$

show that δx is given by

$$\delta x = \frac{V\chi_e}{2A\mu_0 M_s} B_1$$

(c) Let $\theta(x)$ be the angle that the magnetization at point x inside the wall makes with the direction of magnetization of the domains. An ion in this position will have its magnetic moment turned by $\mathbf{B}_1$ through an angle of the order of

$$\delta\theta(x) = \left(\frac{d\theta}{dx}\right)\delta x$$

Show that the rf field at the nucleus in this position of the wall will be

$$B_1' \approx \frac{V\chi_e}{2A\mu_0 M_s}\left(\frac{d\theta}{dx}\right) B_{\mathrm{hf}} B_1$$

where B_{hf} is the hyperfine field, that has the direction of the local magnetization in the wall.

(d) Compute the enhancement factor $\eta(x) = B_1'/B_1$ inside the wall for the case in which $\theta(x) = \operatorname{tg}^{-1}(x/W)$.

8.3 *Inhomogeneous Width of the NMR Line.* Consider an rf pulse of duration τ_p and amplitude B_1'. The angle θ_p through which the nuclear magnetization is rotated by the pulse is given by $\theta_p = \gamma_n B_1' \tau_p$. Assuming that the spin–spin relaxation may be neglected during the application of the pulse, show that for a $\theta_p = \pi/2$ pulse we will have the relation $B_1' \gg 1/(\gamma_n T_2)$. Substitute typical values for γ_n and T_2 in metals in this expression, and estimate the value of B_1 necessary to turn the nuclear spins of $\pi/2$. In the presence of an inhomogeneous linewidth of the order of 5 MHz, what would be the equivalent value of B_1 required to excite all the spins? Comment on your answer.

APPENDIX A

TABLE OF NMR NUCLIDES

Nuclide	Spin	Natural Abundance (%)	Quadrupole Moment ($10^{-28} m^2$)	Sensitivity[a] Relative	Sensitivity[a] Absolute	NMR Frequency (MHz) at a Field of 2.3488 T
^{1}H	$\frac{1}{2}$	99.98	—	1.00	1.00	100.000
^{2}H	1	1.5×10^{-2}	2.73×10^{-3}	9.65×10^{-3}	1.45×10^{-6}	15.351
^{3}H	$\frac{1}{2}$	0	—	1.21	0	106.663
^{3}He	$\frac{1}{2}$	1.3×10^{-4}	—	0.44	5.75×10^{-7}	76.178
^{6}Li	1	7.42	-8.0×10^{-4}	8.50×10^{-3}	6.31×10^{-4}	14.716
^{7}Li	$\frac{3}{2}$	92.58	-4.5×10^{-2}	0.29	0.27	38.863
^{9}Be	$\frac{3}{2}$	100	5.2×10^{-2}	1.39×10^{-2}	1.39×10^{-2}	14.053
^{10}B	3	19.58	7.4×10^{-2}	1.99×10^{-2}	3.90×10^{-3}	10.746
^{11}B	$\frac{3}{2}$	80.42	3.55×10^{-2}	0.17	0.13	32.084
^{13}C	$\frac{1}{2}$	1.108	—	1.59×10^{-2}	1.76×10^{-4}	25.144
^{14}N	1	99.63	1.6×10^{-2}	1.01×10^{-3}	1.01×10^{-3}	7.224
^{15}N	$\frac{1}{2}$	0.37	—	1.04×10^{-3}	3.85×10^{-6}	10.133
^{17}O	$\frac{5}{2}$	3.7×10^{-2}	-2.6×10^{-2}	2.9×10^{-2}	1.08×10^{-5}	13.557
^{19}F	$\frac{1}{2}$	100	—	0.83	0.83	94.077
^{21}Ne	$\frac{3}{2}$	0.257	9.0×10^{-2}	2.50×10^{-3}	6.43×10^{-6}	7.894
^{23}Na	$\frac{3}{2}$	100	0.12	9.25×10^{-2}	9.25×10^{-2}	26.451
^{25}Mg	$\frac{5}{2}$	10.13	0.22	2.67×10^{-3}	2.71×10^{-4}	6.1195
^{27}Al	$\frac{5}{2}$	100	0.149	0.21	0.21	26.057

Nuclide	Spin	Natural Abundance (%)	Quadrupole Moment (10^{-28}m^2)	Sensitivity[a] Relative	Absolute	NMR Frequency (MHz) at a Field of 2.3488 T
^{29}Si	$\frac{1}{2}$	4.7	—	7.84×10^{-3}	3.69×10^{-4}	19.865
^{31}P	$\frac{1}{2}$	100	—	6.63×10^{-2}	6.63×10^{-2}	40.481
^{33}S	$\frac{3}{2}$	0.76	-5.5×10^{-2}	2.26×10^{-3}	1.72×10^{-5}	7.670
^{35}Cl	$\frac{3}{2}$	75.53	-8.0×10^{-2}	4.70×10^{-3}	3.55×10^{-3}	9.798
^{37}Cl	$\frac{3}{2}$	24.47	-6.32×10^{-2}	2.71×10^{-3}	6.63×10^{-4}	8.156
^{39}K	$\frac{3}{2}$	93.1	5.5×10^{-2}	5.08×10^{-4}	4.73×10^{-4}	4.667
^{41}K	$\frac{3}{2}$	6.88	6.7×10^{-2}	8.40×10^{-5}	5.78×10^{-6}	2.561
^{43}Ca	$\frac{7}{2}$	0.145	-0.05	6.40×10^{-3}	9.28×10^{-6}	6.728
^{45}Sc	$\frac{7}{2}$	100	-0.22	0.30	0.30	24.290
^{47}Ti	$\frac{5}{2}$	7.28	0.29	2.09×10^{-3}	1.52×10^{-4}	5.637
^{49}Ti	$\frac{7}{2}$	5.51	0.24	3.76×10^{-3}	2.07×10^{-4}	5.638
^{50}V	6	0.24	± 0.21	5.55×10^{-2}	1.33×10^{-4}	9.970
^{51}V	$\frac{7}{2}$	99.76	-5.2×10^{-2}	0.38	0.38	26.289
^{53}Cr	$\frac{3}{2}$	9.55	$\pm 3.0 \times 10^{-2}$	9.03×10^{-4}	8.62×10^{-3}	5.652
^{55}Mn	$\frac{5}{2}$	100	0.55	0.18	0.18	24.664
^{57}Fe	$\frac{1}{2}$	2.19	—	3.37×10^{-5}	7.38×10^{-7}	3.231
^{59}Co	$\frac{7}{2}$	100	0.40	0.28	0.28	23.614
^{61}Ni	$\frac{3}{2}$	1.19	0.16	3.57×10^{-3}	4.25×10^{-5}	8.936
^{63}Cu	$\frac{3}{2}$	69.09	-0.211	9.31×10^{-2}	6.43×10^{-2}	26.505
^{65}Cu	$\frac{3}{2}$	30.91	-0.195	0.11	3.52×10^{-2}	28.394
^{67}Zn	$\frac{5}{2}$	4.11	0.15	2.85×10^{-3}	1.17×10^{-4}	6.254
^{69}Ga	$\frac{3}{2}$	60.4	0.178	6.91×10^{-2}	4.17×10^{-2}	24.003
^{71}Ga	$\frac{3}{2}$	39.6	0.112	0.14	5.62×10^{-2}	30.495
^{73}Ge	$\frac{9}{2}$	7.76	-0.2	1.4×10^{-3}	1.08×10^{-4}	3.488
^{75}As	$\frac{3}{2}$	100	0.3	2.51×10^{-2}	2.51×10^{-2}	17.126
^{77}Se	$\frac{1}{2}$	7.58	—	6.93×10^{-3}	5.25×10^{-4}	19.067
^{79}Br	$\frac{3}{2}$	50.54	0.33	7.86×10^{-2}	3.97×10^{-2}	25.053
^{81}Br	$\frac{3}{2}$	49.46	0.28	9.85×10^{-2}	4.87×10^{-2}	27.006
^{83}Kr	$\frac{9}{2}$	11.55	0.15	1.88×10^{-3}	2.17×10^{-4}	3.847
^{85}Rb	$\frac{5}{2}$	72.15	0.25	1.05×10^{-2}	7.57×10^{-3}	9.655
^{87}Rb	$\frac{3}{2}$	27.85	0.12	0.17	4.87×10^{-2}	32.721
^{87}Sr	$\frac{9}{2}$	7.02	0.36	2.69×10^{-3}	1.88×10^{-4}	4.333
^{89}Y	$\frac{1}{2}$	100	—	1.18×10^{-4}	1.18×10^{-4}	4.899
^{91}Zr	$\frac{5}{2}$	11.23	-0.21	9.48×10^{-3}	1.06×10^{-3}	9.330
^{93}Nb	$\frac{9}{2}$	100	-0.2	0.48	0.48	24.442
^{95}Mo	$\frac{5}{2}$	15.72	± 0.12	3.23×10^{-3}	5.07×10^{-4}	6.514
^{97}Mo	$\frac{5}{2}$	9.46	± 1.1	3.43×10^{-3}	3.24×10^{-4}	6.652

TABLE OF NMR NUCLIDES

Nuclide	Spin	Natural Abundance (%)	Quadrupole Moment ($10^{-28}\,\text{m}^2$)	Sensitivity[a] Relative	Sensitivity[a] Absolute	NMR Frequency (MHz) at a Field of 2.3488 T
^{99}Ru	$\tfrac{3}{2}$	12.72	−0.19	1.95×10^{-4}	2.48×10^{-5}	3.389
^{101}Ru	$\tfrac{5}{2}$	17.07	7.6×10^{-2}	1.41×10^{-3}	2.40×10^{-4}	4.941
^{103}Rh	$\tfrac{1}{2}$	100	—	3.11×10^{-5}	3.11×10^{-5}	3.147
^{105}Pd	$\tfrac{5}{2}$	22.23	−0.8	1.12×10^{-3}	2.49×10^{-4}	4.576
^{107}Ag	$\tfrac{1}{2}$	51.82	—	6.62×10^{-5}	3.43×10^{-5}	4.046
^{109}Ag	$\tfrac{1}{2}$	48.18	—	1.01×10^{-4}	4.86×10^{-5}	4.652
^{111}Cd	$\tfrac{1}{2}$	12.75	—	9.54×10^{-3}	1.21×10^{-3}	21.205
^{113}Cd	$\tfrac{1}{2}$	12.26	—	1.09×10^{-2}	1.33×10^{-3}	22.182
^{113}In	$\tfrac{9}{2}$	4.28	1.14	0.34	1.47×10^{-2}	21.866
^{115}In	$\tfrac{9}{2}$	95.72	0.83	0.34	0.33	21.914
^{115}Sn	$\tfrac{1}{2}$	0.35	—	3.5×10^{-2}	1.22×10^{-4}	32.699
^{117}Sn	$\tfrac{1}{2}$	7.61	—	4.52×10^{-2}	3.44×10^{-3}	35.625
^{119}Sn	$\tfrac{1}{2}$	8.58	—	5.18×10^{-2}	4.44×10^{-3}	37.272
^{121}Sb	$\tfrac{5}{2}$	57.25	−0.53	0.16	9.16×10^{-2}	23.930
^{123}Sb	$\tfrac{7}{2}$	42.75	−0.68	4.57×10^{-2}	1.95×10^{-2}	12.959
^{123}Te	$\tfrac{1}{2}$	0.87	—	1.80×10^{-2}	1.56×10^{-4}	26.207
^{125}Te	$\tfrac{1}{2}$	6.99	—	3.15×10^{-2}	2.20×10^{-3}	31.596
^{127}I	$\tfrac{5}{2}$	100	−0.79	9.34×10^{-2}	9.34×10^{-2}	20.007
^{129}Xe	$\tfrac{1}{2}$	26.44	—	2.12×10^{-2}	5.60×10^{-3}	27.660
^{131}Xe	$\tfrac{3}{2}$	21.18	−0.12	2.76×10^{-3}	5.84×10^{-4}	8.199
^{133}Cs	$\tfrac{7}{2}$	100	-3.0×10^{-3}	4.74×10^{-2}	4.74×10^{-2}	13.117
^{135}Ba	$\tfrac{3}{2}$	6.59	0.18	4.90×10^{-3}	3.22×10^{-4}	9.934
^{137}Ba	$\tfrac{3}{2}$	11.32	0.28	6.86×10^{-3}	7.76×10^{-4}	11.113
^{138}La	5	0.089	−0.47	9.19×10^{-2}	8.18×10^{-5}	13.193
^{139}La	$\tfrac{7}{2}$	99.91	0.21	5.92×10^{-2}	5.91×10^{-2}	14.126
^{141}Pr	$\tfrac{5}{2}$	100	-5.9×10^{-2}	0.29	0.29	29.291
^{143}Nd	$\tfrac{7}{2}$	12.17	−0.48	3.38×10^{-3}	4.11×10^{-4}	5.437
^{145}Nd	$\tfrac{7}{2}$	8.3	−0.25	7.86×10^{-4}	6.52×10^{-5}	3.345
^{147}Sm	$\tfrac{7}{2}$	14.97	−0.21	1.48×10^{-3}	2.21×10^{-4}	4.128
^{149}Sm	$\tfrac{7}{2}$	13.83	6.0×10^{-2}	7.47×10^{-4}	1.03×10^{-4}	3.289
^{151}Eu	$\tfrac{5}{2}$	47.82	1.16	0.18	8.5×10^{-2}	24.801
^{153}Eu	$\tfrac{5}{2}$	52.18	2.9	1.52×10^{-2}	7.98×10^{-3}	10.951
^{155}Gd	$\tfrac{3}{2}$	14.73	1.6	2.79×10^{-4}	4.11×10^{-5}	3.819
^{157}Gd	$\tfrac{3}{2}$	15.68	2.0	5.44×10^{-4}	8.53×10^{-5}	4.774
^{159}Tb	$\tfrac{3}{2}$	100	1.3	5.83×10^{-2}	5.83×10^{-2}	22.678
^{161}Dy	$\tfrac{5}{2}$	18.88	1.4	4.17×10^{-4}	7.87×10^{-5}	3.294
^{163}Dy	$\tfrac{5}{2}$	24.97	1.6	1.12×10^{-3}	2.79×10^{-4}	4.583

TABLE OF NMR NUCLIDES

Nuclide	Spin	Natural Abundance (%)	Quadrupole Moment ($10^{-28} m^2$)	Sensitivity[a] Relative	Sensitivity[a] Absolute	NMR Frequency (MHz) at a Field of 2.3488 T
^{165}Ho	$\frac{7}{2}$	100	2.82	0.18	0.18	20.513
^{167}Er	$\frac{7}{2}$	22.94	2.83	5.07×10^{-4}	1.16×10^{-4}	2.890
^{169}Tm	$\frac{1}{2}$	100	—	5.66×10^{-4}	5.66×10^{-4}	8.271
^{171}Yb	$\frac{1}{2}$	14.31	—	5.46×10^{-3}	7.81×10^{-4}	17.613
^{173}Yb	$\frac{5}{2}$	16.13	2.8	1.33×10^{-3}	2.14×10^{-4}	4.852
^{174}Lu	1	—	—	—	—	—
^{175}Lu	$\frac{7}{2}$	97.41	5.68	3.12×10^{-2}	3.03×10^{-2}	11.407
^{176}Lu	7	2.59	8.1	3.72×10^{-2}	9.63×10^{-4}	7.928
^{177}Hf	$\frac{7}{2}$	18.5	4.5	6.38×10^{-4}	1.18×10^{-4}	3.120
^{179}Hf	$\frac{9}{2}$	13.75	5.1	2.16×10^{-4}	2.97×10^{-5}	1.869
^{181}Ta	$\frac{7}{2}$	99.98	3.0	3.60×10^{-2}	3.60×10^{-2}	11.970
^{183}W	$\frac{1}{2}$	14.4	—	7.20×10^{-4}	1.03×10^{-5}	4.161
^{185}Re	$\frac{5}{2}$	37.07	2.8	0.13	4.93×10^{-2}	22.513
^{187}Re	$\frac{5}{2}$	62.93	2.6	0.13	8.62×10^{-2}	22.744
^{187}Os	$\frac{1}{2}$	1.64	—	1.22×10^{-5}	2.00×10^{-7}	2.303
^{189}Os	$\frac{3}{2}$	16.1	0.8	2.34×10^{-3}	3.76×10^{-4}	7.758
^{191}Ir	$\frac{3}{2}$	37.3	1.5	2.53×10^{-5}	9.43×10^{-6}	1.718
^{193}Ir	$\frac{3}{2}$	62.7	1.4	3.27×10^{-5}	2.05×10^{-5}	1.871
^{195}Pt	$\frac{1}{2}$	33.8	—	9.94×10^{-3}	3.36×10^{-3}	21.499
^{197}Au	$\frac{3}{2}$	100	0.58	2.51×10^{-5}	2.51×10^{-5}	1.712
^{199}Hg	$\frac{1}{2}$	16.84	—	5.67×10^{-3}	9.54×10^{-4}	17.827
^{201}Hg	$\frac{3}{2}$	13.22	0.5	1.44×10^{-3}	1.90×10^{-4}	6.599
^{203}Ti	$\frac{1}{2}$	29.5	—	0.18	5.51×10^{-2}	57.149
^{205}Ti	$\frac{1}{2}$	70.5	—	0.19	0.13	57.708
^{207}Pb	$\frac{1}{2}$	22.6	—	9.16×10^{-3}	2.07×10^{-3}	20.921
^{209}Bi	$\frac{9}{2}$	100	−0.4	0.13	0.13	16.069
^{235}U	$\frac{7}{2}$	0.72	4.1	1.21×10^{-4}	8.71×10^{-7}	1.790

Source: Reprinted from B. C. Gerstein and C. R. Dybowski, *Transient Techniques in NMR of Solids: An Introduction to Theory and Practice*, Academic Press, Orlando, FL, 1985, p. 8–11.

[a] *Relative sensitivity*—at constant field for equal numbers of nuclei; *Absolute sensitivity*—product of relative sensitivity and natural abundance.

APPENDIX B

SOLUTIONS TO EXERCISES

CHAPTER 1

1.1 The total energy (dipolar + anisotropy) of the sphere is:

$$E_{\text{total}} = -BM_s \cos\theta + K\sin^2\theta$$

Expanding $\cos\theta$ and $\sin^2\theta$ around $\theta = \pi$, we obtain

$$\cos\theta \approx -1 + \tfrac{1}{2}(\theta - \pi)^2 + O(\theta^4) \text{ and } \sin^2(\theta) \approx (\theta - \pi)^2 + O(\theta^3)$$

and the total energy becomes

$$E_{\text{total}} \approx -BM_s\left[-1 + \tfrac{1}{2}(\theta - \pi)^2\right] + K(\theta - \pi)^2$$

Thus, the minimum of E around π will be

$$\frac{dE}{d\theta} = -BM_s(\theta - \pi) + 2K(\theta - \pi) = 0 \text{ or } B = \frac{2K}{M_s}$$

1.2 Consider the gradient of the function $x\mathbf{M}$:

$$\nabla(x\mathbf{M}) = (\nabla x) \cdot \mathbf{M} + x\nabla \cdot \mathbf{M} = \mathbf{i} \cdot \mathbf{M} + x\nabla \cdot \mathbf{M} = M_x + x\nabla \cdot \mathbf{M}$$

Integrating this expression in V, one gets

$$\int_V \nabla(x\mathbf{M})\,dv = \int_V M_x\,dv + \int_V x\nabla\cdot\mathbf{M}\,dv$$

$$\oint_S x\mathbf{M}\cdot\mathbf{n}\,da = \int_V M_x\,dv - \int_V x\rho_m\,dv$$

Multiplying this expression by the unit vector $\mathbf{i}$, we get

$$\int_V \mathbf{i}M_x\,dv = \int_V (\mathbf{i}x)\rho_m\,dv + \oint_S (\mathbf{i}x)\mathbf{M}\cdot\mathbf{n}\,da$$

Doing the same with $\mathbf{j}$ and $\mathbf{k}$ and summing the three expressions we obtain, with $\mathbf{M}\cdot\mathbf{n} = \sigma_m$

$$\int_V (\mathbf{i}M_x + \mathbf{j}M_y + \mathbf{k}M_z)\,dv = \boldsymbol{\mu} = \int_V \mathbf{r}\rho_m\,dv + \oint_S \mathbf{r}\sigma_m\,da$$

1.3 The expression of the total magnetic energy for a distribution of dipoles with magnetization $\mathbf{M}$ is:

$$E = -\tfrac{1}{2}\int_V \mathbf{M}\cdot\mathbf{B}\,dV$$

In the absence of applied field, $\mathbf{B} - -\mu_0 N_d \mathbf{M} = \mathbf{B}_d$. Therefore

$$E = \frac{\mu_0}{2} N_d \int_V M^2\,dV = \frac{\mu_0}{2} N_d M^2 \frac{4\pi}{3} R^3$$

where R is the radius of the sphere. Writing $R = d/2$, where d is the diameter and using for the demagnetizing factor of the sphere $N_d = \tfrac{1}{3}$, we find

$$E = \frac{\pi}{36}\mu_0 M^2 d^3$$

1.4 We use $\nabla\cdot\mathbf{B} = 0$, $\nabla\times\mathbf{H} = 0$, and $\mathbf{H} = -\nabla\Phi^*$:

$$\mathbf{B} = \mu_0(\mathbf{H}+\mathbf{M})$$

$$\nabla\cdot\mathbf{B} = \mu_0(\nabla\cdot\mathbf{H} + \nabla\cdot\mathbf{M}) = \mu_0(-\nabla^2\Phi^* + \nabla\cdot\mathbf{M}) = 0$$

For a uniform magnetization, $\nabla\cdot\mathbf{M} = 0$, and we remain with Laplace equation $\nabla^2\Phi^* = 0$. Taking as solutions inside the sphere and outside the sphere, we have

$$\Phi_1^*(r,\theta) = A_1 r\cos\theta + C_1 r^{-2}\cos\theta \quad \text{and} \quad \Phi_2^*(r,\theta) = A_2 r\cos\theta + C_2 r^{-2}\cos\theta$$

Far from the sphere, we have $\mathbf{H} = H\mathbf{i} = (B/\mu_0)/\mathbf{i}$. Using the boundary conditions: $H_{1\theta} = H_{2\theta}$ and $B_{1r} = B_{2r}$, we finally obtain:

$$\mathbf{H}_{int} = \frac{3B}{\mu + 2\mu_0}\mathbf{i} = \frac{3}{\mu/\mu_0 + 2}H\mathbf{i}; \quad \text{and} \quad \mathbf{B}_{int} = \mu\mathbf{H} = \frac{3}{1 + 2\mu_0/\mu}B\mathbf{i}$$

Since $\mu/\mu_0 > 1$, $H_{int} < H$ and $B_{int} > B$.

CHAPTER 2

2.1 The Lorentz force (in the SI) acting on the electron is:

$$\mathbf{F} = -e(\mathbf{E} + \mathbf{v}\times\mathbf{B}), \quad \text{where} \quad \mathbf{E} = \frac{1}{4\pi\epsilon_0}\frac{e\mathbf{r}}{r^3}$$

is the electric field derived from a central potential. With $\mathbf{B}\|\mathbf{k}$ we find the following system of coupled equations:

$$m_e\frac{dv_x}{dt} + eBv_y + \frac{1}{4\pi\epsilon_0}\frac{e^2 x}{r^3} = 0; \quad m_e\frac{dv_y}{dt} - eBv_x + \frac{1}{4\pi\epsilon_0}\frac{e^2 y}{r^3} = 0$$

Transforming to polar coordinates

$$x(t) = r\cos\omega t \Longrightarrow v_x(t) = -r\omega\sin\theta;$$

$$y(t) = r\sin\omega t \Longrightarrow v_y(t) = r\omega\cos\theta$$

where $\omega = d\theta/dt$. Therefore

$$\frac{dv_x}{dt} = -r\omega^2\cos\theta; \quad \frac{dv_y}{dt} = -r\omega^2\sin\theta$$

Substituting into the equations of motion, we obtain the following equation of the second degree in ω:

$$\omega^2 - \left(\frac{eB}{m_e}\right)\omega - \frac{1}{4\pi\epsilon_0}\frac{e^2}{m_e r^3} = 0$$

whose solutions are

$$\omega = \frac{eB}{2m_e} \pm \sqrt{\left(\frac{eB}{2m_e}\right)^2 + \frac{1}{4\pi\epsilon_0}\frac{e^2}{m_e r^3}}$$

with $e = 1.6 \times 10^{-19}$ C; $m = 9.1 \times 10^{-31}$ kg, and taking $r \sim 10^{-10}$ m and $B \sim 1$ tesla, it follows that $eB/2m \cong 8.8 \times 10^{10}$ s^{-1}. On the other hand, $e^2/(4\pi\epsilon_0 mr^3) \cong 2.5 \times 10^{32}$ s^{-1}. Consequently

$$\left(\frac{eB}{2m}\right)^2 \ll \frac{1}{4\pi\epsilon_0}\frac{e^2}{mr^3} \quad \text{and} \quad \omega \approx \frac{eB}{2m} + \sqrt{\frac{1}{4\pi\epsilon_0}\frac{e^2}{mr^3}} \approx 1.6 \times 10^{16} \text{ s}^{-1}$$

2.2 Starting from the 1s wavefunction of the electron, we obtain

$$\psi_{1s} = \frac{1}{\sqrt{\pi a_0^3}} e^{-r/a_0}$$

We have to compute the expectation value of r^2, $\langle r^2 \rangle$:

$$\langle r^2 \rangle = \int r^2 |\psi_{1s}|^2 d\mathbf{r}$$

In spherical coordinates, we have

$$\langle r^2 \rangle = \frac{1}{\pi a_0^3} \int_0^{2\pi} \int_0^{\pi} \int_0^{\infty} r^4 e^{-2r/a_0} \sin\theta \, dr \, d\theta \, d\phi$$

The angular integrals are trivial. For the integration in r, we use

$$\int_0^{\infty} r^4 e^{-\alpha r} dr = \frac{d^4}{d\alpha^4} \int_0^{\infty} e^{-\alpha r} dr$$

With $\alpha = 2/a_0$, we find $3a_0^5/4$ for this integral. Then $\langle r^2 \rangle = 3a_0^2$. Substituting into Eq. (2.12) we find

$$\chi = -\frac{\mu_0 n Z e^2}{2m_e} a_0^2$$

where $a_0 = 5.3 \times 10^{-11}$ m, $Z = 1$. For ^{1}H, $\rho = 89$ kg m^{-3} and $A = 1$. As a consequence, $n \simeq 5.3 \times 10^{28}$ atoms m^{-3}. Thus $\chi \simeq -2.6 \times 10^{-6}$.

2.3 Saturation magnetization of iron, $M_s = 1.7 \times 10^6$ A m^{-1}; density of iron, $\rho = 7970$ kg m^{-3} and Avogadro constant, $N_A = 6.025 \times 10^{26}$ kg^{-1}; atomic mass of iron, $A = 56$. Then, the number of iron atoms per cubic meter will be $N = 857.5 \times 10^{26}$ m^{-3} The magnetic moment per atom is $m_A = 1.98 \times 10^{-23}$ A m^2. Dividing by the Bohr magneton ($\mu_B = 9.27 \times 10^{-24}$ J T^{-1}) we obtain

$$m_A = 2.14 \, \mu_B$$

2.4 Let $\lambda_{AB} = \lambda_{AB} = -\lambda$ be the (antiferromagnetic) average coupling parameter between the sublattices A and B, and $\lambda_{AA} = \lambda_{BB} = \lambda'$, the (ferromagnetic) coupling parameter within each sublattice. Then, the total fields acting on each sublattice will be

(a) $\mathbf{B}_A = \mathbf{B} - \lambda \mathbf{M}_B + \lambda' \mathbf{M}_A$; $\mathbf{B}_B = \mathbf{B} - \lambda \mathbf{M}_A + \lambda' \mathbf{M}_B$

where $\mathbf{B}$ is the applied field.

(b) Using Eq. (2.88), valid for high temperatures, the magnetization of each sublattice can be written

$$\mathbf{M}_A \approx \frac{C}{T\mu_0}(\mathbf{B} - \lambda \mathbf{M}_B + \lambda' \mathbf{M}_A)$$

and

$$\mathbf{M}_B \approx \frac{C}{T\mu_0}(\mathbf{B} - \lambda \mathbf{M}_A + \lambda' \mathbf{M}_B)$$

where C is the Curie constant.

We thus obtain the following system of equations for $\mathbf{M}_A$ and $\mathbf{M}_B$:

$$\mathbf{M}_A\left(1 - \frac{C\lambda'}{T\mu_0}\right) + \frac{C\lambda}{T\mu_0}\mathbf{M}_B = \frac{C\mathbf{B}}{T\mu_0}$$

$$\frac{C\lambda}{T\mu_0}\mathbf{M}_A + \left(1 - \frac{C\lambda'}{T\mu_0}\right)\mathbf{M}_B = \frac{C\mathbf{B}}{T\mu_0}$$

(c) The Néel temperature is that for which the system has solution $\mathbf{M}_A, \mathbf{M}_B \neq 0$ for $\mathbf{B} = 0$. Therefore the determinant of the coefficients has to be zero

$$\left(1 - \frac{C\lambda'}{T_N\mu_0}\right)^2 = \left(\frac{C\lambda}{T_N\mu_0}\right)^2 = 0, \qquad T_N = \frac{C}{\mu_0}(\lambda \pm \lambda')$$

2.5 The classical partition function is

$$\mathcal{Z} = \int_0^\pi e^{-\mu B \cos\theta/kT} \sin\theta\, d\theta$$

with

$$x = \cos\theta,\ dx = -\sin\theta\, d\theta$$

$$\mathcal{Z} = \int_{-1}^{+1} e^{-(\mu B/kT)x}\, dx = -\frac{kT}{\mu B}\left(e^{-\mu/kT} - e^{+\mu/kT}\right)$$

$$M = \frac{\int_0^\pi e^{-\mu B \cos\theta/kT} \sin\theta\, d\theta}{\mathcal{Z}} = -kT\frac{\partial}{\partial B}\ln\mathcal{Z} = \mu\left[\coth\left(\frac{\mu B}{kT}\right) - \frac{kT}{\mu B}\right]$$

2.6 The magnetic field **B** felt by the electron is

$$-\frac{\mu_0}{4\pi}\mathbf{v}\times\mathbf{E},\ \text{where}\ \mathbf{E} = -\nabla\left(-\frac{V}{e}\right) = \frac{1}{e}\frac{\mathbf{r}}{r}\frac{dV}{dr}$$

Therefore

$$\mathbf{B} = -\frac{\mu_0}{4\pi}\frac{1}{e}\left(\frac{1}{r}\frac{dV}{dr}\right)\mathbf{v}\times\mathbf{r}$$

But, $\mathbf{v} = \mathbf{p}/2m_e$ and $\mathbf{p}\times\mathbf{r} = -\hbar\mathbf{l}$. Therefore

$$\mathbf{B} = \frac{\mu_0}{4\pi}\frac{\hbar}{2m_e}\left(\frac{1}{r}\frac{dV}{dr}\right)\mathbf{l}$$

The interaction of this field with the electronic spin gives the spin–orbit coupling:

$$\mathcal{H}_{so} = -2\mu_B\mathbf{s}\cdot\mathbf{B}\ \text{or}\ \mathcal{H}_{so} = -\frac{\mu_0}{4\pi}2\mu_B\frac{\hbar}{em_e}\left(\frac{1}{r}\frac{dV}{dr}\right)\mathbf{s}\cdot\mathbf{l}$$

Using $\mu_B = -e\hbar/2m_e$, it becomes

$$\mathcal{H}_{so} = \frac{\mu_0}{4\pi}\frac{\hbar^2}{m_e^2}\left(\frac{1}{r}\frac{dV}{dr}\right)\mathbf{s}\cdot\mathbf{l}\ \text{or}\ \mathcal{H}_{so} = \zeta(r)\mathbf{s}\cdot\mathbf{l}$$

2.7 The hamiltonian of the quadrupole interaction (without orthorhombic distortion) is:

$$\mathcal{H}^q_{cf} = B^0_2 O^0_2 = B^0_2(3J^2_c - J^2)$$

where J_c is the component of **J** along the **c** direction of the crystal. If the direction of magnetization (z axis) makes an angle θ with the **c** axis, the relation between the components of **J** in the two systems of coordinates (crystalline and ionic) is:

$$J_c = J_z \cos\theta + J_x \sin\theta$$

and then

$$J^2_c = J^2_z \cos^2\theta + J^2_x \sin^2\theta + (J_z J_x + J_x J_z)\sin\theta\cos\theta$$

Expressing J_x and J_y in terms of the operators J_+ and J_- we obtain:

$$J^2_c = J^2_z \cos^2\theta + [J^2 + \tfrac{1}{4}(J^2_+ + J^2_- - 2J_- J_+ - 2J_z) - J^2_z]\sin^2\theta$$
$$+ \tfrac{1}{2}(J_z J_+ + J_z J_- + J_+ J_z + J_- J_z)$$

Substituting this expression in $\mathcal{H}^q_{cf}$ and computing $\langle JJ|\mathcal{H}^q_{cf}|JJ\rangle$, all the nondiagonal terms cancel, remaining only the terms in J_z, J^2_z and J^2:

$$\langle \mathcal{H}^q_{cf}\rangle = \langle JJ|\mathcal{H}^q_{cf}|JJ\rangle = B^0_2\left[3J^2\cos^2\theta + \frac{3J}{2}\sin^2\theta - J(J+1)\right]$$

Rearranging the terms, we find

$$\langle \mathcal{H}^q_{cf}\rangle = B^0_2\left[J\underbrace{\left(\frac{1-3\cos^2\theta}{2}\right)}_{-\mathcal{P}_2(\cos\theta)}(1-2J)\right]$$

Therefore,

$$\langle \mathcal{H}^q_{cf}\rangle = B^0_2 J(2J-1)\mathcal{P}_2(\cos\theta)$$

2.8 (a) The potential acting at the origin due to 6 charges is:

$$V = \frac{q}{4\pi\epsilon_0}\left(\frac{1}{|\mathbf{r}-a\mathbf{i}|} + \frac{1}{|\mathbf{r}+a\mathbf{i}|} + \frac{1}{|\mathbf{r}-a\mathbf{j}|} + \frac{1}{|\mathbf{r}+a\mathbf{j}|} + \frac{1}{|\mathbf{r}-b\mathbf{k}|} + \frac{1}{|\mathbf{r}+b\mathbf{k}|}\right)$$

For $r \ll a, b$, each term of this expression can be expanded in a power series. For example:

$$\frac{1}{|r \pm ai|} \approx \frac{1}{a} \mp \frac{x}{a^2} - \frac{r^2}{2a^3} + \frac{3x^2}{2a^3} + \ldots$$

with similar expressions for the other terms. One then finds for the potential:

$$V \simeq \text{constant} + \frac{q}{4\pi\epsilon_0}\left(\frac{1}{a^3} - \frac{1}{b^3}\right)r^2 + 3\left(\frac{1}{b^3} - \frac{1}{a^3}\right)z^2$$

that is,

$$V = \frac{q}{4\pi\epsilon_0}\left(\frac{1}{b^3} - \frac{1}{a^3}\right)(3z^2 - r^2) + \text{constant}$$

Thus, we may write $\mathcal{H} = A(3z^2 - r^2)$ with $A > 0$ (since $b < a$). Note that if the symmetry were cubic ($b = a$), we would have $A = 0$.

(b) The wavefunctions p of the electron are

$$p_x = xf(r) = rf(r)\sin\theta\cos\phi = R(r)\sin\theta\cos\phi;$$
$$p_y = yf(r) = rf(r)\sin\theta\sin\phi = R(r)\sin\theta\sin\phi$$
$$p_z = zf(r) = rf(r)\cos\theta = R(r)\cos\theta$$

Thus, the interaction energy of one electron in the orbital p_x in the crystal field will be

$$\langle p_x|\mathcal{H}|p_x\rangle = A\int r^2|R(r)|^2 d\tau_r \int \sin^2\theta\cos^2\phi(3\cos^2\theta - 1)d\Omega$$

The integral in r is simply the expectation value $<r^2>$. The angular integral is

$$\int \sin^2\theta\cos^2\phi(3\cos^2\theta - 1)d\Omega = -\tfrac{8}{15}\pi$$

Therefore

$$\langle p_x|\mathcal{H}|p_x\rangle = -\tfrac{8}{15}\pi A\langle r^2\rangle = \langle p_y|\mathcal{H}|p_y\rangle$$

For p_z, one finds $\langle p_z|\mathcal{H}|p_z\rangle = \tfrac{16}{15}\pi A\langle r^2\rangle$. Consequently, the p_x and p_y

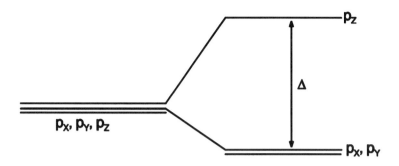

Figure B.1 Crystal field energy levels.

levels remain degenerate, while the p_z level is "lifted" by a quantity $\Delta = \frac{8}{5}\pi A \langle r^2 \rangle$ as shown in Fig. B.1.

(c) With a field applied along the z direction the total hamiltonian becomes (considering only the orbital magnetic moment)

$$\mathcal{H} = A(3z^2 - r^2) + \mu_B B L_z$$

In order to compute the matrix elements of this hamiltonian between the states p_x, p_y and p_z, we need only $\langle p_x, p_y, p_z | L_z | p_x, p_y, p_z \rangle$ and the nondiagonal terms of $\mathcal{H}_{cf}$ since the diagonal terms have been computed in the previous item. However, we may show that L_z has no diagonal elements in this representation. For example:

$$\langle p_x | L_z | p_x \rangle = \frac{\hbar}{i} \int xf(r) \left(x \frac{\partial}{\partial y} - y \frac{\partial}{\partial x} \right) xf(r) d^3 r$$

or, in spherical coordinates

$$\langle p_x | L_z | p_x \rangle = -\frac{\hbar}{i} \int |rf(r)|^2 (r^2 dr) \int_0^\pi \sin^3 \theta d\theta \int_0^{2\pi} \sin\phi \cos\phi \, d\phi = 0$$

The same is found for the other diagonal elements. Outside the diagonal, the only nonzero elements are

$$\langle p_y | L_z | p_x \rangle = i\hbar = -\langle p_x | L_z | p_y \rangle$$

The hamiltonian matrix then becomes

$$\mathcal{H} = \begin{pmatrix} E_0 & -i\mu_B B & 0 \\ i\mu_B B & E_0 & 0 \\ 0 & 0 & E_1 \end{pmatrix}$$

where $E_0 = -\frac{8}{15}\pi^2 A\langle r^2\rangle$ and $E_1 = \frac{16}{15}\pi^2 A\langle r^2\rangle$.

(d) The eigenvalues of this hamiltonian are obtained from the equation:

$$\begin{vmatrix} E_0 - \epsilon & -i\mu_B B & 0 \\ i\mu_B B & E_0 - \epsilon & 0 \\ 0 & 0 & E_1 - \epsilon \end{vmatrix} = 0$$

which has the roots

$$\epsilon_1 = E_1; \quad \epsilon_2 = E_0 + \mu_B B; \quad \epsilon_3 = E_0 - \mu_B B$$

that is, the degeneracy in x and y is lifted by the field, while E_1 is not altered.

CHAPTER 3

3.1 First part: we need the expression (3.86); second part—as follows:

$$\hbar\omega = 2JS\left[z - \sum_\delta \cos(\mathbf{k} \cdot \boldsymbol{\delta})\right]$$

For $\mathbf{k} \cdot \boldsymbol{\delta}$ small:

$$\cos(\mathbf{k} \cdot \boldsymbol{\delta}) \approx 1 + \frac{(\mathbf{k} \cdot \boldsymbol{\delta})^2}{2} = 1 - \frac{(ka)^2}{2};$$

$$\hbar\omega \approx 2JS\left[z - \sum_\delta \left(1 - \frac{(ka)^2}{2}\right)\right] \approx (2JSa^2)k^2$$

3.2 $\mathcal{U} = \dfrac{V}{(2\pi)^3}\displaystyle\int d^3k \dfrac{\hbar\omega(k)}{e^{\hbar\omega/kT} - 1}; \quad \omega(k) = Ak^2; \quad d^3k = 4\pi k^2 dk$

$$\mathcal{U} = \frac{V}{(2\pi)^3} \int 4\pi k^2 dk \frac{\hbar A k^2}{e^{\hbar A k^2/kT} - 1}$$

Let $y = \hbar A k^2/kT$; then

$$dy = \frac{\hbar A}{kT} 2k\, dk = 2\left(\frac{\hbar A}{kT}\right)^{1/2} y^{1/2} dk; \quad dk = \frac{1}{2} y^{-1/2} \left(\frac{kT}{\hbar A}\right)^{1/2} dy$$

$$\mathcal{U} = \frac{V}{(2\pi)^3} 4\pi \hbar A \int \left(\frac{kT}{\hbar A}\right) y \frac{1}{2} \left(\frac{kT}{\hbar A}\right)^{1/2} y^{-1/2} dy \left(\frac{kT}{\hbar A}\right) y \frac{1}{e^y - 1}$$

$$\mathcal{U} = \frac{2\pi V}{(2\pi)^3} \hbar A \left(\frac{kT}{\hbar A}\right)^{5/2} \int \frac{y^{3/2} dy}{e^y - 1}$$

$$\mathcal{U} = \text{constant} \times T^{5/2}; \quad \left(\frac{\partial \mathcal{U}}{\partial T}\right)_v = C_v \propto T^{3/2}$$

3.3 $M(T) = M(0)\left[1 - \frac{1}{NS}\sum_k n_k\right]; \quad \sum_k n_k \longrightarrow \frac{V}{(2\pi)^3} \int \frac{d^3k}{e^{\hbar\omega/kT} - 1}$

At low temperatures, $\omega(k) \approx (2JSa^2)k^2$ (Exercise 3.1)

$$\int \frac{d^3k}{e^{\hbar\omega/kT} - 1} = \int \frac{4\pi k^2\, dk}{e^{\hbar A k^2/kT} - 1}$$

Substituting $x = \hbar A k^2/kT$, $dk = \frac{1}{2}(kT/\hbar A)^{1/2} x^{-1/2} dx$, the integral becomes

$$\int 4\pi \left(\frac{kT}{\hbar A}\right) \frac{1}{2} \left(\frac{kT}{\hbar A}\right)^{1/2} x^{-1/2} \frac{dx}{e^x - 1} = \text{constant} \times T^{3/2}$$

Thus

$$\frac{M(T) - M(0)}{M(0)} = \text{constant} \times T^{3/2}$$

CHAPTER 4

4.1 The susceptibility of the gas is given by

$$\chi(T) = \frac{\partial M(T)}{\partial H},$$

where $M(T) = -\mu_B(n_+ - n_-)$ and $n_\pm(T) = \int n_\pm(\epsilon) f(\epsilon) d\epsilon$ with $f(\epsilon)$ the Fermi-Dirac distribution and

$$n_\pm(\epsilon) = \tfrac{1}{2}n(\epsilon \pm \mu_B B)$$

For $\mu_B B \ll E_F$ we may approximate

$$n_\pm(\epsilon) \approx \tfrac{1}{2}n(\epsilon) \pm \tfrac{1}{2}\mu_B B n'(\epsilon)$$

where $n'(\epsilon)$ is the first derivative of n in relation to ϵ. With this approximation we obtain

$$M(T) = \mu_B^2 B \int n'(\epsilon) f(\epsilon) d\epsilon \quad \text{and} \quad \chi(T) = \mu_B^2 \int n'(\epsilon) f(\epsilon) d\epsilon$$

To evaluate this integral, we have to make a Sommerfeld expansion (e.g., Ashcroft and Mermin 1976):

$$\int_{-\infty}^{+\infty} n'(\epsilon) f(\epsilon) d\epsilon \approx \int_{-\infty}^{\mu} n'(\epsilon) d\epsilon + \frac{\pi^2}{6}(kT)^2 n''(E_F)$$

where the integral on the right hand side may be written as

$$\int_{-\infty}^{\mu} n'(\epsilon) d\epsilon = \int_0^{E_F} n'(\epsilon) d\epsilon + (\mu - E_F) n'(E_F) = n(E_F) + (\mu - E_F) n'(E_F)$$

where μ is the chemical potential. Consequently

$$\chi(T) \approx \mu_B^2 \left[n(E_F) + (\mu - E_F) n'(E_F) + \frac{\pi^2}{6}(kT)^2 n''(E_F) \right]$$

Now, the condition for the conservation of the number of electrons with T is (e.g., Ashcroft and Mermin 1976).

$$\mu(T) = E_F - \frac{\pi^2}{6}(kT)^2 \frac{n'(E_F)}{n(E_F)}$$

With this, we obtain

$$\chi(T) \approx \chi_0 \left\{ 1 - \frac{\pi^2}{6}(kT)^2 \left[\left(\frac{n'(E_F)}{n(E_F)} \right)^2 - \left(\frac{n''(E_F)}{n(E_F)} \right) \right] \right\}$$

where $\chi_0 = \mu_B^2 n(E_F)$. For a gas of free electrons, the density of states may be written as

$$n(\epsilon) = \frac{3}{2}\frac{n}{E_F}\left(\frac{\epsilon}{E_F}\right)^{1/2}$$

Therefore

$$\chi(T) = \chi_0\left[1 - \frac{\pi^2}{12}\left(\frac{kT}{E_F}\right)^2\right]$$

4.2 The total number of electrons with moment up is given by

$$N_\uparrow = \frac{1}{2}\int_{-\mu_B B}^{E_F} n(\epsilon + \mu_B)d\epsilon$$

and the energy of these electrons is

$$E_\uparrow = \frac{1}{2}\int_{-\mu_B B}^{E_F} \epsilon n(\epsilon + \mu_B)d\epsilon$$

where for the free-electron gas $n(\epsilon) = A\epsilon^{1/2}$. Substituting this expression in the preceding integrals one finds

$$N_\uparrow = \tfrac{1}{3}A(E_F + \mu_B B)^{2/3}$$

and

$$E_\uparrow = \tfrac{6}{10}N_\uparrow(E_F + \mu_B B) - \mu_B B N_\uparrow$$

But

$$E_F + \mu_B B = \left(\frac{3N_\uparrow}{2A}\right)^{2/3} \quad \text{with} \quad N_\uparrow = \tfrac{1}{2}N(1+\zeta)$$

one finds

$$E_\uparrow = E_0(1+\zeta)^{5/3} - \tfrac{1}{2}N\mu_B B(1+\zeta)$$

with $E_0 = (\tfrac{3}{10})NE_F$. The computation for $N_\downarrow$ is identical, changing only $\epsilon + \mu_B$ for $\epsilon - \mu_B$ in the argument of n.

The total energy $E = E_\uparrow + E_\downarrow$ will then be equal to

$$E = E_0[(1+\zeta)^{5/3} + (1-\zeta)^{5/3}] - N\mu_B B\zeta$$

In the limit $\zeta \ll 1$ we may approximate

$$(1 \pm \zeta)^{5/3} \approx 1 \pm \frac{5}{3}\zeta + \frac{5}{9}\zeta^2$$

Therefore,

$$E \approx \frac{10}{9} E_0 \zeta^2 - N\mu_B B\zeta, \quad \text{and} \quad \frac{dE}{d\zeta} \cong \frac{20}{9} E_0 \zeta - N\mu_B B$$

The value of ζ for which the derivative is annulled is

$$\zeta_0 = \frac{9}{20} \frac{N\mu_B B}{E_0}$$

The magnetization will be

$$M = \mu_B(N_\uparrow - N_\downarrow) = \mu_B N \zeta_0 = \frac{9}{20} \frac{\mu_B^2 N^2 B}{E_0}$$

and

$$M = \frac{3}{2} \frac{N\mu_B^2}{E_F} B$$

4.3
$$\frac{2}{3} \approx \frac{k\theta'}{E_F} \quad \text{and} \quad \theta' = \frac{\lambda n \mu_B^2}{k}$$

$$\frac{2}{3} - \frac{\lambda \mu_B^2 n}{E_F} \approx 0$$

We have to find a relation between n, E_F, and $n(E_F)$:

$$n(E_F) = \frac{1}{\pi^2} \frac{m_e^{3/2}}{\hbar^3} (2E_F)^{1/2}$$

We know that

$$k_F = (3\pi^2 n)^{1/3}$$

$$E_F = \frac{\hbar^2}{2m_e}(3\pi^2 n)^{2/3}$$

then

$$\frac{m_e}{\hbar^2} = \frac{(3\pi^2 n)^{2/3}}{2E_F}$$

substituting into $n(E_F)$, we obtain

$$n(E_F) = \frac{3n}{2E_F}$$

replacing above

$$\tfrac{2}{3} - \tfrac{2}{3}\lambda \mu_B^2 n(E_F) = 0$$

then

$$1 - \lambda \mu_B^2 n(E_F) = 0$$

4.4 With $N_\uparrow$ equal to the total number of moments up, the total number of pairs of distinct moments up electrons will be equal to

$$\tfrac{1}{2}(N_\uparrow^2 - N_\uparrow) = \tfrac{1}{2}N_\uparrow(N_\uparrow - 1) \approx \tfrac{1}{2}N_\uparrow^2$$

since $N_\uparrow \approx 10^{23} \gg 1$. Since each pair of parallel spins interacts with a constant exchange energy equal to $-V$, the total exchange energy will be equal to

$$E_{\text{exch}} = -\tfrac{1}{2}VN_\uparrow^2 = -\tfrac{1}{8}VN^2(1+\zeta)^2$$

Adding to the interaction energy the interaction with the field B computed in exercise 4.2, we have

$$E^+ = E_0(1+\zeta)^{5/3} - \tfrac{1}{8}VN^2(1+\zeta)^2 - \tfrac{1}{2}N\mu_B B(1+\zeta)$$

In the same way we find the energy for the moment down band and the total energy will be

$$E = E_0[(1+\zeta)^{5/3} + (1-\zeta)^{5/3}] - \tfrac{1}{8}VN^2[(1+\zeta)^2 + (1-\zeta)^2] - N\mu_B B\zeta$$

Computing $dE/d\zeta$ and taking the limit $\zeta \ll 1$, we find that the value of ζ corresponding to the extreme of E, ζ_0, will be

$$\zeta_0 = \frac{3\mu_B B}{2E_F - (3/2)VN}$$

and the magnetization of the gas with exchange interactions will be

$$M = \mu_B(N_\uparrow - N_\downarrow) = \mu_B N \zeta_0 = \frac{3N\mu_B^2 B}{2E_F - \tfrac{3}{2}VN}$$

with the susceptibility given by

$$\chi = \frac{\partial M}{\partial H} = \frac{3N\mu_0 \mu_B^2}{2E_F - \tfrac{3}{2}VN}$$

We thus see that for $V > 0$ the exchange interaction increases the magnetic susceptibility of the gas. To ensure that there is spontaneous magnetization in the gas ($B = 0$), the denominator of the expression for M must be zero. That is, $V = 4E_F/3N$. In order to assure stability to such state of spontaneous magnetization, the extreme given by ζ_0 above must represent a minimum of E. For this to occur, we have to impose that its second derivative in ζ_0 is positive, which implies $V < 4E_F/3N$.

4.5 The number of particles with momentum between k and $k + dk$ is

$$dN = 2 \times \frac{4\pi k^2}{8\pi^3/V} dk = n(E)dE$$

Replacing k^2 and dk from $E = \hbar^2 k^2/2m_e$, one obtains

$$n(E)dE = \frac{V}{2\pi^2}\left(\frac{2m_e}{\hbar^2}\right)^{3/2} E^{1/2} \, dE$$

CHAPTER 5

5.1 The anisotropy energy is written under the form

$$U = K_1(\alpha_1^2\alpha_2^2 + \alpha_1^2\alpha_3^2 + \alpha_2^2\alpha_3^2) + K_2\alpha_1^2\alpha_2^2\alpha_3^2$$

We know that $\alpha_1^2 + \alpha_2^2 + \alpha_3^2 = 1$. Therefore

$$(\alpha_1^2 + \alpha_2^2 + \alpha_3^2)^2 = \alpha_1^4 + \alpha_2^4 + \alpha_3^4 + 2(\alpha_1^2\alpha_2^2 + \alpha_1^2\alpha_3^2 + \alpha_2^2\alpha_3^2) = 1$$

Thus,

$$(\alpha_1^2\alpha_2^2 + \alpha_1^2\alpha_3^2 + \alpha_2^2\alpha_3^2) = -\frac{\alpha_1^4 + \alpha_2^4 + \alpha_3^4}{2}$$

which shows that it is not necessary to include the term of the fourth power in the expression of U.

5.2

$$U_{\text{total}} = U_{el} + U_K + U_a \quad \text{and} \quad \frac{\partial U_{\text{total}}}{\partial \epsilon_{xy}} = C_{44}\epsilon_{xy} + B_2\alpha_1\alpha_2$$

Therefore

$$\frac{\partial U_{\text{total}}}{\partial \epsilon_{xy}} = 0 \Longrightarrow \epsilon_{xy} = -\frac{B_2\alpha_1\alpha_2}{C_{44}}$$

In the same way one finds

$$\epsilon_{xz} = -\frac{B_2\alpha_1\alpha_3}{C_{44}}; \quad \epsilon_{yz} = -\frac{B_2\alpha_2\alpha_3}{C_{44}}$$

The diagonal terms are

$$\frac{\partial U_{\text{total}}}{\partial \epsilon_{xx}} = C_{11}\epsilon_{xx} + C_{12}(\epsilon_{yy} + \epsilon_{zz}) + B_1\alpha_1^2 = 0$$

$$\frac{\partial U_{\text{total}}}{\partial \epsilon_{yy}} = C_{11}\epsilon_{yy} + C_{12}(\epsilon_{xx} + \epsilon_{zz}) + B_1\alpha_2^2 = 0$$

$$\frac{\partial U_{\text{total}}}{\partial \epsilon_{zz}} = C_{11}\epsilon_{zz} + C_{12}(\epsilon_{yy} + \epsilon_{xx}) + B_1\alpha_3^2 = 0$$

SOLUTIONS TO EXERCISES

Computing the principal determinant Δ and the determinants Δ_{xx}, Δ_{yy}, and Δ_{zz}, we can solve the system of equations

$$\epsilon_{xx} = \frac{\Delta_{xx}}{\Delta} = \frac{B_1[C_{12} - \alpha_1^2(C_{11} + 2C_{12})]}{(C_{11} - C_{12})(C_{11} + 2C_{12})}$$

and similar expressions for ϵ_{yy} and ϵ_{zz}.

5.3

$$J = \int_0^{\phi_0} [U_K + U_{\text{exch}}(\phi, \phi')] dx$$

(a)

$$\delta J = \int_0^{\phi_0} \left[\frac{dU_K}{d\phi} \delta\phi + \frac{dU_{\text{exch}}}{d\phi} \delta\phi + \frac{dU_{\text{exch}}}{d\phi'} \delta\phi' \right] dx$$

But $dU_{\text{ex}}/d\phi = dh/d\phi\, \phi'^2$; $dU_{\text{exch}}/d\phi' = 2h(\phi)\phi'$. Thus

$$\delta J = \int_0^{\phi_0} \left[\frac{dU_K}{d\phi} \delta\phi + \frac{dh}{d\phi} \phi'^2\, \delta\phi + 2h\phi'\, \delta\phi' \right] dx$$

Substituting $\phi'\, \delta\phi' = d/dx(\phi'\, \delta\phi) - \phi''\, \delta\phi$ leads to

$$\delta J = \int_0^{\phi_0} \left[\frac{dU_K}{d\phi} \delta\phi + \frac{dh}{d\phi} \phi'^2\, \delta\phi + 2h \frac{d}{dx}(\phi'\, \delta\phi) - 2h\phi''\, \delta\phi \right] dx$$

Thus

$$\delta J = \int_0^{\phi_0} \left[\left(\frac{dU_K}{d\phi} + \frac{dh}{d\phi} \phi'^2 - 2h\phi'' \right) \delta\phi + 2h \frac{d}{dx}(\phi'\, \delta\phi) \right] dx$$

(b) Integration of the last term yields

$$\int_0^{\phi_0} 2h \frac{d}{dx}(\phi'\, \delta\phi) dx$$

Calling $dv = (d/dx)(\phi' \, \delta\phi) \Rightarrow v = \phi' \, \delta\phi;\ u = 2h \Rightarrow du = 2dh/dx = 2(dh/d\phi)\, \phi'$. This results in

$$\int_0^{\phi_0} 2h \frac{d}{dx}(\phi' \, \delta\phi)dx = 2h(\phi)\phi' \, \delta\phi \Big]_0^{\phi_0}$$

$$-\int_0^{\phi_0} \phi' \, \delta\phi \, 2\frac{dh}{d\phi}\phi' dx = -2\int_0^{\phi_0} \phi'^2 \frac{dh}{d\phi} \delta\phi \, dx$$

(c) Substituting this result in the expression of δJ, we obtain

$$\delta J = \int_0^{\phi_0} \left[\left(\frac{dU_K}{d\phi} + \frac{dh}{d\phi}\phi'^2 - 2h\phi'' - 2\frac{dh}{d\phi}\phi'^2 \right) \delta\phi \, dx \right]$$

$\delta J = 0$ implies

$$\frac{dU_K}{d\phi} - \frac{dh}{d\phi}\phi'^2 - 2h\phi'' = 0$$

Multiplying by $\phi' = d\phi/dx$, we obtain

$$\frac{dU_K}{dx} - \frac{dh}{dx}\phi'^2 - 2h\phi'\phi'' = \frac{dU_K}{dx} - \frac{dh}{dx}[\phi'^2 h(\phi)] = 0 \text{ or}$$

$$U_K = \phi'^2 h(\phi) = U_{\text{exch}}$$

CHAPTER 6

6.1 The probability of occupation of a state with energy E_m is given by: $p_m = \exp[-(E_m/kT)]/Z$ where Z is the partition function

$$Z = \sum_{m=-I}^{+I} e^{-(E_m/kT)}$$

With the interaction with the magnetic field given by $\mathcal{H} = -\boldsymbol{\mu} \cdot \mathbf{B}$ and using $\mu' = gI$, the nuclear moment in units of μ_N

$$\frac{E_m}{k} = \frac{\mu' \mu_N B}{kI} m = 0.366 \left(\frac{\mu B}{I} \right) m$$

in millikelvins per tesla. Therefore, at 300, 4 and 0.01 K, we will have, respectively, $E_m/kT = 0.458 \times 10^{-5}$ m, 0.343×10^{-3} m, and 0.137 m. For

$I = 2$, we have $m = -2, -1, 0, 1, 2$, which, for $T = 10$ mK, gives

$$Z = e^{0.274} + e^{0.137} + 1 + e^{-0.137} + e^{-0.274} = 5.094$$

Thus, the probabilities of occupation of the states at 10 mK will be $p_{-2} \cong 0.258$ $p_{-1} \cong 0.255$, $p_0 \cong 0.196$, $p_1 \cong 0.171$, and $p_2 \cong 0.149$. The cases for $T = 300$ K and $T = 4$ K may be computed in a similar way.

6.2 The interaction energy of the nuclear moment with the magnetic field generated by an electron of momentum J can be written as

$$E = \frac{A}{\hbar} \langle \mathbf{I} \cdot \mathbf{J} \rangle$$

With the total angular momentum of the atom $\mathbf{F} = \mathbf{I} + \mathbf{J}$, we obtain

$$\langle \mathbf{I} \cdot \mathbf{J} \rangle = \tfrac{1}{2} \hbar [F(F+1) - I(I+1) - J(J+1)]$$

In the case of ^{87}Rb we have $I = \tfrac{5}{2}$ and a single electron in the state s, and therefore $J = S = \tfrac{1}{2}$. Thus

$$F = |I - J|, |I - J + 1|, ..., |I + J| = 2, 3, 4, 5, 6$$

and we obtain the following values for $2E/A\hbar$: -3.5, 2.5, 10.5, 20.5, and 32.5, corresponding to $F = 2, 3, 4, 5, 6$, respectively. If there is an applied field sufficiently strong to break the coupling I—J, the hyperfine energy takes the form

$$E_{m_I, m_J} = A m_I m_J - g_I \mu_N B m_I$$

There will then be two groups of nuclear energy levels, corresponding to $m_J = \tfrac{1}{2}$ and $m_J = -\tfrac{1}{2}$ with $m_I = -\tfrac{5}{2}, -\tfrac{3}{2}, -\tfrac{1}{2}, \tfrac{1}{2}, \tfrac{3}{2}, \tfrac{5}{2}$. For example, for $m_I = \tfrac{3}{2}$ we will have $E_{3/2} = 3A/4 - (\tfrac{3}{2}) g_I \mu_N B$ if $m_J = \tfrac{1}{2}$, and $E_{3/2} = -3A/4 - (\tfrac{3}{2}) g_I \mu_N B$ if $m_J = -\tfrac{1}{2}$. The other levels can be easily computed from the preceding expression.

6.3

$$B_i = \frac{a}{N} \sum_{j=1}^{N} I_j^z; \qquad B_i^2 = \left(\frac{a}{N}\right)^2 \sum_{j=1}^{N} \sum_{j'=1}^{N} I_j^z I_{j'}^z$$

If there is no interaction between the spins, the total wavefunction will be given by

$$|M\rangle = |m_1, m_2, \ldots, m_j, \ldots, m_{j'}, \ldots, m_N\rangle = |m_1\rangle \ldots |m_N\rangle$$

Therefore

$$\langle B_i^2 \rangle = \left(\frac{a}{N}\right)^2 \sum_{j=1}^{N} \sum_{j'=1}^{N} \frac{1}{2} m_j m_{j'} \delta_{j,j'}$$

$$\langle B_i^2 \rangle = \left(\frac{a}{N}\right)^2 \sum_{j=1}^{N} \frac{1}{2}\left(\frac{1}{2} \times \frac{1}{2} + \frac{-1}{2} \times \frac{-1}{2}\right) = \left(\frac{a}{2N}\right)^2 N$$

In the same way, we may compute

$$B_i^4 = \left(\frac{a}{N}\right)^4 \sum_{j=1}^{N} \sum_{k=1}^{N} \sum_{l=1}^{N} \sum_{m=1}^{N} I_j^z I_k^z I_l^z I_m^z$$

Computing the expectation value with the previous wavefunction, there will appear the following possibilities for the nonzero terms:

$$j = k;\ l = m; \qquad j = l;\ k = m; \qquad j = m;\ k = l$$

These three possibilities are equivalent to the multiplication of the sum by three. We also have that $m_j m_k m_l m_m = \frac{1}{16}$. It follows that

$$B_i^4 = \left(\frac{a}{N}\right)^4 \times 3 \times N^2 \times \frac{1}{16} = 3\left(\frac{a}{2N}\right)^4 N^2$$

6.4

$$\mathcal{H}_{cf} = \tfrac{3}{4} B_2^0 J_-^2$$

$$C^2 = \frac{1}{(g-1)^2 J} \sum_{M=-J}^{J-2} \frac{|\langle J; M|\mathcal{H}_{cf}|J; J\rangle|^2}{J - M}$$

$$J_-^2 |J; M\rangle = \sqrt{J(J+1) - M(M-1)}$$
$$\times \sqrt{J(J+1) - M(M-1)(M-2)}|J;\ M-2\rangle$$

$$\langle J; M|\mathcal{H}_{cf}|J;\ J\rangle = \tfrac{3}{4} B_2^0 \langle J; M|J_-^2|J;\ J\rangle$$

$$= \tfrac{3}{4}B_2^0\sqrt{J(J+1) - M(M-1)}$$
$$\times \sqrt{J(J+1) - M(M-1)(M-2)}|J; \quad M-2\rangle\delta_{M,J-2}$$

Substituting the values of J, g, and B_2^0, we obtain $C^2 = 66$ K. Thus:

$$\langle J_z \rangle = J\left(1 - \frac{66}{10000}\right) = 0.9934\ J$$

CHAPTER 7

7.1

$$\mathcal{H} = -\boldsymbol{\mu} \cdot \mathbf{B}_{hf} = \gamma\hbar\mathbf{I} \cdot \mathbf{B}_{hf}; \quad E_m = \hbar\omega_0 m; \quad \omega_0 = \gamma B_{hf}; \quad g\mu_N = \gamma\hbar$$

But $\mu' = gI$. It then follows that, $\gamma = (\mu_N/h)(\mu'/I(B_{hf}))$ and $\omega_0 = 2\pi\nu_0$. Then,

$$\nu_0 = \frac{\mu_N}{h}\left(\frac{\mu'}{I}B_{hf}\right)$$

Substituting $\mu_N = 5.05 \times 10^{-27}$ J T^{-1}, $h = 6.63 \times 10^{-34}$ J s^{-1} and the values of μ' and I, one obtains $\nu_0 = 17.3$ MHz.

7.2

$$\mathbf{A} = A_x\mathbf{x} + A_y\mathbf{y} + A_z\mathbf{z}$$

The components of **F** in the rotating system seen by an observer *in the laboratory system* are

$$\mathbf{A}(t) = A_{x'}(t)\mathbf{x}'(t) + A_{y'}(t)\mathbf{y}'(t) + A_{z'}(t)\mathbf{z}'(t)$$

Then

$$\frac{d\mathbf{A}}{dt} = \frac{dA_{x'}}{dt}\mathbf{x}' + \frac{dA_{y'}}{dt}\mathbf{y}' + \frac{dA_{z'}}{dt}\mathbf{z}' + A_{x'}\frac{d\mathbf{x}'}{dt} + A_{y'}\frac{d\mathbf{y}'}{dt} + A_{z'}\frac{d\mathbf{z}'}{dt}$$

Now take the particular case in which z and z' coincide. In this case we can deduce

$$\mathbf{x}' = \cos\omega t\,\mathbf{x} + \sin\omega t\,\mathbf{y}; \quad \mathbf{y}' = -\sin\omega t\,\mathbf{x} + \cos\omega t\,\mathbf{y}; \quad \mathbf{z}' = \mathbf{z}$$

Therefore

$$\frac{d\mathbf{x}'}{dt} = \omega(-\sin\omega t\mathbf{x} + \cos\omega t\mathbf{y})$$

$$\frac{d\mathbf{x}'}{dt} = \omega\mathbf{z} \times \sin\omega t\mathbf{y} + \cos\omega t\mathbf{x}$$

$$\frac{d\mathbf{x}'}{dt} = \boldsymbol{\omega} \times \mathbf{x}'$$

This result was computed for a particular direction of $\boldsymbol{\omega}$, the z direction. For an arbitrary direction $\boldsymbol{\omega} = \omega_{x'}\mathbf{x}' + \omega_{y'}\mathbf{y}' + \omega_{z'}\mathbf{z}'$, we will have

$$\boldsymbol{\omega} \times \mathbf{x}' = -\omega_{y'}\mathbf{z}' + \omega_{z'}\mathbf{y}'; \quad \boldsymbol{\omega} \times \mathbf{y}' = \omega_{x'}\mathbf{z}' - \omega_{z'}\mathbf{x}'; \quad \boldsymbol{\omega} \times \mathbf{z}' = -\omega_{x'}\mathbf{y}' + \omega_{y'}\mathbf{x}'$$

Thus

$$\frac{d\mathbf{x}'}{dt} = \omega_{z'}\mathbf{y}' - \omega_{y'}\mathbf{z}'$$

$$\frac{d\mathbf{y}'}{dt} = -\omega_{z'}\mathbf{x}' + \omega_{x'}\mathbf{z}'$$

$$\frac{d\mathbf{z}'}{dt} = \omega_{z'}\mathbf{x}' - \omega_{x'}\mathbf{y}'$$

Then

$$\frac{d\mathbf{A}}{dt} = \left.\frac{d\mathbf{A}}{dt}\right|_G + \boldsymbol{\omega} \times \mathbf{A}$$

where the subscript G refers to the *rotating system*. Under the action of a field $\mathbf{B}_0$ in the laboratory system, the magnetization follows the equation (neglecting relaxation effects)

$$\frac{d\mathbf{M}}{dt} = \gamma \mathbf{M} \times \mathbf{B}_0$$

In the rotating system we will then have:

$$\left.\frac{d\mathbf{M}}{dt}\right|_G = \gamma \mathbf{M} \times \mathbf{B}_0 + \mathbf{M} \times \boldsymbol{\omega}$$

$$\left.\frac{d\mathbf{M}}{dt}\right|_G = \gamma \mathbf{M} \times \left(\mathbf{B}_0 + \frac{\boldsymbol{\omega}}{\gamma}\right)$$

$$\left.\frac{d\mathbf{M}}{dt}\right|_G = \gamma \mathbf{M} \times \mathbf{B}_{\text{eff}}$$

where $\mathbf{B}_{\text{eff}}$ is the *effective field* in the rotating system. We see that if $\omega = -\gamma B_0$, the effective field will be zero and the magnetization in the rotating system will remain stationary. If then a field $\mathbf{B}_1 = B_1 \mathbf{x}'$ is applied, the motion of the magnetization will be

$$\left.\frac{d\mathbf{M}}{dt}\right|_G = \gamma \mathbf{M} \times \mathbf{B}_1 = -B_1 M_{y'} \mathbf{z}' + B_1 M_{z'} \mathbf{y}'$$

We will then have (omitting, to simplify the notation, the G and the prime)

$$\frac{dM_x}{dt} = 0; \quad \frac{dM_y}{dt} = \omega_1 M_z; \quad \frac{dM_z}{dt} = -\omega_1 M_y$$

where $\omega_1 = \gamma B_1$. This set of equations has the following solutions:

$$M_x(t) = \text{constant} = 0; \quad M_y(t) = M_0 \sin \omega_1 t; \quad M_z(t) = M_0 \cos \omega_1 t$$

where we have used the initial conditions $M_x(0) = 0, M_y(0) = 0$ and $M_z(0) = M_0$. The time τ required for the field B_1 to turn the magnetization of 180°, that is, from M_0 to $-M_0$, will be given by

$$\omega_1 \tau = \pi, \quad \tau = \frac{\pi}{\omega_1} = \frac{\pi}{\gamma B_1}$$

7.3 The equation of motion of **M**, neglecting relaxation is:

$$\frac{d\mathbf{M}}{dt} = \gamma \mathbf{M} \times \mathbf{B}$$

In the presence of an anisotropy field $\mathbf{B}_a$, assumed in the z direction, the total field is $\mathbf{B} = (B_0 + B_a)\mathbf{k}$ and the resonance frequency is

$$\omega_0 = \gamma (B_0 + B_a)$$

7.4 The magnetization will be proportional to $n = N_2 - N_1$; with $N = N_2 + N_1$, we will have

$$N_1 = \frac{N-n}{2}; \quad N_2 = \frac{N+n}{2}$$

Then

$$\frac{dN_1}{dt} = W_{21}N_2 - W_{12}N_1 + W_{rf}N_2 - W_{rf}N_1$$

$$\frac{dN_2}{dt} = W_{12}N_1 - W_{21}N_2 + W_{rf}N_1 - W_{rf}N_2$$

Subtracting the second from the first, we arrive at

$$\frac{d}{dt}(N_1 - N_2) = \frac{dn}{dt} = -2W_{rf}n + \left[N\frac{W_{12} - W_{21}}{W_{12} + W_{21}} - n\right](W_{12} + W_{21})$$

or

$$\frac{dn}{dt} = -2W_{rf}n + \frac{n_0 - n}{T_1}$$

where $n_0 \equiv N(W_{12} - W_{21})/(W_{12} + W_{21})$ and $1/T_1 = (W_{12} + W_{21})$. In the stationary state, $dn/dt = 0$, and

$$n = \frac{n_0}{1 + 2W_{rf}T_1}$$

When $2W_{rf}T_1 \ll 1$, $n \approx n_0$, and M_z does not change much. The power absorbed by the field will be

$$\frac{dE}{dt} = N_2 W_{rf}\hbar\omega - N_1 W_{rf}\hbar\omega = nW_{rf}\hbar\omega$$

or

$$\frac{dE}{dt} = n_0\hbar\omega\frac{W_{rf}}{1 + 2W_{rf}T_1}$$

We know that $W_{rf} \propto B_{rf}^2$. Therefore, as long as $2W_{rf}T_1 \ll 1$, the absorbed power may be increased by increasing B_1. However, when $2W_{rf}T_1 \approx 1$, W_{rf} in the numerator tends to cancel with the denominator, dE/dt does not depend any more on B_1^2, and the system does not absorb more energy from the field. This is the phenomenon of saturation.

CHAPTER 8

8.1

$$B_1(\omega) = \frac{1}{\sqrt{2\pi}}\int_{-\infty}^{+\infty} B_1(t)e^{i\omega t}dt = \frac{1}{\sqrt{2\pi}}\int_{-T/2}^{+T/2} B_1(t)e^{i\omega t}dt = \frac{2}{\sqrt{2\pi}}\frac{B_1}{\omega}\sin\left(\frac{\omega T}{2}\right)$$

The absorbed power is proportional to $B_1(\omega)^2$:

$$P(\omega) \propto \frac{2B_1^2}{\omega^2} \sin^2\left(\frac{\omega T}{2}\right)$$

The function $P(\omega)$ may be written as

$$P(\omega) = P_0 \frac{\sin^2(\pi\nu T)}{(\pi\nu T)^2} = P_0 \frac{\sin^2\theta}{\theta^2}$$

where $\theta = \pi\nu T$. For $\theta = 0$, applying l'Hospital's rule one finds: $P(0) = P_0$. For $\theta = \pi/3$ we will have $P(\pi/3) \simeq 0.68 P_0$. For $\theta = \pi/2$ we find $P(\pi/2) \simeq 0.41 P_0$. Therefore, the half-height of the curve will occur around $\theta \sim \pi/2 \sim \pi\nu T$ or $\nu \sim 1/2T$, and the width of the curve (bandwidth) will be

$$\Delta\nu \sim 2 \times \frac{1}{2T} = \frac{1}{T}$$

For $T = 10\,\mu$s the width will be $\Delta\nu \sim 1$ MHz.

8.2 (a) Let M_s = the magnetization and δx = the displacement of the wall in the x direction. The fractional increase of the volume is

$$\delta V = \frac{A\,\delta x}{V}$$

The increase in the magnetization in the direction of B_1 is

$$\delta M_1 \mathbf{k} = M_s\,\delta V \mathbf{k} - M_s\,\delta V(-\mathbf{k}) = 2M_s\,\delta V\,\mathbf{k} = \frac{2M_s A\,\delta x \mathbf{k}}{V}$$

(b) But $\delta M_1 \equiv (\chi_e/\mu_0) B_1$. Thus:

$$\frac{\chi_e}{\mu_0} B_1 = \frac{2M_s A\delta x}{V}, \quad \text{or} \quad \delta x = \frac{V\chi_e B_1}{2M_s A\mu_0}$$

(c)
$$\delta\theta(x) = \left(\frac{d\theta}{dx}\right)\delta x$$

But the hyperfine field is parallel to the moment of the ion, and is rotated to the same amount. We see that

$$\delta B_1 \approx B_{hf}\delta\theta$$

and

$$B_1' = B_1(1+\eta) \approx \eta B_1 = \delta\theta \; B_1 B_{hf} = \frac{V\chi_e}{2A\mu_0 M_s}\left(\frac{d\theta}{dx}\right) B_{hf} B_1$$

Therefore

$$\eta(x) = \frac{B_1'}{B_1} = \frac{V\chi_e}{2A\mu_0 M_s}\left(\frac{d\theta}{dx}\right) B_{hf}$$

8.3

$$\theta_p = \gamma_n B_1' \tau_p$$

The condition of negligible relaxation during the application of the pulse is $T_2 \gg \tau_p$. For $\theta_p = \pi/2$, we will have

$$T_2 \gg \frac{\pi}{2}\frac{1}{\gamma_n B_1'} \quad \text{or} \quad \frac{1}{B_1'} = \frac{2\gamma_n T_2}{\pi}$$

With $2/\pi \approx 1$, we will have

$$B_1' \gg \frac{1}{\gamma_n T_2}$$

For $\gamma_n \approx 1$ MHz T^{-1} and $T_2 \approx 100$ μs we find $B_1' \gg 0.01$ T. Fields generated in laboratory coils are of the order of 10 G or 0.001 T. However, with amplification factors of the order of 10^3, we will have $B_1' \approx 1$ T. A linewidth of 5 MHz is equivalent to $T_2 = 0.2$ μs, which corresponds to a field of the order of $(1$ MHz T$^{-1} \times 0.2$ μs$)^{-1} = 5$ T, an extremely large value to be generated by an rf coil.

REFERENCES

Ashcroft, N. W. and N. D. Mermin, *Solid State Physics*, Holt Rinehart and Winston, New York, 1976.

APPENDIX C

PHYSICAL CONSTANTS

Quantity	Symbol	Value	CGS	SI
Speed of light in vacuum	c	2.997925	10^{10} cm s^{-1}	10^{8} m s^{-1}
Elementary charge	e	1.60218	—	10^{-19} C
Planck constant	h	6.62607	10^{-27} erg s	10^{-34} J s
	$\hbar = h/2\pi$	1.054572	10^{-27} erg s	10^{-34} J s
Avogadro constant	N_A	6.02214 $\times 10^{23}$ mol^{-1}		
Atomic mass constant	m_u	1.66054	10^{-24} g	10^{-27} kg
Electron mass	m_e	9.10939	10^{-28} g	10^{-31} kg
Proton mass	m_p	1.67262	10^{-24} g	10^{-27} kg
Ratio of proton and electron masses	m_p/m_e	1836.153	—	—
Proton gyromagnetic ratio	γ_p	2.67522128	10^{4} s^{-1} G^{-1}	10^{8} s^{-1} T^{-1}
Electron Compton wavelength	λ_c	2.42631	10^{-10} cm	10^{-12} m
Bohr radius	a_0	0.529177	10^{-8} cm	10^{-10} m
Bohr magneton	μ_B	9.2740154	10^{-21} erg G^{-1}	10^{-24} J T^{-1}
Nuclear magneton	μ_N	5.0507866	10^{-24} erg G^{-1}	10^{-27} J T^{-1}
Boltzmann constant	k	1.380658	10^{-16} erg K^{-1}	10^{-23} J K^{-1}
Reciprocal of fine-structure constant	$1/\alpha$	137.036	—	—
Rydberg constant	$R_\infty hc$	2.179874	10^{-11} erg	10^{-18} J

PHYSICAL CONSTANTS

Quantity	Symbol	Value	CGS	SI
Molar gas constant	R	8.31451	10^7 erg mol^{-1} K^{-1}	J mol^{-1} K^{-1}
Vacuum permittivity	ϵ_0	—	1	$10^7/4\pi c^2$
Vacuum permeability	μ_0	—	1	$4\pi \times 10^{-7}$ H m^{-1}
Electronvolt	eV	1.60218	10^{-12} erg	10^{-19} J
Electronvolt/h	eV h^{-1}	2.41797×10^{14} Hz	—	—
Electronvolt/hc	eV h^{-1}c^{-1}	8.06546	10^3 cm^{-1}	10^5 m^{-1}
Electronvolt/k	eV k^{-1}	1.16044×10^4 K	—	—
Electronvolt/cm^{-1}	eV cm^{-1}	1.986	10^{-16} erg	10^{-23} J
Ångström	Å	1	10^{-8} cm	10^{-10} m

AUTHOR INDEX

Bold page number indicates reference citation.

Abe et al. (1966), 221, **224**
Abragam (1961), 161, **186**, **187**, **223**
Abragam and Bleaney (1970), 163, **186**, **187**, **223**
Ailion and Ohlsen (1983), **223**, **246**
Aleksandrov (1966), **223**
Allen (1976), **186**
Anderson and Clogston (1961), 73
Arif et al. (1975), 161, **187**
Ashcroft and Mermin (1976), 35, 87, 266, **281**

Baker and Williams (1962), 174, **187**
Barata and Guimarães (1985), 223, **224**
Barbara (1988), **155**
Bleaney (1967), 165, 169, **186**, **187**
Bleaney (1972), **186**
Bloch (1946), 195, **224**
Bloom (1955), 220, **224**, 240, **247**
Bobek et al. (1993), 212, **224**
Boll (1994), 142, **155**, **156**
Borovik-Romanov et al. (1984), 217, **224**
Buschow (1994), 143, **156**
Butterworth (1965), 221, **224**

Cahn et al. (1994), 143, **155**
Carr (1969), **155**
Carr and Purcell (1954), 214, **224**
Carrington and McLachlan (1967), 205, **223**, **224**
Chikazumi (1997), **155**
Coehoorn (1990), 70
Coey (1996), 135, **155**, **156**
Cohen and Giacomo (1987), 1, **24**
Crangle (1991), 36, 49, **59**, **112**, **155**
Craik (1995), **59**, **89**, **246**
Culken (1994), **155**
Crooks (1979), XII, 2, **24**
Cullity (1971), **155**
Cullity (1972), 132, **155**, **156**

de Gennes et al. (1963), 232, 233, **247**
de Gronckel et al. (1991), 230, **247**
Dormann (1991), 231, **246**, **247**

Elliott (1972), 47, **59**
Elliott and Stevens (1953), 165, **187**
Evetts (1992), 10, 23, **24**, 143, **156**

Farrar and Becker (1971), **223**
Figiel (1991), **246**
Foner (1963), **246**
Fukushima and Roeder (1981), **223**

Gerl (1973), 94, **112**
Gerstein and Dybowski (1985), **224**, 254
Gignoux (1992), **112**
Givord (1996), 148, 149, **155**, **156**
Goldanskii (1968), **186**
Gradman (1993), 22, **24**, 126, **156**
Grandjean and Long (1990), 10, **24**, **155**
Grant and Phillips (1990), 1, 2, **24**
Greenwood and Gibb (1971), **186**
Guimarães (1971), unpublished, 231, **247**

Hagn (1986), 183, **186**, **187**
Hahn (1950), 219, **224**
Henry (1952), 44, **59**
Herring (1964), **89**
Herrmann (1991), 115, 116, 117, 118, **156**
Hurd (1982), 13, 14, 15, 16, 17, 18, **24**
Hutchings (1966), 57, **59**

Iannarella et al. (1982), 108, 109, 110, **112**
Ingram (1979), **155**

Jackson (1975), 166, 167, 169, **186**, **187**
Jaynes (1995), 220, **224**
Jensen and Mackintosh (1991), 58, **59**
Jiles (1991), **155**

Kapusta et al. (1996), 229, 231, **247**
Kittel (1949), 119, 127, 129, 130, **155**, **156**
Kittel (1986), 80, 84, **89**, **156**
Kronmüller (1995), 143, **156**
Kündig (1967), 186, **187**

Landau and Lifshitz (1959), 128, **155**, **156**
Landau and Lifshitz (1968), 109, **112**
Landolt-Börnstein (1962), 6, **24**
Landolt-Börnstein (1986), 3, 21, **24**, 73, **89**, 182, **187**, 237, **247**
Landolt-Börnstein (1991), 39, **59**, 102, **112**
Lea et al. (1962), 58, **59**
Lee (1979), **155**

Martin (1967), 71, 80, **89**, **155**
Matthias et al. (1962), **186**, **187**
McCausland and Mackenzie (1980), 58, **59**, 174, 180, **186**, **187**, 212, **224**, 231, **246**, **247**
McCurrie (1994), **155**
McMorrow et al. (1989), 174, 177, **187**
Meyers (1987), 39, 40, **59**
Mims (1972), **224**
Morrish (1965), 59, 153, **155**, **156**, 245, **246**, **247**
Morup (1983), 17, **24**
Mydosh (1996), 18, **24**

Narath (1967), 181, **187**, 237, **246**, **247**
Néel (1954), 126, **156**
Netz (1986), 170, **187**

Oguchi (1955), 74, 75, **89**

Parker (1990), **155**, 186, **187**
Patterson (1971), 63, **89**
Poole and Farah (1971), **224**
Portis and Gossard (1960), 239, **247**
Portis and Lindquist (1965), **246**

Raghavan et al. (1975), **187**

Rathenau (1969), **155**
Rhyne (1972), 54, **59**
Ruderman and Kittel (1954), 72, **89**, 181, **187**

Seeger and Kronmüller (1989), **112**
Shadowitz (1975), XII, 2, **24**
Shimizu (1981), **112**
Slichter (1990), 163, **187**, 204, **224**
Smart (1966), 75, 77, 78, 79, **89**
Sparks (1964), **247**
Stancil (1994), **247**
Stearns (1966), 181, **187**
Stearns (1967), 240, 241, **247**
Stevens ((1952), 56, **59**
Stone (1986), **187**
Stoner (1938), 100, **112**
Swartzendruber (1991), 54, **59**
Schultz (1973), **155**

Taylor (1971), **187**
Taylor (1995), 1, **24**
Taylor and Darby (1972), **59**, **187**
Tribuzy and Guimarães (1997), unpublished, 231, **247**
Turov and Petrov (1972), 232, **246**

Vonsovskii (1946), 181, **187**
Vonsovskii (1966), **247**

Watson (1967), 74, **89**
Weisman et al. (1973), 207, 213, **224**, 231, **246**, **247**
Weiss and Forrer (1926), 20, **24**
Williams (1986), **89**, **112**, **155**
Wohlfarth (1976), 111, **112**
Wohlfarth (1980), 88, **89**, 96, 102, **112**
Wohlfarth (1982), **24**

Zener (1951), 70, **89**, 181, **187**
Zhang et al. (1991), 228, 231, **247**
Zinn (1974), **246**

SUBJECT INDEX

a (lattice parameter), 90, 151
a (quadrupole interaction parameter), 221–222
A (mass number), 160
A (hyperfine interaction constant), 173
A' (exchange stiffness), 151
$\mathbf{A}(\mathbf{r})$ (vector potential), 166–167
AB_2, 108
Absolute sensitivity, see NMR absolute sensitivity
Absorbed power, 202–203, 236–237, 279, 280
Absorption, 199
Absorptive part of the susceptibility, 202
Actinides, 40, 91, 93, 170, 180
Activation energy, 17, 207
AFMR, see Antiferromagnetic resonance
After-effect, 148–149
Aging, 149
Al, 181
Alkali metals, 91, 188
α_i (direction cosines), 123
α_J (2nd order crystal field constant), 56, 57
α-Mn, 14
Aluminum, 40, 180
Amorphous, 13, 228
Am^{2+}, 170, 183
Ampère law, generalized, 2
Amperes per meter, 3
Amplification, 227, 232, 238
Amplification factor, 233, 237, 281. See also Enhancement factor
An (actinides), 93
Angular frequency, 150, 154, 182, 189, 194, 197–199, 204, 215
Anhysteretic curve, 141
Anisotropy
  axis, 144–145
  constant, 123–124, 143
  energy, 25, 123–124, 127–128, 131, 133, 225, 255, 271

  field, 124, 135, 141, 145, 153, 157, 232, 237, 243–245, 278
  mechanisms, 132
  tensor, 244
Antiferromagnet, 15, 60, 178, 227
Antiferromagnetic resonance (AFMR), 153, 245–246
Antiferromagnetism, 14–15
Antisymmetry, 65–66
Arrott plot, 109, 111–112
Asperomagnetism, 17
Asymmetry parameter, 163
Atomic radius, 37, 40
Avogadro constant, 60, 259
Axial anisotropy constants, 132
Azbel'-Kaner resonance, 153

$\mathbf{B}$, B (magnetic induction), 2, 7, 149, 165, 167, 169, 181, 185, 194–195, 260, 269
$\mathbf{B}_a$ (anisotropy field), 153, 225, 232, 237, 244
$\mathbf{B}_c$ (contact hyperfine field), 169–170
$\mathbf{B}_{cp}$ (core polarization hyperfine field), 170, 174–175, 179
$\mathbf{B}_{dip}$ (dipole field), 174–175
$\mathbf{B}_{hf}$ (hyperfine field), 173–174, 179–180, 185, 232, 239, 248
B_i (magnetoelastic coupling constants), 130
$\mathbf{B}_{orb}$ (orbital hyperfine field), 175
B_1 (time-dependent field), 203–204, 214–215, 220, 224, 227–228, 232, 237–240, 242, 248–249, 278
B_2 (effective rf field), 236–237
$\mathbf{B}_s$ (spin magnetic dipole field), 168
B_n^m parameters, 57–58
$BaFe_{12}O_{19}$, 124
Band, 91–93, 100–102, 107, 113, 270
Barium ferrite, 124
Barkhausen effect, 137
Barkhausen noise, 137
Basal plane, 124

SUBJECT INDEX

β_J (4th order crystal field constant), 57
Bethe-Peierls-Weiss method, 79
(BH) (energy product), 12, 23
$B - H$ curve, 23, 140-141
$B \times H$ curve, 121, 138
$B - H$ plot, 115
$(BH)_{max}$, 23, 141, 148
Bloch equation, 153, 191, 195-196, 198-199, 207, 210, 227, 230, 236
Bloch functions, 168
Bloch $T^{3/2}$ law, 87, 90
Bloch wall, 134-135, 157, 240, 245
Blocking temperature, 17
Bohr magneton, 28, 48, 60, 160, 172, 192, 259
Boltzmann distribution, 35, 41, 190, 207-209
Boltzmann population, 187
Bose statistics, 65
Bosons, 65
Boundary conditions for H and B, 4
Bowing, 136
Brillouin function, 20, 43-44, 48, 60

C (Curie constant), 78, 259
c_{ij} (elastic moduli), 128, 130
C_p (specific heat), 75
c-axis, 124
Canted magnetism, 17
Centimeter-gram-second (CGS), 2-5, 9-11, 28, 124, 177
CESR, see Conduction electron spin resonance
CF, see Crystal field
CGS, see Centimeter-gram-second
Chemical potential, 95, 266
χ (susceptibility), 118
χ (susceptibility tensor), 244-245
χ_0 (Pauli susceptibility), 97
χ_α (spin wavefunction), 81
χ_β (spin wavefunction), 81
χ_{CGS} (χ in the CGS system), 5
χ_m (susceptibility per mole), 77-78
χ_n (nuclear susceptibility), 48, 200-201
χ_P (Pauli susceptibility), 71
$\chi(\mathbf{R})$ (nonuniform susceptibility), 71
χ', 200, 202
χ'', 200, 202-203
Closed shells, 92, 170
Closure domain, 133
Cluster glass, 17
Co, 54, 91, 229-230
Co^{2+}, 33-34, 36
^{59}Co, 213, 223, 229-230
Cobalt, 3, 135, 213
Cobalt metal, 124
Co/Cu multilayer, 229-230

Coercive field, 18, 140-141, 148
Coercive force, 24, 140
Coercivity, 24, 115, 140-141, 143, 148
Coherence, 204, 215
Combined electric and magnetic interactions, 183
Complex susceptibility, 236
Conduction band, 74, 100
Conduction electron, 70-72, 92-93, 108, 113, 153, 161, 170, 177, 181, 211-213
Conduction electron spin resonance (CESR), 153
Contact hyperfine field, 173
Continuous wave (CW) technique, 199, 220
Convection currents, 2
Coordinate system, 196, 198, 261
Copper, 40, 154, 182
Copper oxide, 13
Core correction factor, 183
Core polarization, 170, 175, 179
Correlation function, 78, 204-205
Correlation time, 204-206
Coulomb energy, 65
Coulomb interaction, 72, 159
Coulomb potential, 64
Coulomb term, 161, 167
Coulomb's law, 119
Coupled two-spin system, 227, 230
Cr, 44
Cr^{3+}, 44
Critical exponents, 112
Critical field, 118
Cr_2O_3, 18
Crystal
  anisotropy, 123
  field, 55-56, 58, 61, 129, 160, 175, 179, 188, 262
  field interaction, 55-57
Crystalline
  anisotropy, 68, 132, 144, 243-244
  field (CF), 46-47, 123, 188
  materials, 13
Crystallites, 132
Cs, 91
Cu-Ni alloys, 104-105
Curie constant, 45, 78, 259
Curie law, 13-14, 45, 48, 201
Curie temperature, 13, 22, 51-55, 74, 130
Curie-Weiss dependence, 107
Curie-Weiss law, 13-14, 55, 76
Current density, 2, 166-167, 171
CW, see Continuous wave
Cyclotron, 153

SUBJECT INDEX **289**

Cyclotron resonance, 153

D (diffusion coefficient), 214
D (stiffness constant), 80, 84, 88–89, 154
D (displacement vector), 2
d electrons, 57, 92, 108
d group, 107
d series, 46
d transition metal, 24, 107
$d - d$ interaction, 107–108
Damping, 245
Defects, 136–137
δ_{ij} (Kronecker delta), 161
Δ (hyperfine anomaly), 173
Demagnetization curve, 120–121
Demagnetizing factor, 8–10, 25, 116, 125, 177, 239, 242–243, 256
Demagnetizing field, 8, 10, 119–121, 125, 140, 152–153, 177, 228, 232, 242
Demagnetizing tensor, 9, 242
Density, 5, 104, 164
Density matrix, 193
Density of free poles, 8
Density of states, 92–96, 98, 101, 103, 113–114, 211–212, 266
Detailed balance, 209
Diamagnet, 1, 13–14, 18
Diamagnetic susceptibility, 49, 60, 99
Diamagnetic volume susceptibility, 28–29
Diamagnetism, 13, 28
Differential susceptibility, 28
Diffusion, 149, 214
Diffusion coefficient, 214, 219
Dipole-dipole interactions, 211
Dipolar field, 168, 175–177, 205, 210
Dirac's equation, 61
Dirac's theory, 31
Direction cosines, 123, 129–132
Direction of easy magnetization, 122, 124
Direction of magnetization, 61, 115, 124, 130, 136, 145–146, 213, 240, 248, 261
Disaccommodation, 149
Disordered materials, 13
Dispersion, 199
Dispersion relation, 84, 89, 93
Dispersive part of the susceptibility, 202
Displacement vector, 2
Domain, 13, 49, 115, 119, 123, 132–135, 137, 141, 144–147, 151–152, 157, 212, 214, 227, 237–238, 240, 248
Domain wall, 132–133, 135–137, 141, 144, 149–150, 152, 212, 214, 228, 237, 238, 240, 259
Domain wall center (DWC), 239

Domain wall edge (DWE), 238
Domain wall motion, 115, 150
Domain wall thickness, 133–134
Dynamic effects, 148
Dynamic frequency shift, 229, 233

E (Young's modulus), 128
E (electric field), 7
E_a (activation energy), 207
E_{el} (elastic energy), 128
E_F (Fermi energy), 74, 94–95, 109
E_K (magnetic anisotropy energy), 123–124, 128, 133
E_{ME} (magnetoelastic energy), 128, 131
E_{ms} (magnetostatic energy), 120–121
$E(k)$ (dispersion curve), 94
Easy axis, 125
Easy direction, 122, 126, 132, 137
Easy magnetization, 124–125, 133, 135, 145
Easy plane, 122, 126
Eddy currents, 149, 150, 245
Effective mass, 133
Effective paramagnetic moment, 45–46
EFG, see Electric field gradient
ε_{ijk} (Levi-Civita symbol), 167
Elastic energy, 128
Elastic moduli, 128, 130
Electric charge density, 2, 162, 164
Electric field gradient (EFG), 162–163, 165, 182–184, 203, 221
Electric field gradient (EFG) tensor, 163, 183
Electric monopole moment, 159
Electric quadrupole interaction, 162, 221, 224, 228
Electric quadrupole moment, 159–160, 162–163
Electric quadrupole moment tensor, 162
Electron-electron interaction, 72–73, 100, 102
Electron gas, 101, 114, 211, 213, 266–267
Electronic susceptibilies, 235
Electron paramagnetic resonance (EPR), 153–154
Electron spin resonance (ESR), 153
Ellipsoid, 9, 125
Emu, 3–4
Energy product, 12, 23, 141, 147–148
maximum, 148
Enhanced Pauli paramagnetism, 15
Enhanced susceptibility, 73, 235
Enhancement, 227, 232
Enhancement factor, 73, 220, 232, 237, 240, 248
ε_{ij} (strain), 127, 128
EPR, see Electron paramagnetic resonance
Equilibrium magnetization, 121

Er, 58
η (asymmetry parameter), 163
η (enhancement factor), 220, 233, 237, 239–240, 248
η_d (domain enhancement factor), 237–238
η_w (wall enhancement factor), 238–241
Eu, 46, 174
Eu^{2+}, 174–175
Eu^{3+}, 46
Exchange
 constant, 70–71, 89
 energy, 127, 133, 135, 157–158, 269
 integral, 65, 70, 72–73
 interaction, 13, 63, 69, 113, 119, 123, 126–127, 135, 160, 175, 188, 270
 parameter, 127
 stiffness, 151
Expectation value, 67, 193, 258, 275
Extraionic field, 56, 177–178
Extraionic interactions, 175

F (correlation function), 204
F (enhancement factor), 73
F (total angular momentum), 48, 174, 274
$f(E)$ (Fermi-Dirac distribution), 98
f electrons, 57
f states, 73
Faraday law, 2, 150
Fe, 18, 44, 54, 70, 91–93, 102, 122, 170, 180–181, 212, 242
^{57}Fe, 179, 181, 212, 228
Fe^{3+}, 44
FeB, 229
$Fe_{86}B_{14}$, 228
Fe-Al, 181
Fe-C alloy, 149
FeO, 14
Fe_3Mn, 14
Fe_2O_3, 18, 21
Fermi contact, 169, 175, 212
Fermi-Dirac distribution, 98, 211, 266
Fermi-Dirac function, 95, 105
Fermi-Dirac statistics, 65
Fermi distribution, 97
Fermi energy, 94–95
Fermi level, 93, 96, 98, 102, 181, 212
Fermi surface, 113
Fermi temperature, 112
Fermions, 65
Ferrimagnet, 16
Ferrimagnetism, 15–16
Ferrite, 49, 150
Ferromagnet, 1, 15, 18, 53, 55, 74, 79–80, 86, 178, 225, 227, 235–236, 242, 244–245, 248
Ferromagnetic material, 132, 144, 148, 150, 242
Ferromagnetic nuclear resonance (FNR), 227
Ferromagnetic resonance (FMR), 153–154, 225, 234, 242, 244–245
Ferromagnetism, 13, 15, 49, 63, 91, 100, 102–103, 113–114
FeSi alloy, 150
FID, see Free induction decay
Filling factor, 220
$5d$ series, 56
$5f$ electrons, 93
$5f$ shell, 92
Fluctuations, 69, 203–206, 210–211, 213, 215, 220
Flux density, 150
FMR, see Ferromagnetic resonance
FNR, see Ferromagnetic nuclear resonance
Forced magnetization, 129, 144
$4f$ electron, 47, 70, 91, 93
$4f$ ion, 58
$4f$ orbital, 70, 108
$4f$ series, 56
$4f$ shell, 13, 46, 92, 175
$4f$ state, 73
$4s$ electrons, 22, 91–92
Fourier intensity, 206
Fourier spectrum, 203, 220, 247
Fourier transform, 205, 217–218
Fourier transform NMR, 220
Free electrons, 93–96, 98–99, 101, 104, 113–114
Free energy, 109
Free induction, 217, 220
Free induction decay (FID), 215, 218–220, 240–242
Free ion, 70, 165, 174–175, 179, 182
Free poles, 7, 10, 177
Frequency diffusion, 214
Frequency pulling, 229, 232
Frustration, 18

g (g-factor), 28, 35, 154, 189
G (gauss), 2, 4
g_I (nuclear g-factor), 160
Gadolinium, 13, 53, 70
γ (gyromagnetic ratio), 28, 48, 80, 151, 153, 160, 182, 189, 191, 193, 197, 214, 231, 239
γ_∞ (Sternheimer factor), 183
γ_J (6th order crystal field constant), 57
Gauss, 2, 4
Gaussian, 219
Gauss law, 2
Gd, 44, 54, 74, 87

SUBJECT INDEX

Gd^{3+}, 33, 44, 163, 170, 175, 180
^{157}Gd, 224
GdCo$_2$, 223
GdFe$_2$, 15, 108
GdNi$_2$, 108
Giant magnetoimpedance, 154
Gilbert equation, 153, 245
Gold, 40
Gyromagnetic ratio, 28, 48, 80, 151, 153, 160, 182, 189, 191, 193, 197, 211, 214, 231, 239

H, H (magnetic field intensity), 2, 7, 115, 200
H_a (anisotropy field), 124, 145
H_c (coercivity), 148
H_c (critical field), 118
H_{c1} (critical field), 118
$\mathcal{H}_{cf}$, 56–58, 61, 160, 188, 261, 263, 275
$\mathcal{H}_{coul}$, 55–56
H_d (demagnetizing field), 8, 120–121, 140, 242
$\mathcal{H}_{el}$, 175
$\mathcal{H}_{exch}$, 160
$\mathcal{H}_{hf}$, 160, 173–175, 183
$\mathcal{H}_{LS}$, 55–56, 160, 175
$\mathcal{H}_{mag}$, 56, 183
$\mathcal{H}_Q$ (electric quadrupole hamiltonian), 162–164, 183
$_BH_c$ (coercivity), 140
$_HH_c$ (coercivity), 140
Hahn, 218
Hahn echo, 221–222
Hard magnetic material, 116, 141, 147
Heisenberg hamiltonian, 63, 68–69, 81, 87–88
Heisenberg representation, 192
Helimagnet, 18
Ho, 58
^{165}Ho, 180
Holmium, 180
Homogeneity, 221
Homogeneous broadening, 216, 220
Hooke's law, 128
Hund's rules, 33–34
Hybridization, 73, 93
Hydrogen, 60
Hyperfine anomaly, 173–174
Hyperfine field, 57, 165, 167–168, 170, 172–175, 177–181, 185, 221, 224, 227–228, 230, 232–233, 237, 239, 248, 280
Hyperfine interaction, 159–160, 173, 175, 182, 188, 227, 229–230
Hyperfine interaction constant, 173
Hyperfine quantum number, 174
Hyperfine splitting, 174
Hysteresis, 17, 141, 150
    curve, 140–141, 145, 147

cycle, 137, 141
loop, 137, 141, 147–148
loss, 120, 141, 150

I, I (nuclear angular momentum), 160, 200–201, 213, 221
I (magnetization intensity), 5
Ideal demagnetized state, 145
Ideal diamagnet, 115, 118
Ideal diamagnetic material, 117, 119
Ideal hard magnetic material, 115–116
Ideal magnetic materials, 115
Ideal nonmagnetic material, 115–116
Ideal soft magnetic material, 115–117
Imaginary part of the susceptibility, 236
Impurity, 116, 136–137, 180–181
Incipient ferromagnetism, 15
Incomplete shell, 12–13, 55, 93, 170, 174–175
Indirect interaction, 70, 72, 229
Induction, magnetic, 1
Inhomogeneity, 215–216, 220–221, 242
Inhomogeneous broadening, 248
Initial permeability, 138–139, 147
Intermediate magnetic material, 141, 143
Intermetallic compound, 107–108, 124, 177–178, 230
Intraionic interactions, 175
Intrinsic susceptibility, 244
Inverse magnetostrictive effect, 129
Ion-electron system, 110
Iron, 3, 13, 20–21, 60, 70, 92, 95–96, 102, 121–122, 132, 135, 154, 178–180, 241, 259
Iron carbon, 21
Iron group, 13, 40, 91
Irreversible process, 135–136, 150
Ising hamiltonian, 68
Isochromat, 215, 217, 219
Isochrone, 215, 217–218
Isomer shift, 164
Itinerant antiferromagnet, 14
Itinerant electron, 71, 91, 93, 100, 105
Itinerant ferromagnet, 102
Itinerant magnetism, 18, 107
Itinerant moment, 107

j (current density), 2
J (polarization), 3, 5, 140
J (total angular momentum), 69, 75, 165, 173–174, 261
$\mathcal{J}$ (exchange integral), 65–68, 73–74, 77, 82, 89, 105, 108–109, 127
j$_t$ (total current density), 2
$J(\omega)$ (spectral density function), 205

jj coupling, 33

k (Boltzmann constant), 103
K (potassium), 91
K_0 (anisotropy constant), 123–124
K_1 (anisotropy constant), 123–124, 143
K_2 (anisotropy constant), 123–124
K_s (surface anisotropy constant), 126
K_u (uniaxial anisotropy constant), 126, 131, 135
Kennelly convention, 5
Knight shift, 177, 181–182, 212
Korringa relation, 212
Kronecker delta, 161

L (total orbital angular momentum), 32, 172–173, 175
La, 40
λ (magnetostriction), 129
λ (spin-orbit coupling constant), 34
λ_m (molecular field parameter), 50, 53, 103, 105, 259
λ_s (saturation magnetostriction), 130–131
Landau diamagnetic susceptibility, 29
Landau diamagnetism, 13
Landau expansion, 109
Landau–Lifshitz equations, 153, 245
Landé g-factor, 35, 47, 68
Langevin, 17, 48
Langevin function, 48–49
Langevin magnetism, 60
Langevin paramagnetism, 47
Lanthanide, 40, 93
Laplace's equation, 163, 256
Laplacian, 80, 164
Larmor diamagnetic susceptibility, 29
Larmor frequency, 29, 59, 196–199, 203–206, 215
Legendre polynomial, 30, 61
Lenz's law, 28, 150
Levi–Civita symbol, 167
Li, 91
Lifetime, 215
Linear chain, 83, 88
Lines of force, 7, 117–119
Linewidth, 178, 211, 221, 224, 228
Ln (lanthanides), 93
Load line, 121
Local field, 206
Local magnetization, 177, 248
Local order, 74–77, 79
Local technique, 19
Localization, 70, 92
Localized-itinerant systems, 107

Localized magnetism, 18, 105
Localized spin fluctuations, 16
Longitudinal relaxation, 196, 203, 207, 211–212
Lorentz field, 177
Lorentz force, 2, 29, 257
Lorentzian, 200, 218
Lorentz sphere, 177
Loss factor, 151
Loss of memory, 214, 220
LS coupling, 32–33, 172
Lu, 40
$LuFe_2$, 108
$LuNi_2$, 108

m (magnetic dipole moment), 119
M (magnetization), 104–105, 120, 149, 154
M, M (magnetization), 69, 104–105, 115, 120, 125–126, 135, 149, 154, 168, 177, 189, 192, 194–195, 201, 225, 228, 237, 242–243, 256, 270, 278
m_e (electron mass), 27, 48, 94
M_e (equilibrium magnetization), 120–121
M_l (local magnetization), 177
m_p (proton mass), 48, 160, 192
M_r (remanence), 148
M_s (saturation magnetization), 120–121, 124, 141, 148–149, 243, 248
$M - H$ curve, 115, 118, 140, 144–145
$M - H$ loop, 138, 140, 148
Magnet, 2, 7, 22, 132
Magnetically hard material, 18
Magnetically soft material, 18
Magnetic anisotropy, 119, 121, 123, 135
Magnetic anisotropy energy, 125
Magnetic circuit, 10
Magnetic dipole interaction, 63, 165
Magnetic dipole moment, 3, 27, 119, 159–160, 165, 167–168, 170, 189
Magnetic field, 1, 69, 123, 129, 132, 135, 144, 149–150, 152–154, 167, 181–182, 188–192, 195, 198, 200, 203, 205–207, 213–215, 220–221, 225, 227, 231, 244–245, 260, 273–274
definition, 1
Magnetic field gradient, 214
Magnetic field intensity, 2–3
definition, 3
Magnetic flux, 150, 220
Magnetic flux density, 2
Magnetic induction, 1–3, 216
Magnetic material, 19–20, 119–120, 122, 127, 129, 137, 143, 145, 150, 160, 165, 178, 212, 214, 216, 227–228, 237, 240, 243

SUBJECT INDEX **293**

Magnetic moment, 2–3, 25, 27–28, 32, 60, 69–70, 75–76, 91–92, 96, 98, 107, 122, 132, 134, 151, 153–154, 165, 167, 171, 175, 178, 181, 191–193, 210, 213–214, 217, 227, 229–230, 238–239, 242, 248, 259, 263
   definition, 3
Magnetic orbital moment, 27, 32, 34
Magnetic permeability, 5, 6, 118, 147, 149–150, 154
Magnetic pole density, 25
Magnetic poles, 119
Magnetic quantities, 1–2, 11
Magnetic quantum number, 31
Magnetic recording, 19
Magnetic recording media, 18
Magnetic resonance, 153, 189–192, 200, 211, 230, 232, 242
Magnetic saturation, 130
Magnetic self-energy, 25
Magnetic shield, 117
Magnetic units, 1–2, 11
Magnetic susceptibility, 5, 244
Magnetic susceptibility tensor, 244, 245
Magnetism, 1, 229
Magnetite, 15, 21
Magnetization, 2–3, 69, 72, 74–75, 77, 79–80, 82, 86–87, 89–90, 97–100, 102–103, 105–106, 109, 113, 115, 118, 120–122, 124–125, 127, 129, 131–133, 135, 137–138, 141, 144–145, 148–150, 152, 154, 157, 168, 170, 175, 179, 181, 189–190, 193, 195, 197, 199–204, 214–220, 222, 225, 227–228, 230, 232, 234, 236–237, 239, 241–243, 246, 248, 256, 259, 268, 270, 277–278, 280
   definition, 3
Magnetization curve, 115, 118, 123, 137–141, 146–147
Magnetization intensity (**I**), 5
Magnetization loop, 120
Magnetization process, 137
Magnetocrystalline anisotropy, 123
Magnetoelastic coupling, 156
Magnetoelastic coupling constants, 129–130, 157
Magnetoelastic energy, 127–128, 131–132
Magnetoelastic interaction, 119
Magnetogyric ratio, 28, 191
Magnetostatic energy, 119–121, 125, 133
Magnetostriction, 119, 127, 129, 131–132
Magnetostriction constant, 130
Magnon, 19, 84, 86, 89–90, 211–213, 229
Maximum permeability, 138–139

Maxwell equations, 2
Mean field, 50, 69, 74, 246
Mean field approximation, 63, 100
Measurement time, 178
Medium crystal field, 56
Metal, 13, 46, 70, 72, 91–92, 95, 107, 124, 153, 161, 165, 175, 181–183, 188, 207, 211–213, 249
Metallic film, 22
Metamagnetism, 15–16
Microstruture, 24
Mictomagnetism, 17
Minor loop, 138–139
Mixing, 74
Mn^{2+}, 233
Molecular field, 50, 53, 63, 72–74, 79, 100, 103, 188, 246
   approximation, 69, 112, 232
   coefficient, 50, 60, 73
   constant, 50, 70, 77
   parameter, 53, 232
Moment rotation, 141
Mössbauer effect, 164
Mössbauer spectroscopy (MS), 19, 164, 174, 179, 181
Motional narrowing, 205
μ (chemical potential), 95, 266
$\boldsymbol{\mu}$ (magnetic dipole moment), 189
μ (permeability), 5, 118, 140, 154
μ_0 (vacuum permeability), 2–3, 73, 119, 121, 140, 166, 235
μ_B (Bohr magneton), 160
μ_i (initial permeability), 138, 147
μ_m (maximum permeability), 138
μ_N (nuclear magneton), 160, 174, 192, 276
μ_r (relative permeability), 5–7
Multidomain sample, 123, 240, 245
Multiplet, 33, 36–37, 45–46
Muon, 19
Muon spin rotation, 19

N (north), 7
N_d (demagnetizing factor), 8–10, 116, 121, 239
$\mathbf{N}_d$ (demagnetizing tensor), 9, 242
$n(E)$ (density of states per volume), 92, 94, 96, 98
$N(E)$ (density of states), 93, 95–96
NdFeB, 18, 148
Néel temperature, 14, 60, 259
Néel wall, 135
Neutron, 19, 79, 84
Neutron diffraction, 19, 92
Ni, 18, 20, 54, 70, 76, 91, 102, 122, 213, 242

Nickel, 3, 13, 20, 92, 102, 107, 121, 129
^{61}Ni, 212
90° domain wall, 135
NMR, see Nuclear magnetic resonance
Noble gases, 37, 40
Noble metals, 40, 92-94
Nonsecular broadening, 204, 206
Normal metals, 40, 92, 94
Normal modes, 233
North pole, 7, 118
ν (Poisson ratio), 128
ν (Rayleigh coefficient), 147
Nuclear g-factor, 48, 160, 173, 182, 191
Nuclear magnetic relaxation, 229
Nuclear magnetic resonance (NMR), 19, 153, 174, 177-180, 204, 227, 237
  absolute sensitivity, 201, 254
  experiment, 201, 212, 214, 221-222
  frequency, 181-182, 224, 233, 242
  line, 214, 218, 248
  in metals, 242
  nuclide, 229
  pulsed, 214, 218, 248
  relative sensitivity, 201, 254
  sensitivity, 201, 254
  signal, 201, 229
  spectrum, 217, 221, 223-224, 228, 230
  spectrum shape, 217
  technique, 214, 220, 229
Nuclear magnetism, 48, 160, 165
Nuclear magneton, 48, 160, 174, 192, 273
Nuclear magnetic susceptibility, 48-49, 200-201
Nuclear radius, 164, 173
Nuclear resonance frequency, 235-236
Nucleation, 137
Nutation, 214

O_n^m operators, 57
Oersted (Oe), 3, 4
Oguchi method, 73
Oguchi model, 77, 79
ω (angular frequency), 150, 154, 189, 198-199
$\omega(k)$ (dispersion relation), 84
ω_0 (Larmor frequency), 199, 205-206, 235
ω_L (Larmor frequency), 29
One-dimensional chain, 127
180° wall, 134-136, 152
Operator equivalents, 56
Orbital angular momentum, 27, 28, 31-32, 123, 168, 172, 175
Orbital contribution, 170
Orbital quantum number, 31

Order parameter, 74-75, 78-79
Oscillating field, 192, 199, 201, 211, 227
Oscillations, 221, 223

p (pole strength), 119
$\mathcal{P}$ (absorbed power), 202-203
$p_{1/2}$ electron, 164, 169
p orbital, 61
PAC, 174
Pair, 75-77, 79, 85, 127, 133, 269
Palladium, 73
Palladium group, 40
Paramagnet, 1, 19, 53, 103
Paramagnetic Curie temperature, 55, 78
Paramagnetic material, 55, 178
Paramagnetic moment, effective, 45-46
Paramagnetic sample, 45
Paramagnetic susceptibility, 93
Paramagnetism, 13
  Van Vleck, 19
Paramagnon, 16
Parity, 162
Partition function, 260, 273
Pauli matrices, 81
Pauli paramagnetism, 13, 91
Pauli principle, 34
Pauli susceptibility, 29, 71, 98-99, 112-113, 181
Pd, 102
p_{eff} (effective paramagnetic moment), 45-46
Penetration depth, 154
Periodic table, 37-38, 40
Permalloy, 18, 129
Permanent magnet, 3, 18, 21, 23-24, 108, 116, 141, 148
Permanent magnet material, 143, 147-148
Permeability, 150-151
Permeance coefficient, 121
Perturbation theory, 185
Perturbed angular correlations, 174
Phonons, 211
π pulse, 218
$\pi/2$ pulse, 214-216, 218
Pinning, 136-137
Planck distribution, 86
Platinum group, 40
Poisson equation, 164
Poisson ratio, 128
Polarization, 3, 5, 140, 175, 177, 181-182
Pole strength, 119
Positron, 19
Positron annihilation, 19
Power transformers, 150
Principal quantum number, 31

SUBJECT INDEX   295

Pulsed magnetic resonance, 203, 214
Pulsed resonance, 199, 203

Q (electric quadrupole moment), 160, 165, 221
Q (quality factor), 220
Quadrupole interaction, 162, 165, 221, 223–224, 261
Quadrupole interaction parameter, 221–222
Quadrupole oscillation, 221–223
Quality factor, 220
Quantum mechanics, 63, 162, 229
Quantum number, 170
Quenching, 57, 61, 188

R (EFG correction factor), 183
Radiofrequency, 190, 214
RAl_2, 177–178
Rare earth, 13, 24, 40, 46–47, 56, 58, 68, 70–71, 73, 91, 93, 107–108, 160, 165, 170, 175, 177–178, 180, 183, 188
Rayleigh coefficient, 147
Rayleigh curve, 146–147
Rb, 91
^{87}Rb, 274
$RbMnFe_3$, 84
Reduced magnetization, 104–105
Reduced mass, 30
Relative magnetic permeability, 6–7
Relative magnetization, 100
Relative permeability, 5
Relative permeability tensor, 245
Relative sensitivity, NMR, 201, 254
Relaxation, 190, 203, 211, 214, 224, 231, 236, 245, 277
Relaxation mechanisms, 212
Relaxation process, 193–194
Relaxation time, 195, 203–204, 206–207, 211–212
Remanence, 140–141, 147–148
Remanent magnetization, 140
Resonance, 148, 152–154, 220
Retentivity, 115, 140–141, 147
rf, 201, 214
rf field, 203, 228, 236–242, 244–245, 248
RFe_2, 107
R_2Fe_{17}, 107
R_2O_3 oxide, 46
RNi_2, 108
ρ (electric charge density), 2
Rotating axes, 196, 199, 203
Rotating field, 196, 201
Rotating coordinate system, 197–198, 214, 216, 219, 224–225, 276–278

Rubidium, 188
Ruderman-Kittel, 213
Ruderman-Kittel-Kasuya-Yoshida (RKKY) interaction, 213
Ruderman-Kittel-Kasuya-Yoshida (RKKY) model, 72
Russell-Saunders coupling, 32

s (penetration depth), 154
s (conduction electron spin), 71
S (south), 7
S (total spin angular momentum), 32, 172, 173
s electron, 73, 164–165, 169–170, 175
S state, 163, 170, 180
$s - f$ hybridization, 73
Saturation, 131–132, 138, 140, 144–145, 148, 196, 203, 225, 279
Saturation magnetization, 115, 121, 123–124, 140–141, 148, 243, 248, 259
Saturation magnetostriction, 130, 132
Saturation magnetostriction constant, 130
Sc, 40
Schrödinger equation, 29–30, 63, 82
Secular broadening, 204, 206
Self-correlation, 204
Self-energy, 120
Semiconductor, 153, 161, 165
Shape anisotropy, 125–126, 144
Shape anisotropy energy, 125
Shape effect, 132
Shell, 30, 47, 180, 182
Shielding, 46
Short range correlation, 74
Short range order, 75
SI, 2–3, 5, 9–11, 25, 28, 50, 119, 124, 163, 177, 257
σ (magnetization per unit mass), 104
σ (projection of **S** along **J**), 68–69, 177–178, 180
σ (stress), 127–128
Silver, 40
Single-crystal, 121–122
Single domain, 123, 135–136, 149
Singlet, 65
Skin effect, 154, 242
Slater-Pauling curve, 20, 22
Slater-Pauling dependence, 107
Slow passage, 199
Sm, 46
^{147}Sm, 229
Sm^{3+}, 46
Sm_2Fe_{17}, 229
Sodium chloride, 13

Soft magnetic material, 116–117, 141–142
Solenoid, 3
Sommerfeld convention, 5
Sommerfeld expansion, 266
South pole, 7
Space quantization, 31
Spatial diffusion, 214
Specific heat, 75, 90
Spectral density function, 205
Spectral diffusion, 214
Spectroscopic splitting factor, 192
Sperimagnetism, 17
Speromagnet, 17
Speromagnetism, 17, 18
Spherical harmonics, 161
Spin, 28
Spin angular momentum, 28, 32, 57, 168
Spin density, 71, 80, 87–88, 168–170, 173
Spin echo, 214, 218, 220–223, 229, 240
Spin function, 65
Spin-glass, 17–18, 148
Spin-lattice interaction, 212
Spin-lattice relaxation, 190, 213, 215
Spin-orbit, 46
Spin-orbit coupling, 34, 123, 260
Spin-orbit coupling constant, 34
Spin-orbit interaction, 55, 57, 60, 160
Spin-orbit splitting, 47
Spin magnetic moment, 34
Spin operator, 81–82, 127, 185
Spin packet, 217
Spin quantum number, 31, 85
Spin-spin interaction, 190, 212–213, 218, 220
Spin-spin process, 215
Spin-spin relaxation, 215, 249
Spin-spin relaxation time, 190, 215
Spin temperature, 208, 210
Spin wave, 79–81, 83–86, 90, 228–229, 245
Spin wavefunction, 81
Spin wave resonance, 153–154
Static nuclear susceptibility, 4
Sternheimer factor, 183
Steven's operators, 4, 56
Stiffness constant, 80, 84, 88–89
Stimulated echo, 221
Stoner criterion, 100, 102, 107, 113
Stoner-Hubbard parameter, 100
Stoner model, 102–107, 109, 111
Stoner parameter, 100
Strain, 127–129, 229–230
Stress, 127–129, 131–132
Strong crystal field, 56
Strong ferromagnet, 103
Strong itinerant magnet, 18, 102

Subshell, 40
Suhl–Nakamura (SN) effect, 229
Suhl–Nakamura (SN) interaction, 213
Superconductor, type II, 118
Superferromagnetism, 17
Superparamagnetism, 16–17, 48–49
Surface anisotropy, 126
Susceptibility, 53–54, 71, 76, 78, 97, 99, 104, 107, 109, 113, 118, 135, 235, 248, 265, 270
  differential, 5
  extrinsic, 244
  intrinsic, 244
  mass, 5–6
  specific, 5
  volume, 5–7, 45
Système International d'Unités, see SI

T (tesla), 2
T'_2, 204, 206
T_1, 194–195, 199, 203–207, 210–211, 213–214, 225
T_2, 190, 194–196, 199, 203–207, 210–211, 213–214, 216, 220, 222, 249, 281
T_2^*, 215–216, 218–219, 220
T_c (Curie temperature), 13, 52–55, 74–79, 105, 108–109, 111, 124, 129, 178
T_f (freezing temperature), 17
T_L (lattice temperature), 208, 210
T_N (Néel temperature), 14, 259
T_s (spin temperature), 208
$T^{3/2}$ law, 87, 90
τ (order parameter), 74, 78–79
τ (pulse separation), 214, 218–219, 221, 223
τ (torque), 124
$τ_0$ (correlation time), 204–206
Taylor series, 87, 128, 161
Tb^{3+}, 188
Technical saturation, 144
Tesla, 2, 5, 178, 200
Tesseral harmonics, 161
Texture, 132
Thermal average, 74, 173, 177, 193
Thermal reservoir, 189, 207, 209, 211
Thermal vibrations, 203
$θ_p$ (paramagnetic Curie temperature), 55, 78
$θ'$ (Stoner molecular field parameter), 103, 105
Thin film, 125–126, 243
3d, 34
3d electrons, 22, 31, 91–93, 103
3d elements, 73, 108
3d metals, 88
3d series, 22, 56–57, 91, 180
3d shell, 13, 92
3d transition elements, 20

Time effects, 148
Tm, 58
Toroid, 10, 12
Total angular momentum, 69, 165, 170, 188–189, 274
Total magnetic moment, 34
Transferred field, 177, 178
Transition elements, 40, 46, 55
Transition metals, 40, 92–94, 100, 102, 108
Transition probability, 209
Transition rate, 208
Transverse magnetization, 190, 204–205, 214–219, 232, 234
Transverse relaxation, 196, 203, 205, 210, 213–215
Transverse susceptibility, 232, 234–235
Triplet, 65
Types of magnetism, 12

Uniaxial anisotropy, 23, 124, 126, 131, 135, 144
Uniform mode, 229
Uniform mode precession frequency, 243
Unit vector, 196
Unpaired electrons, 27

Vacuum magnetic permeability, 2, 29
Vacuum permeability, 3, 119, 235
Vacuum permittivity, 164
Van Vleck induced paramagnetism, 19
Vector model, 31

Vector potential, 166–167, 170–171
Virgin curve, 137–138, 140
Viscosity, magnetic, 149
V_{zz}, 163, 185, 221

W (crystal field scaling parameter), 58
Wall mobility, 144
Weak crystal field, 56
Weak itinerant magnet, 18, 111
Weak ferromagnet, 102–103
Webers per square meter, 3
Weger process, 212
Weiss approximation, 74
Weiss model, 52–55, 73–75, 77–79, 105
Weiss molecular field, 69
Weiss, P., 49, 51, 63
Wigner–Eckart theorem, 163
Work, 119–120, 131
Working point, 121

Y, 40
Young's modulus, 128

Z (atomic number), 160, 182
z (number of nearest neighbors), 77, 79, 82
Zeeman levels, 190, 216, 221
Zeeman term, 83
Zener model, 71
Zero-field NMR, 216, 229
ζ (relative magnetization), 100, 103–104
$ZrZn_2$, 111